SAFFRON
Biology and Biotechnology

Editors:
Deepu Pandita
Government Department of School Education
Jammu, Jammu and Kashmir, India

Anu Pandita
Vatsalya Clinic, New Delhi, India

Tanushri Kaul
International Centre for Genetic Engineering & Biotechnology
New Delhi, India

CRC Press
Taylor & Francis Group
Boca Raton London New York

CRC Press is an imprint of the
Taylor & Francis Group, an **informa** business
A SCIENCE PUBLISHERS BOOK

First edition published 2026
by CRC Press
2385 NW Executive Center Drive, Suite 320, Boca Raton FL 33431

and by CRC Press
4 Park Square, Milton Park, Abingdon, Oxon, OX14 4RN

CRC Press is an imprint of Taylor & Francis Group, LLC

Library of Congress Cataloging-in-Publication Data (applied for)

ISBN: 978-1-032-51744-5 (hbk)
ISBN: 978-1-032-51748-3 (pbk)
ISBN: 978-1-003-40376-0 (ebk)

DOI: 10.1201/9781003403760

Typeset in Times New Roman
by Prime Publishing Services

Preface

Crocus sativus L. (family-Iridaceae) is generally known as the Saffron or "Red Gold or Gold Condiment" due to its taste, aroma and being most expensive spice globally (€30,000 /kg). The word Saffron originated from an Arabic term Zafaran which indicates to become yellow because of usage of Saffron as a bright yellow colorant. As stated by some researchers the Latin name *crocus* is derivative of a Sanskrit word for saffron. Saffron geophyte flowering exotic crop belongs to family Iridaceae and genus Crocus which possess about 85 species. Saffron plant is small with maximum height of 10–30 cm, autumn flowering, triploid ($2n = 3x = 24$), male-sterile, medicinal and aromatic plant and propagates only vegetatively via corms, and native of Asia Minor. Saffron is a male-sterile triploid flower crop plant. The spice is derived from dry stigmas of saffron plant. The plant fetches very high values, especially as the growth of Saffron plant is highly inadequate.

Saffron flower has a perigonium made up of six tepals which are purple/violet in colour and measuring 20 mm to 47 mm long and 11 mm to 23 mm wide. Flowers have an underground tubular ovary and a thin style of 9–10 cm length divided into three dark red branched stigmas, each one up to 30–40 mm long, characteristic reddish or orange red in colour, and subclavate. Three stigmas are the main red part of the saffron plant as distal ends of the plant's tricarpellate gynoecium in a flower. Dried red stigmas of flower of Saffron form spice. About 2 lakh stigmas hand-picked from approximately 70,000 *C. sativus* flowers produces 1 pound of commercial saffron and takes round 200 hours.

The growth pattern of saffron is different from other crops and it can be divided into three stages, namely, flowering, vegetative stage, and formation of corms. The flower usually appears in autumn (October–November), followed by vegetative stage throughout winter, and the formation of replacement corms at the base of the shoots. Saffron cultivation, manual harvesting, flower picking and stigma separation makes this golden spice costly, so even little boost in the yield and/or quality can enhance profitability.

Saffron is primarily cultivated in a variety of countries including Iran, India, Japan, France, Switzerland, Morocco, Spain, Greece, Italy, Turkey, Pakistan, China, and Australia. India is the second-largest saffron importer in world, and the nation's swift growth in economy has made saffron more affordable to a rising figure of middle- and upper-income households. In India, saffron is cultivated under temperate climatic conditions in Kashmir valley on uplands with lacustrine deposits ("Karewas" at an altitude of 1585-1677 m above sea level) and Kishtwar region of Jammu and Kashmir State. Notably, Iran produces the highest amount of saffron, meeting nearly 80% of global demand. Saffron has been cultivated as a spice in Egypt and the Middle East for at least 5,000 years.

Saffron is a medicinal and an aromatic flower. Saffron is full of treasured natural metabolites, 150 volatile and aroma-yielding compounds and various non-volatile active compounds. The apocarotenoids such as, picrocrocin (responsible for bitter taste), crocin, crocetin (responsible for bright natural color), and the odor-active safranal, terpenoids (Saponins) and flavonoids are bestowed by Nature with various pharmacological and health benefits. Saffron is expectorant, and has aphrodisiac, anti-diabetic, anti-inflammatory, anti-depressant, anti-spasmodic, anti-hypertensive, anti-oxidant, hepatoprotective, neuroprotective anti-tumor, anti-mutagenic, anti-cancerous and immune-modulating properties.

This volume covers the detailed introduction of Saffron, its Cultivation and Ecology, Genetics and Cytogenetics, Role of Saffron Bioactives in Nutrition and Medicine, Economic Importance, Beneficial Effects, Applications, Safety Evaluation and Limitations of Saffron. Further, Transcriptomics, Post transcriptomics, Metabolomics, Computational Biology, Molecular Markers, Genome Editing, Microbiome and Endophytic interactions, Micropropagation and Phytochemistry and uses of Saffron in Human Health and Diseases have been deliberated in detail. Genomics and Transcriptomics may provide deeper insights into the genes responsible for the production of saffron metabolites. Further, the integratome omics frameworks give an insight into the systemic and comprehensive saffron biology and future genetic progress by tools of genome editing, epigenetics, breeding and micro-propagation.

This book is essential for the farmers, industries, pharmaceutical industries, readers, scholars, researchers and faculty members familiar with the field and for newcomers in the area of saffron. The area of saffron biology is an important research area and the audience will get a comprehensive updated knowledge about it in the form of this book. It will be most recent updated compilation on saffron biology and biotechnology.

Deepu Pandita
Anu Pandita
Tanushri Kaul

June 2025

Contents

1 Saffron (*Crocus sativus* L.): The Red Gold

Deepu Pandita[1*] and Anu Pandita[2]

Crocus sativus L., commonly known as "Red Gold or the Gold Condiment", is revered for its taste, aroma and status as the world's most expensive spice. The word *Saffron* originates from the Arabic term Zafaran, meaning "*to become yellow*", due to its historical use as a bright yellow colorant. Some researchers suggest that its Latin name, *Crocus,* derives from a Sanskrit word for saffron. Saffron is an exotic, geophytic flowering crop belonging to the Iridaceae family and the Crocus genus, which *comprises* about 85 species. The saffron plant is small, reaching a maximum height of 10–30 cm. It is a sterile triploid ($2n = 3x = 24$), propagated via corms and native to Asia Minor. Its flowers feature a perigonium composed of six violet or purple tepals measuring 20–47 mm in length and 11–23 mm in width. The three red stigmas, situated at the distal ends of the flower's tricarpellate gynoecium, are the most commercially valuable part of the plant, forming the saffron spice when dried. Notably, the stigmas make up only about 7% of the total flower. Saffron is cultivated in numerous countries, including Iran, India, Japan, France, Switzerland, Morocco, Spain, Greece, Italy, Turkey, Pakistan, China, and Australia. In India, its primary cultivation occurs in the Pampore district of Kashmir.Although India is the second-largest saffron importer, in world, rapid economic growth has made saffron increasingly accessible to middle- and upper-income households, Iran dominates global saffron production, supplying nearly 80% of total demand. Historically, saffron has been cultivated as a spice in Egypt and the Middle East for over 5,000 years. Genetic studies using genome-wide single-nucleotide polymorphism (SNP) analysis have confirmed that *C. sativus* originated in Attica, Greece, from *Crocus cartwrightianus*.The flowers are hand-harvested from late October to early November, shortly after dawn, as exposure to sunlight quickly diminishes their color and aroma. Approximately 200,000 stigmas, meticulously gathered from about 70,000 flowers. yield just 0.453 kg of commercial saffron—requiring around 200 hours of labor. Crocin imparts a red to reddish-brown hue to saffron stigmas, safranal enhances its aroma, and picrocrocin adds bitterness. These compounds—along with crocetin—are non-volatile and contribute to saffron's

[1] Government Department of School Education, Jammu, Jammu and Kashmir, India.
ORCID iD -https://orcid.org/0000-0002-8361-1872
[2] Vatsalya Clinic, Krishna Nagar, New Delhi, India.
* Corresponding author: deepupandita@gmail.com

DOI: 10.1201/9781003403760-1

medicinal properties. Saffron exhibits expectorant, aphrodisiac, anti-inflammatory, antioxidant, neuroprotective, anti-cancer, and immune-modulating effects.

1. Biological Classification of Saffron

Kingdom	Plantae
Phylum	Magnoliophyta
Class	Liliopsida
Order	Liliales
Family	Iridaceae
Genus	*Crocus*
Species	*sativus*
Botanical Name	*Crocus sativus* L.
Synonyms	*Safran officinarum* Medik. *Crocus pendulus* Stokes *Crocus autumnalis* Sm. nom. illeg. *Crocus orsinii* Parl. *Crocus officinalis* (L.) Honck. *Geanthus autumnalis* Raf. *Crocus setifolius* Stokes
Common Names (*http://gernot-katzers-spice-pages.com*)	English: Saffron Arabic & Hebrew: *Zafran* Farsi: *Za'afaran* Chinese Mandarin: *Fan hong hua* French, German: *Safran* Italian: *Zaffarano* Spanish: *Azafrán* Greek: *Krokos* Hindi: *Kesar*

2. Origin, History & Folklore of Saffron

Crocus sativus L. has been cultivated in Egypt and the Middle East for over 5,000 years, first mentioned during the Akkadian empire (2300 BC). Minoan frescoes from Knossos (1700–1600 BC) depict saffron harvesting.The ancient Greek term Krokos appears in Homer's Iliad and is linked to mythology, where Hermes transformed Krokos into the Saffron plant. Theophrastus, Hippocrates and Galen praised saffron for its digestive benefits. In Egypt, it was cited in the Ebers Papyrus for kidney ailments, while the Bible references it in Song of Songs (4:14). Persian rulers introduced saffron to India, where Ayurveda recognized its role in treating arthritis, infertility, and impotence. The Romans brought it to Britain, and Arabians spread it to Spain (http://gernot-katzers-spice-pages.com/engl/Croc_sat.html;Velasco-Negueruela 2001; Kumar et al., 2008).

The origin of saffron remains debated. Vavilov attributes it to the Middle East, while others trace its roots to Asia Minor or the southwestern Greek islands (Winterhalter and Straubinger, 2000). The Minoan civilization (c. 1550 B.C.) documented saffron in the Papyrus Ebers, though cultivation predates this. References to saffron appear in the Old Testament (12th century B.C.) and the Hebrew Bible (Tanakh) (Halvorson, 2008). In ancient Persia (10th century B.C.), C. sativus adorned royal carpets and burial veils before Persians introduced it to Kashmir. They pioneered saffron-infused baths, a practice later adopted by Alexander the Great's Macedonian troops (356–323 B.C.) (Magesh et al., 2006). Safranbolu, Turkey, hosts an annual saffron harvest festival, celebrating its cultivation. Since the late 20th century, Consuegra, Spain has held the "Saffron Rose Festival on the last weekend of October to honor saffron's cultural significance (Melnyk et al., 2010; Dai et al., 2021).

Saffron-dyed fabrics were rare, luxurious, and symbols of royalty in ancient times. Irish kings wore saffron mantles, while Assyrian rulers received saffron-dyed textiles from Phoenicians (Velasco-Negueruela, 2001). Artifacts from Castilla-La Mancha, Spain, indicate saffron was used as a dye as early as 2400 B.C. It later became a condiment during Hammurabi's reign (1800–1700 BC) (Joukar et al., 2013).

The triploid C. sativus, which reproduces only via corms, has an uncertain evolutionary origin (Rashed-Mohassel, 2006; Aghayev et al., 2009). Some theories suggest it resulted from autotriploidy, while others propose it is an allotriploid with *Crocus cartwrightianus* as one parent (Mathew, 1999; Negbi & Negbi, 2002). Genetic studies using SNP analysis confirmed C. sativus originated from C. cartwrightianus in Attica, Greece (Nemati et al., 2019).

Botanical Description of Saffron

Crocus sativus, known as Saffron or "Red Gold", is prized for its taste, aroma, and status as the world's costliest spice. A geophytic plant of the Iridaceae family and Crocus genus, it includes about 85 species. Saffron is a small, sterile triploid ($2n = 3x = 24$), propagated via corms, native to Asia Minor, and primarily cultivated in Mediterranean regions, especially Spain and Iran (Kumar et al., 2008; Fiore et al., 2010; Lopez-Corcoles et al., 2015; Leone et al., 2018; Shokrpour, 2019; Tahiri et al., 2023). Its flowers feature six purple-violet petals, three stamens and a pistil, with three red-orange stigmas forming a saffron thread . Saffron planting varies by climate zone, typically occurring between May and August. In Khorasan, Iran, cultivation spans late May to early October, while timing differs in Spain and India. Major saffron-producing countries include India, China, Spain, Iran, Greece and Morocco, with Spain and Iran leading global production. La Mancha, Spain is renowned for high-quality saffron (*http://gernot-katzers-spice-pages.com*; Velasco-Negueruela 2001). *C. sativus* is the only species used as a spice and is the world's most expensive. The term saffron originates from the Arabic Zafaran (meaning "to become yellow"), the Persian sahafaran, and possibly the Sanskrit root of Crocus (http://gernot-katzers-spice-pages.com; Winterhalter and Straubinger, 2000; Shokrpour, 2019).

Saffron stigmas are carefully separated and dried after harvest, with Spanish producers traditionally roasting them over charcoal embers. The labor-intensive harvesting and stigma segregation contribute to saffron's high price, exceeding $4,000 per pound (https://www.businessinsider.com/why-real-saffron-is-so-expensive-2020-6). Around 200,000 stigmas, hand-picked from approximately 70,000 *C. sativus* flowers, yield just one pound of saffron, requiring about 200 hours of work (Velasco-Negueruela 2001; Kumar et al., 2008). Harvesting occurs by hand from late October to early November, shortly after dawn,as exposure to sunlight quickly diminishes saffron's color and aroma (Velasco-Negueruela 2001; Kumar et al., 2008). Figure 1 illustrates the annual publication count on saffron research, peaking at 332 papers in 2022.

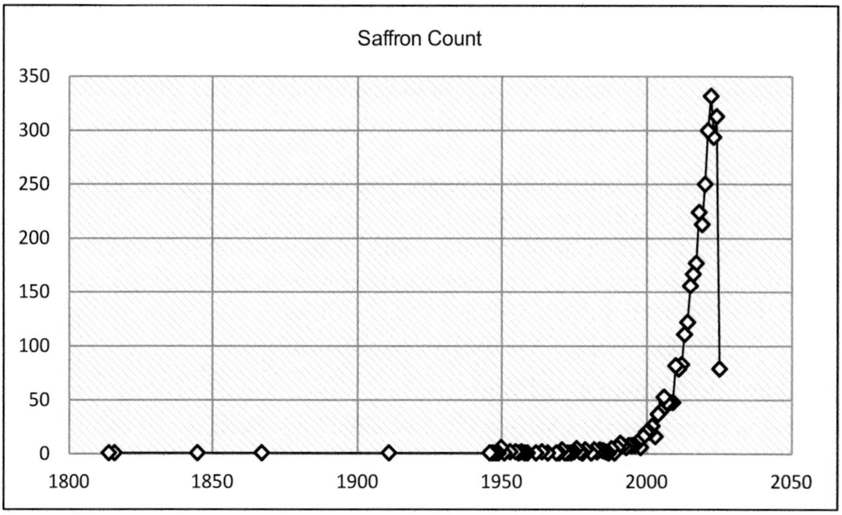

Fig. 1 Results of search word Saffron by year in PubMed with highest 332 papers in 2022.

2.1 Saffron Plant

Crocus sativus, is a sub-hysteranthous, autumnal herbaceousus geophyte of the Iridaceae family. It thrives in Mediterranean Europe and Western Asia, spanning 10°W to 80°E longitude and 30°N to 50°N latitude (Molina et al., 2005; Kumar et al., 2009; Namayandeh et al., 2013; Yildirim et al., 2017; Leone et al., 2018; Cardone, et al., 2019; Mykhailenko et al., 2020; Mohtashami et al., 2021; Ma et al., 2024). A perennial plant, saffron reaches 20–30 cm in height.

The corm of *Crocus sativus* is flattened to ovoid or sub-globose, compact, and flat at the base, resembling an onion bulb. It ranges from 0.5 cm to 6.5 cm in diameter and weighs between 0.5 g and 70 g. A minimum size of 2.5 cm is required for flowering. Corms are encased in fine reticulate fibrous tunics extending about 5 cm above the plant's neck (Rechinger, 1975; Wendelbo, 1977; Kumar et al., 2009). Corms measuring 3–3.5 cm typically develop two or three apical buds, producing leaves, a floral axis, and daughter corms. Additionally, four to seven secondary buds, arranged in a spiral pattern,contribute to leaf growth. Saffron corms are planted annually from July to September, with parent corms (brownish-white) buried 15-20 centimetres underground. Each corm generates one to three flowers with purple petals. The starchy corm consists mainly of parenchymatous tissue and a basal node that forms adventitious roots (Kumar et al., 2009). Its biochemical composition includes glucose, amino acids (threonine, lysine, aspartic acid, glutamic acid, cysteine, glycine, serine, alanine, valine, tyrosine, arginine, leucine, histidine, methionine, proline, phenylalanine), triterpenic and steroid saponins, and high molecular weight proteins (Rubio Moraga et al., 2013).

Crocus sativus has three types of adventitious roots: absorbing, contractile, and contractile-absorbing roots (Negbi, 1999). Absorbing roots, emerging from the corm base, are long, fibrous, and slender. Contractile roots, small and abundant, develop at the base of sprouting buds, helping to anchor newly formed corms

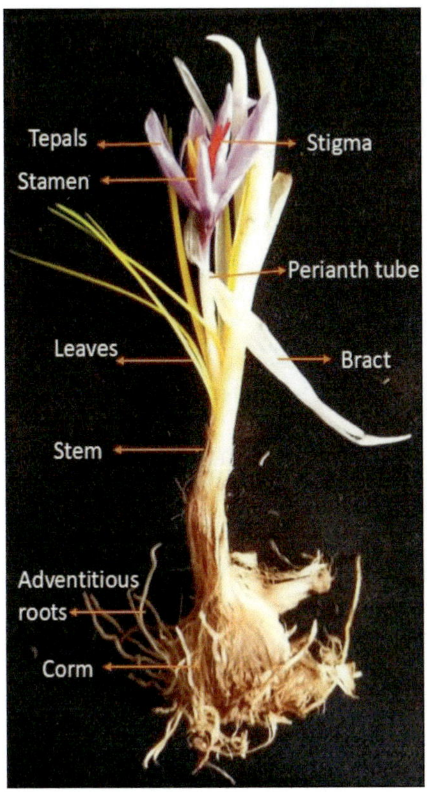

Fig. 2 Saffron Plant

deeper into the soil while also aiding nutrient absorption. Contractile-absorbing roots grow near sprouting buds on the parent corm and exhibit both contractile and absorbing functions (Negbi, 1999). Kumar et al. (2009) classify saffron roots into two structural types: fibrous and contractile. Fibrous roots, thin and linear (1 mm thick, 15–20 cm long), emerge from the corm base as shoots grow, efficiently absorbing water and nutrients. Contractile roots, large and colorless, form in layers as daughter corms develop, pulling them deeper underground in response to soil temperature and light variations (Kumar et al., 2009).

Crocus sativus leaves are slender, linear, and grass-like, with a pointed apex and entire margins. They range from 40–70 cm in length and 2–3 cm in width, displaying a dark green hue with a white central stripe and two darker lateral stripes. Leaf count depends on bud health, corm diameter, and weight, varying from 29 leaves in 6–12 g corms to 52 leaves in 36–42 g corms (Renau-Morata et al., 2012). During emergence, leaves stand upright (orthotopic), later shifting to a horizontal position (plagiotopic) as they develop (Kafi et al., 2006). Photosynthesis is crucial for daughter corm multiplication, with leaves growing from September to May before drying in the senescence phase (Renau-Morata et al., 2012). The flower stem produces white membranous bracteoles with long-tapering tips, which are narrower and shorter than bracts.

2.2 Saffron Flowers

In *Crocus sativus*, flowers can bloom either before or after leaf emergence, depending on environmental conditions (Molina et al., 2005).

Fig. 3 Saffron Flower

Petals: The saffron flower has a perigonium with six violet-purple tepals, measuring 20–47 mm in length and 11–23 mm in width. Its petals envelop the style, stigmas, and stamens, containing polyphenols with antioxidant properties beneficial to pharmaceuticals, food, and cosmetics.

Stamen: The androecium of *Crocus sativus* consists of three yellow stamens with short filaments and 15 mm-long yellow anthers containing pollen (Kumar et al., 2009). Saffron stamens contribute to yellow saffron, which retains the spice's characteristic scent. While widely used in pharmaceuticals, they have minimal culinary applications. Due to abnormal pollen grains and low viability, saffron is self-sterile and reproduces exclusively through corms (Shokrpour, 2019).

Carpel: The gynoecium, or pistil, features an inferior ovary with three locules, forming a tricarpellate and trilocular structure. Stigmas are the main red part of the saffron plant at distal ends of the plant's carpels. Each saffron flower has three red stigmas at the distal ends of its carpels, collectively forming a thread with the style. Ovule placentation is axial, with 18–20 ovules per locule arranged in two rows. The gynoecium develops into a 9 cm-long yellow-green filiform stylus, extending through the perigonium tube and terminating in a red stigma with three threads. The stigma, 30–40 mm long, has a trumpet-shaped apex and elongates beyond the anthers and tepals during anthesis, bending downward as the flower opens. A physiological, non-genetic anomaly in flower formation can increase the number of stigmas, stamens and violet tepals (Gresta et al., 2008). A single fresh saffron flower weighs 300–500 mg, with fresh stigmas ranging from 25–47 mg and dried stigmas weighing 6–9 mg. Producing 1 kg of saffron spice requires 110,000–160,000 flowers (Cardone et al., 2019). After handpicking, stigmas are separated, dried, and processed into saffron (Roshanravan & Ghaffari, 2022). The dried red stigmas, comprising only 7% of the flower, are prized in the food industry

Fig. 4 Dried Saffron Stigmas

for their coloring, flavoring, and aromatic properties in dishes and alcoholic beverages (Mzabri et al., 2019; Serrano-Díaz et al., 2013; Bakshi et al., 2022).

3. Saffron Chromosomes

Saffron (*Crocus sativus*) is an autotriploid that originated in Attica, Greece, through the fusion of two distinct *C. cartwrightianus* genotypes (Nemati et al., 2019). Karasawa (1932, 1933, 1942) first identified cultivated saffron as an autotriploid

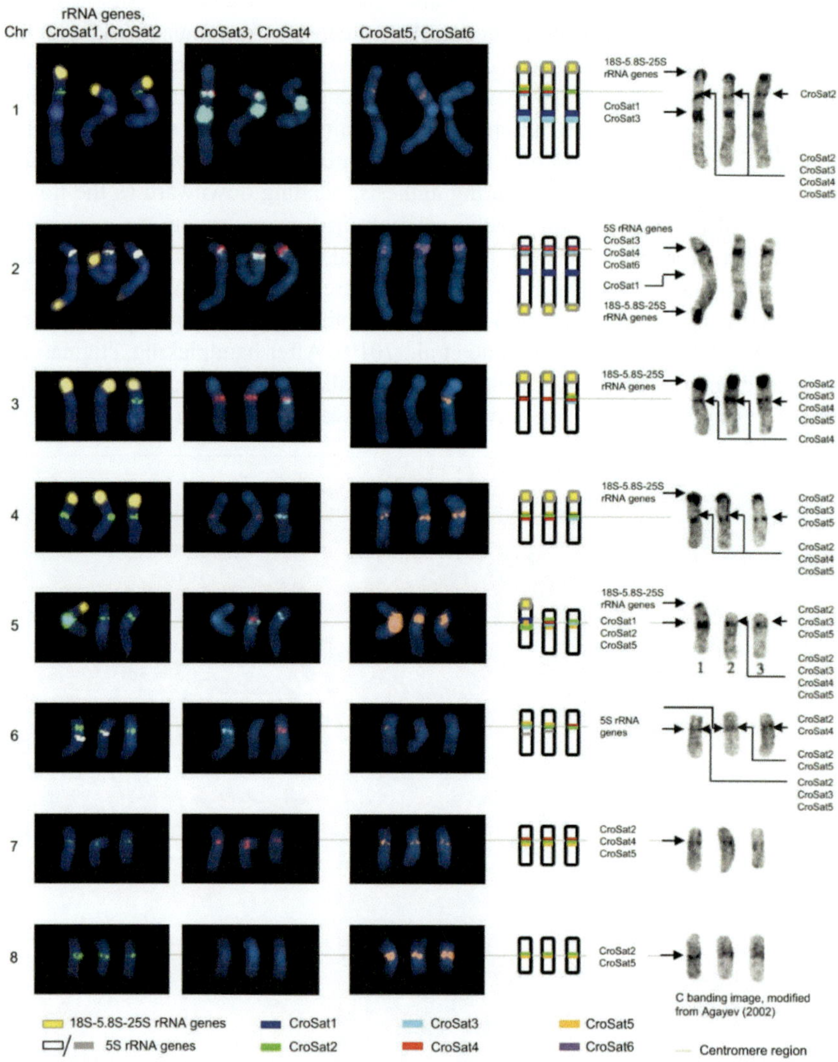

Fig. 5 The FISH of CroSat satellites and assignment to heterochromatic C-bands (*Source:* Agayev, 2002; Schmidt et al., 2019)

with 2n = 24 chromosomes. Its triploid genome (x = 8; 2n = 3x = 24) disrupts chromosome pairing during meiosis, leading to irregular distribution at metaphase and resulting in sterility (Shokrpour, 2019; Hajyzadeh et al., 2020). Due to abnormal pollen grains and low viability, saffron is self-sterile and reproduces exclusively through corms (Shokrpour, 2019).

Fluorescent *in situ* hybridization (FISH) was used to analyze CroSat satellites and their association with heterochromatic C-bands in *Crocus sativus*. DAPI-stained mitotic chromosomes appeared blue. The left panels displayed signal patterns for CroSat clones (CroSat1.3–CroSat6.4), rRNA genes (18S-5.8S-25S in yellow, 5S in white), across all eight saffron triplets. The middle panels illustrated ideograms of 24 chromosomes, while the right panels correlated satellite arrays with C-bands. Centromere positions were marked by thin lines (Agayev, 2002; Schmidt et al., 2019).

4. Saffron Production

The precise origin of saffron remains debated, with some attributing it to Iran (Middle East) and others to Greece, particularly Crete and Santorini. It later spread to India, China, the Mediterranean basin, and Eastern Europe (Shokrpour, 2019). India, the second-largest saffron importer, has seen increased accessibility due to economic growth. Major saffron-producing countries include Iran, Spain, France, India, Greece, Morocco, Italy, Japan, Turkey, Switzerland, China, Pakistan, and Australia. Iran dominates global production, supplying nearly 80% of total demand (Menia et al., 2018). India contributes only 5% to global saffron output, with 90% of its production concentrated in Jammu and Kashmir, primarily in Pulwama and Budgam districts (Tanra et al., 2017).

Table 1 Top Exporters and Importers of Saffron.
(https://oec.world/en/visualize/tree_map/hs92/export/ind/all/2091020/2023)

Top Saffron exporting countries		
Country	*Percentage*	*Exports value*
Iran	(43.4%)	$115M
Spain	(23.3%)	$62.1M
Afghanistan	(21.4%)	$56.9M
Greece	(1.56%)	$4.14M
Portugal	(1.25%)	$3.31M
Netherlands	(1.24%)	$3.31M
France	(1.12%)	$2.99 M
Poland	(0.81%)	$2.17M
Austria	(0.6%)	$1.6M

Contd.

Table 1 *Contd.*

The top importers of Saffron		
Country	*Percentage*	*Exports value*
Spain	(17.1%)	$45.5M
India	(13.6%)	$36.1M
China	(10%)	$26.7M
Saudi Arabia	(7.37%)	$19.6M
United States	(8.02%)	$21.3M
United Arab Emirates	(7.26%)	$19.3M
Italy	(5.43%)	$14.4M
France	(4.08%)	$10.9M
Kuwait	(3.11%)	$8.28M
Sweden	(2.89 %)	$7.7M
Germany	(1.99%)	$5.28M
United Kingdom	(1.97%)	$5.25M
Portugal	(1.54%)	$4.09M

5. Chemical Constituents of Saffron

Crocus sativus comprises over 150 volatile aromatic chemical compounds, along with non-volatile active carotenoids such as zeaxanthin, carotenes, and lycopene (Liakopoulou-Kyriakides and Kyriakidis, 2002).

Saffron's volatiles include over 34 compounds, such as terpenes, terpene alcohols and esters, contributing to its strong aroma (Abe and Saito, 2000). Key bioactive elements include crocin, responsible for the red to reddish-brown stigma color (10% in dried saffron), safranal, which imparts aroma, and picrocrocin, a glycosidic precursor of safranal that adds bitterness (Maggi et al., 2010; Shahi et al., 2016; Zhang et al., 2019). Non-volatile compounds in saffron stigmas include crocins, carotenes, crocetin, picrocrocin, and safranal (Wallis, 2005). Crocin ($C_{44}H_{64}O_{24}$), a natural cooking colorant, acts as an antioxidant and may inhibit cancer cell production by reducing RNA, DNA, and protein synthesis in tumor cells (Kabiri et al., 2017; Kothari et al., 2021). Crocetin ($C_{20}H_{24}O_4$), a lipophilic carotenoid, exhibits anti-cancer, anti-diabetic, anti-inflammatory, cardiovascular-protective, and antidepressant properties while also supporting liver health (Samarghandian et al., 2013; Wang et al., 2014; Ohba et al., 2016; Kabiri et al., 2017). Safranal ($C_{10}H_{14}O$) offers cardioprotective benefits and aids in treating neurological disorders (Maggi et al., 2010; Kabiri et al., 2017). Picrocrocin is recognized for its potential in Alzheimer's treatment (Samarghandian and Borji, 2014).

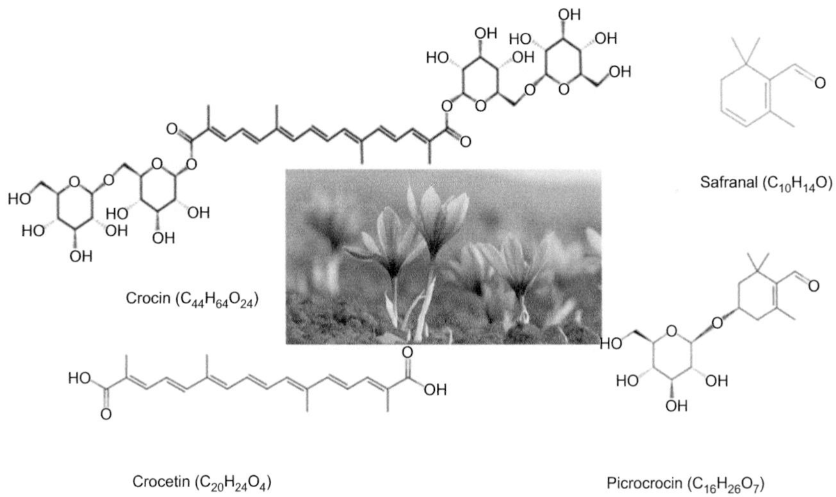

Fig. 6 Chemical structure of Crocin, Safranal, Crocetin and Picrocrocin

6. Nutritional Profile of Saffron

Saffron is a nutrient-rich spice containing proteins, sugars, vitamins B1 and B2, and essential minerals. Its key bioactive compounds include crocin, picrocrocin, crocetin and safranal (Al-Snafi, 2016; Figure 3). The full nutritional profile, as per the USDA database, is detailed in Table 2.

Table 2 Nutritional value per 100 grams of Saffron.
https://fdc.nal.usda.gov/food-details/170934/nutrients

Component Name	*Amount*
Water	11.9 g
Energy	310 kcal
Protein	11.4 g
Total lipid (fat)	5.85 g
Ash	5.45 g
Carbohydrate, by difference	65.4 g
Fiber, total dietary	3.9 g
Calcium, Ca	111 mg
Iron, Fe	11.1 mg
Magnesium, Mg	264 mg
Phosphorus, P	252 mg
Sodium, Na	148 mg

Contd.

Table 2 *Contd.*

Potassium, K	1720 mg
Manganese, Mn	28.4 mg
Zinc, Zn	1.09 mg
Copper, Cu	0.328 mg
Vitamin C, total ascorbic acid	80.8 mg
Selenium, Se	5.6 µg
Niacin	1.46 mg
Thiamin	0.115 mg
Riboflavin	0.267 mg
Vitamin B-6	1.01 mg
Vitamin A, RAE	27 µg
Vitamin A, IU	530 IU
Folate, total	93 µg
Folate, food	93 µg
Folate, DFE	93 µg
Fatty acids, total saturated	1.59 g
SFA 14:0	0.006 g
SFA 16:0	1.16 g
SFA 18:0	0.247 g
Fatty acids, total monounsaturated	0.429 g
MUFA 16:1	0.003 g
MUFA 18:1	0.39 g
MUFA 20:1	0.006 g
Fatty acids, total polyunsaturated	2.07 g
PUFA 18:2	0.754 g
PUFA 18:3	1.24 g
PUFA 20:4	0.013 g
PUFA 22:5 n-3 (DPA)	0.006 g

7. Medicinal Properties of Saffron

Saffron is a medicinal plant with diverse therapeutic properties (Anaeigoudari et al., 2023) Its key bioactive compounds—crocins, crocetin, and safranal—help inhibit inflammatory responses and neutralize free radicals, exhibiting strong antioxidant effects (Abe and Saito, 2000; Ashktorab et al., 2019; Cardone et al., 2020; Azami et al., 2021; Anaeigoudari et al., 2023).

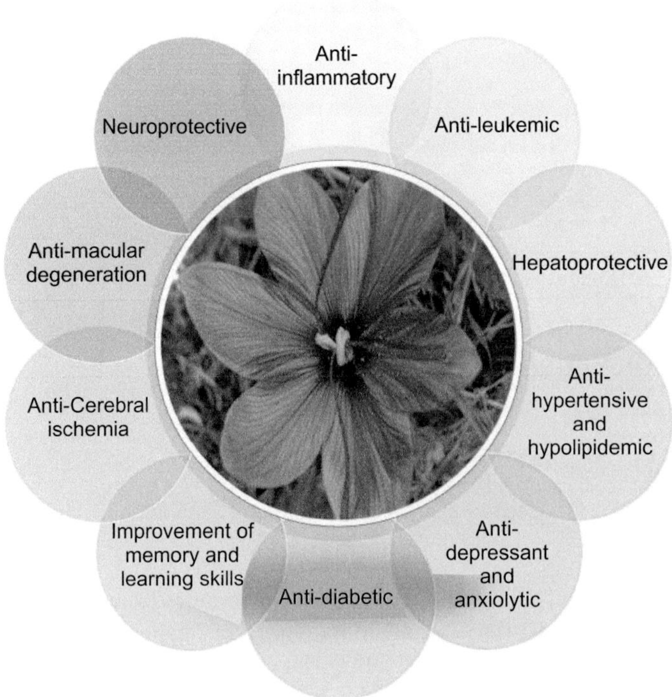

Fig. 7 Medicinal Uses of Saffron

Crocin, safranal and crocetin exhibit cytotoxic effects on breast, lung, and liver cancer cell lines (Bhandari, 2015). They induce apoptosis, inhibit cell proliferation, and block angiogenesis in lung, ovarian, breast, and tumor cancers (Aung et al., 2007). Additionally, they downregulate the NF-κB signaling pathway, which plays a role in inflammation and cancer progression (Zeinali et al., 2019; Anaeigoudari et al., 2023). Saffron enhances the expression of sirtuin 1 (SIRT1) and nuclear factor erythroid 2-related factor 2 (Nrf2), improving organ function while inhibiting inducible nitric oxide synthase (iNOS) and cyclooxygenase-2 (COX2) (Anaeigoudari et al., 2023). SIRT1 and Nrf2 regulate inflammatory responses and oxidative stress, with studies linking saffron's protective effects to their upregulation. The inhibition of iNOS and COX2 further contributes to crocin's anti-inflammatory and antioxidant effects of crocin.

Abbreviations: CAT—catalase; SOD—superoxide dismutase; COX2—cyclooxygenase-2; IL—interleukin; NF-κB—nuclear factor kappa B; iNOS—inducible nitric oxide synthase; Nrf2—nuclear factor erythroid 2-related factor 2; TNF-α—tumor necrosis factor alpha. Pandita (2020) provides a comprehensive overview of saffron's phytochemistry, therapeutic significance and omics-based biology. Saffron exhibits expectorant, aphrodisiac, anti-diabetic, anti-inflammatory,

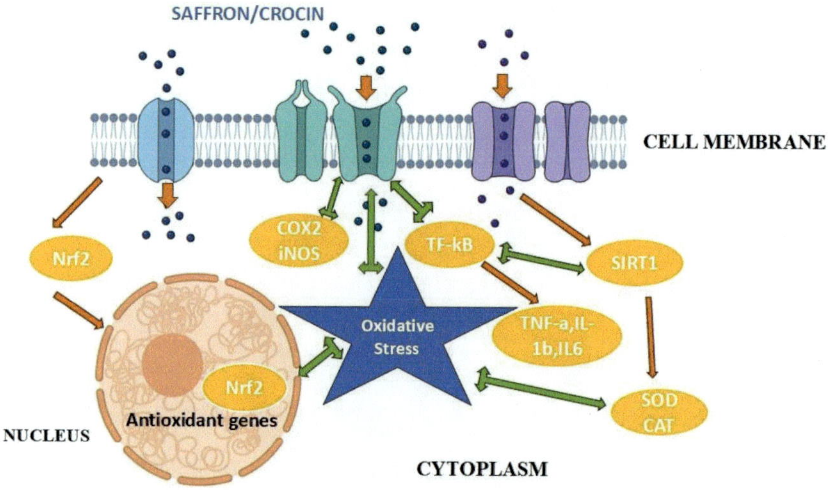

Fig. 8 Mechanism of Medicinal Effects of Saffron and Crocin. (Adapted from Anaeigoudari et al., 2023)

anti-depressant, anti-spasmodic, anti-hypertensive, antioxidant, hepatoprotective, neuroprotective, anti-tumor, anti-mutagenic, anti-cancerous and immune-modulating properties (Abdullaev, 2003; Magesh et al., 2006; Lechtenberg et al., 2008; Jan et al., 2014; Bhandari, 2015; Rangarajan et al., 2015 Cardone et al., 2020; Azami et al., 2021; Lambrianidou et al., 2021). It aids in treating coughs, wounds, gastrointestinal disorders, smallpox, sleeplessness, scarlet fever, chronic uterine bleeding, colds, respiratory issues, colic, and cardiovascular conditions, while also soothing teething infants (Abdullaev, 2003; Abdullaev and Espinosa-Aguirre, 2004).

The anti-inflammatory potential of saffron stigmas was evaluated using various concentrations of aqueous and ethanolic extracts, attributed to crocin and crocetin (Hosseinzadeh and Younesi, 2002). These compounds also exhibit antileukaemic effects by inhibiting nucleic acid synthesis, enhancing antioxidation, and promoting promyelocytic leukaemia differentiation (Moradzadeh et al., 2018).

Safranal, crocin, and kaempferol contribute to saffron's neuroprotective effects (Zheng et al., 2007). Compared to donepezil, saffron improved cognitive function and reduced vomiting in patients with mild to moderate Alzheimer's disease (Akhondzadeh et al., 2010; Hatziagapiou et al., 2019). In Wistar rats pre-treated with crocetin, neuromodulatory effects suggested its potential for Parkinson's prevention by lowering thiobarbituric acid reactive content and preserving glutathione and dopamine levels (Ahmad et al., 2005; Hatziagapiou et al., 2019). Saffron is widely used to treat neurodegenerative disorders, particularly depression, by modulating neurotransmitter levels and opioid activity. Clinical studies show it is more effective than placebos and synthetic antidepressants, with fewer side effects like agitation, pain, and nausea (Jalali-Heravi et al., 2009; Khazdair et al.,

2015; Kell et al., 2017). It enhances memory by altering neurotransmitter systems, increasing cerebral blood flow, and reducing oxidative stress (Kell et al., 2017). Additionally, saffron extracts act as hepatoprotective antioxidants, improving liver biomarkers and diabetic hepatopathy (Rahbani et al., 2012). The vascular endothelial growth factor further supports its neuroprotective role in cerebral ischemia/reperfusion injury (Abdel-Rahman et al., 2020).

Saffron enhances insulin sensitivity by improving β-cell function, which regulates insulin release. It also inhibits pathways leading to β-cell failure and promotes GLUT-4 a translocation to the cell membrane, facilitating glucose uptake in insulin-dependent tissues (Yaribeygi et al., 2019). Studies show it lowers fasting blood glucose and HbA1c levels in type 2 diabetic patients while preventing neuropathy, retinopathy, and nephropathy (Abu-Izneid et al., 2022; Sanaie et al., 2023). Its key bioactive compounds—crocin, crocetin, and safranal—further enhance insulin sensitivity (Kianbakht and Hajiaghaee, 2011). Additionally, saffron boosts glucose absorption by adipocytes and improves insulin responsiveness (Sepahi et al., 2018).

Conclusion

Saffron, known as "Red Gold," is a geophyte flowering crop from the Iridaceae family and Crocus genus. It is a sterile triploid ($2n = 3x = 24$) propagated via corms. Its flower features a perigonium with six violet tepals and a tricarpellate gynoecium with three red stigmas, which form the spice when dried. Crocin inhibits cancer cell production by reducing RNA, DNA, and protein synthesis in tumor cells. Crocetin offers anti-cancer, anti-diabetic, anti-inflammatory, cardiovascular-protective, antidepressant, and hepatoprotective benefits (Samarghandian et al., 2013; Wang et al., 2014; Ohba et al., 2016; Kabiri et al., 2017). Safranal supports heart health and neurological function, while picrocrocin aids in Alzheimer's treatment.

References

Abdel-Rahman, R.F., El Awdan, S.A., Hegazy, R.R., Mansour, D.F., Ogaly, H.A et al. (2020). Neuroprotective effect of Crocus sativus against cerebral ischemia in rats. *Metab. Brain Dis.*, *35*(3): 427–439.

Abdullaev, F. (2003). Crocus sativus against cancer. *Archives of Medical Research*, *4*(34): 354.

Abdullaev, F.I. and Espinosa-Aguirre, J.J. (2004). Biomedical properties of saffron and its potential use in cancer therapy and chemoprevention trials. *Cancer Detection and Prevention*, *28*(6): 426–432.

Abe, K. and Saito, H. (2000). Effects of saffron extract and its constituent crocin on learning behavior and long-term potentiation. *Phytotherapy Research*, *14*(3): 149–152.

Abu-Izneid, T., Rauf, A., Khalil, A.A., Olatunde, A., Khalid, A. et al. (2022). Nutritional and health beneficial properties of saffron (Crocus sativus L): A comprehensive review. *Critical Reviews in Food Science and Nutrition*, *62*(10): 2683–2706.

Aghayev, Y.M., Fernandez, J.A. and Zarif, E. (2009). Clonal section of saffron (Crocus sativus L.): the first optimistic experimental results. *Euphytica*, *169*: 81–99.

Ahmad, A.S., Ansari, M.A., Ahmad, M., Saleem, S., Yousuf, S. et al. (2005). Neuroprotection by crocetin in a hemi-parkinsonian rat model. *Pharmacol. Biochem. Behav.*, *81*(4): 805–813.

Akhondzadeh, S., Sabet, M.S., Harirchian, M.H., Togha, M., Cheraghmakani, H. et al. (2010). Saffron in the treatment of patients with mild to moderate Alzheimer's disease: a 16-week, randomized and placebo-controlled trial. *J. Clin. Pharm. Ther.*, *35*(5): 581–588.

Al-Snafi, A.E. (2016). The pharmacology of Crocus sativus—A review. *IOSR Journal of Pharmacy*, *6*: 8–38.

Anaeigoudari, F., Anaeigoudari, A. and Kheirkhah-Vakilabad, A. (2023). A review of therapeutic impacts of saffron (Crocus sativus L.) and its constituents. *Physiological reports*, *11*(15): e15785. https://doi.org/10.14814/phy2.15785

Ashktorab, H., Soleimani, A., Singh, G., Amin, A., Tabtabaei, S. et al. (2019). Saffron: the golden spice with therapeutic properties on digestive diseases. *Nutrients*. *11*(943): 1–16.

Aung, H.H., Wang, C.Z., Ni, M., Fishbein, A., Mehendale, S.R. et al. (2007). Crocin from Crocus sativus possesses significant anti-proliferation effects on human colorectal cancer cells. *Experimental Oncology*, *29*(3): 175–180.

Azami, S., Shahriari, Z., Asgharzade, S., Farkhondeh, T., Sadeghi, M. et al. (2021). Therapeutic potential of saffron (Crocus sativus L.) in ischemia stroke. *Evidence-Based Complementary and Alternative Medicine*, 1–8.

Bakshi, R.A., Sodhi, N.S., Wani, I.A., Khan, Z.S., Dhillon, B. and Gani, A. (2022). Bioactive constituents of saffron plant: Extraction, encapsulation and their food and pharmaceutical applications. *Applied Food Research*, *2*(1): 100076.

Bhandari, P.R. (2015). Crocus sativus L.(saffron) for cancer chemoprevention: A mini review. *Journal of traditional and complementary medicine*, *5*(2): 81–87.

Cardone, L., Castonuovo, D., Perniola, M., Cicco, N., Candido, V. (2019). Evaluation of corm origin and climatic conditions on saffron (Crocus sativus L.) yield and quality. *J. Sci. Food Agric. 99*: 5858–5869.

Cardone, L., Castronuovo, D., Perniola, M., Cicco, N. and Candido, V. et al. (2020). Saffron (Crocus sativus L.), the king of spices: An overview. *Scientia Horticulturae*, *272*: 109560.

Dai, R.C., Nabil, W.N.N. and Xu, H.X. (2021). The history of saffron in China: From its origin to applications. Chinese Medicine and Culture, *4*(4): 228–234.

Fiore, A., Pizzichini, D., Diretto, G., Scossa, F., Spano, L. et al. (2010). Genomic transcriptomic of saffron, new tools to unravel the secret of an attractive spice. In: *Husaini, A.M. (Ed.), Saffron. Global Science Books*, UK, pp. 25–30.

Gresta, F., Lombardo, G.M., Siracusa, L. and Ruberto, G. (2008). Effect of mother corm dimension and sowing time on stigmas yield, daughter corms and qualitative aspects of saffron (Crocus sativus L.) in a Mediterranean environment. *J. Sci. Food Agric. 88*: 1144–1150.

Hajyzadeh, M., Olmez, F. and Khawar, K.M. (2020). Molecular approaches to determine phylogeny in saffron. In *Saffron* (pp. 57–68). Elsevier.

Halvorson, S. (2008). Saffron cultivation and culture in Central Spain. *FOCUS on Geography*, *51*(1): 17–24.

Hatziagapiou, K., Kakouri, E., Lambrou, G.I., Bethanis, K., Tarantilis, P.A. et al. (2019). Antioxidant Properties of Crocus sativus L. and its constituents and relevance to neurodegenerative diseases; focus on Alzheimer's and Parkinson's disease. *Curr. Neuropharmacol. 17*(4): 377–402.

Hosseinzadeh, H. and Younesi, H.M. (2002). Antinociceptive and anti-inflammatory effects of Crocus sativus L. stigma and petal extracts in mice. *BMC Pharmacol. 2*(1): 2–7.

https://fdc.nal.usda.gov/food-details/170934/nutrients. Database accessed on 11 March 2025.

Jalali-Heravi, M., Parastar, H. and Ebrahimi-Najafabadi, H. (2009). Characterization of vola tile components of Iranian saffron using factorial-based response surface modeling of ultra sonic extraction combined with gas chromatography-mass spectrometry analysis. *Journal of Chromatography 1216*(33): 6088–6097.

Jan, S., Wani, A.A., Kamili, A.N. and Kashtwari, M. (2014). Distribution, chemical com position and medicinal importance of saffron (Crocus sativus L.). *African Journal of Plant Science 8*(12): 537–545.

Joukar, S., Ghasemipour-Afshar, E., Sheibani, M., Naghsh, N. and Bashiri, A. et al. (2013). Protective effects of saffron (Crocus sativus) against lethal ventricular arrhythmias induced by heart reperfusion in the rat: A potential anti-arrhythmic agent. *Pharmaceutical Biology. 51*(7): 836–843.

Kabiri, M., Rezadoost, H. and Ghassempour, A. (2017). A comparative quality study of saffron constituents through HPLC and HPTLC methods followed by isolation of crocins and picrocrocin. *LWT-Food Science and Technology. 84*: 1–9.

Kafi, M., Koocheki, A., Rashed, M.H. and Nassiri, M. (2006). *Saffron (Crocus Sativus) Production and Processing*. Science Publishers, United States of America, pp. 1–221.

Karasawa, K. (1932). Karyological studies of some flowering Crocus.—*Bot. Mag. Tokyo 46*: 800–802.

Karasawa, K. (1933). On the triploidy of Crocus sativus L. and its high sterility.—*Jap. J. Genet*. 9: 6–8.

Karasawa, K. (1942). Karyological studies in Crocus III.—*Jap. J. Bot. 12*: 475

Kell, G., Rao, A., Beccaria, G., Clayton, P., Inarejos-García, A.M. et al. (2017). Saffron a novel saffron extract (Crocus sativus L.) improves mood in healthy adults over 4 weeks in a double-blind, parallel, randomized, placebo-controlled clinical trial. *Complementary Therapies in Medicine. 33*: 58–64.

Khazdair, M.R., Boskabady, M.H., Hosseini, M., Rezaee, R. and Tsatsakis, A.M. et al. (2015). The effects of Crocus sativus (saffron) and its constituents on nervous system: A review. *Avicenna Journal of Phytomedicine. 5*(5): 376–391.

Kianbakht, S. and Hajiaghaee, R. (2011). Anti-hyperglycemic effects of saffron and its active constituents, crocin and safranal, in alloxan-induced diabetic rats. *Journal of Medicinal Plants. 10*(39): 82–89.

Kothari, D., Thakur, R. and Kumar, R. (2021). Saffron (Crocus sativus L.): Gold of the spices— a comprehensive review. *Horticulture, Environment, and Biotechnology. 62*(5): 661–677.

Kumar, R., Singh, V., Devi, K., Sharma, M., Singh, M.K. et al. (2008). State of Art of Saffron (Crocus sativus L.) Agronomy: A Comprehensive Review. *Food Reviews International. 25*(1): 44–85. https://doi.org/10.1080/87559120802458503

Kumar, R., Singh, V., Devi, K., Sharma, M., Singh, M.K. et al. (2009). State of art of saffron (Crocus sativus L.). Agronomy: a comprehensive review. *Food Rev. Int. 25*: 44–85.

Lambrianidou, A., Koutsougianni, F., Papapostolou, I. and Dimas, K. (2021). Recent Advances on the Anticancer Properties of Saffron (Crocus sativus L.) and Its Major Constituents. *Molecules.* 26:86. https://dx.doi.org/10.3390/ molecules26010086

Lechtenberg, M., Schepmann, D., Niehues, M., Hellenbrand, N., Wünsch, B. et al. (2008). Quality and functionality of saffron: Quality control, species assortment and affinity of extract and isolated saffron compounds to NMDA and σ1 (sigma-1) receptors. *Planta Medica. 74*(07): 764–772.

Leone, S., Recinella, L., Chiavaroli, A., Orlando, G., Ferrante, C. et al. (2018). Phytotherapic use of the Crocus sativus L. (Saffron) and its potential applications: a brief overview. *Phytother. Res. 32*(12): 2364–2375.

Liakopoulou-Kyriakides, M. and Kyriakidis, D.A. (2002). Croscus sativus-biological active constitutents. *Studies in Natural Products Chemistry. 26*: 293–312.

Lopez-Corcoles, H., BrasaRamos, A., Montero Garcia, F., Romero Valverde, M., Montero Riqueime, F. et al. (2015). Phenological growth stages of saffron plants (Crocus sativus L.) according to the BBCH scale. *Span. J. Agric. Res. 13*(3): 1–6.

Ma, Y., Pan, Y. and Mao, B. (2024). Regenerating Saffron (Crocus sativus L.) from Corm Lateral Buds via Indirect Somatic Embryogenesis. *Plants 13*: 10. https://doi.org/ 10.3390/plants13010010

Magesh, V., Singh, J.P.V., Selvendiran, K., Ekambaram, G. and Sakthisekaran, D. et al. (2006). Antitumour activity of crocetin in accordance to tumor incidence, antioxidant status, drug metabolizing enzymes and histopathological studies. *Molecular and cellular biochemistry. 287*: 127–135.

Maggi, L., Carmona, M., Zalacain, A., Kanakis, C.D., Anastasaki, E. et al. (2010). Changes in saffron volatile profle according to its storage time. *Food Research International Journal. 43*: 1329–1334.

Mathew, B. (1999). Botany, taxonomy, and cytology of Crocus sativus L. and its allies. In: Negbi, M. (Ed.), *Saffron: Crocus sativus L.* Harwood Academic Publishers, Amsterdam, pp. 19–30.

Melnyk, J.P., Wang, S. and Marcone, M.F. (2010). Chemical and biological properties of the world's most expensive spice: Saffron. *Food Research International Journal. 43*: 1981–1989.

Menia, M. Sadaf, I., Zahida, R., Tahir, S., Kanth, R.H. et al. (2018). Production technology of saffron for enhancing productivity. *J. Pharmacognosy Phytochem. 7*(1): 1033–1039

Mohtashami, L., Amiri, M.S., Ramezani, M., Emami, S.A. and Simal-Gandara, J. et al. (2021). The genus crocus l.: A review of ethnobotanical uses, phytochemistry and pharmacology. *Indian Crops Production. 171*: 113923.

Molina, R.V., Valero, M., Navarro, Y., Guardiola, J.L., Garcia-Luis, A. et al. (2005). Temperature effects on flower formation in saffron (Crocus sativus L.). *Sci. Hort. 103*: 361–379.

Moradzadeh, M., Reza Kalani, M., Avan, A. (2018). The antileukaemic effects of saffron (Crocus sativus L.) and its related molecular targets: a mini review. *J. Cell. Biochem. 120*(2): 1–7.

Mykhailenko, O., Desenko, V., Ivanauskas, L. and Georgiyants, V. (2020). Standard operating procedure of Ukrainian saffron cultivation according to good agriculture and collection prac tices to assure quality and traceability. *India Crops Production. 151*: 112376.

Mzabri, I., Addi, M. and Berrichi, A. (2019). Traditional and modern uses of saffron (Crocus Sativus). *Cosmetics. 6*(4): 63.

Namayandeh, A., Nemati, Z., Kamelmanesh, M.M., Mokhtari, M., Mardi, M. et al. (2013). Genetic relationships among species of Iranian crocus (Crocus spp.). *Crop Breeding. J. 3*: 61–67.

Negbi, M. (1999). Saffron cultivation, past, present and future prospects. In: Negbi, M. (Ed.), Saffron: Crocus sativus L. Harwood Academic Publishers, Amsterdam, pp. 1–18.

Negbi, M. and Negbi, O. (2002). Saffron domestication in bronze age crete. World island in prehistory. In: Waldern, W.H., Ensayat, J.A. (Eds.), World Island in Prehistory. International Insular Investigations. British Archeological Reports. *Archeopress, International series* 1095.

Nemati, Z., Harpke, D., Gemicioglu, A., Kerndorff, H. and Blattner, F.R. et al. (2019). Saffron (Crocus sativus) is an autotriploid that evolved in Attica (Greece) from wild Crocus cart wrightianus. *Molecular Phylogenetics and Evolution. 136*: 14–20.

Ohba, T., Ishisaka, M., Tsujii, S., Tsuruma, K., Shimazawa, M. et al. (2016). Crocetin protects ultraviolet A-induced oxidative stress and cell death in skin in vitro and in vivo. *European Journal Pharmacology. 789*: 244–253.

Pandita, D. (2020). Saffron (Crocus sativus L.): It's Phytochemistry, Therapeutic Significance and Omics-based Biology. In: T. Aftab, K. R. Hakeem (eds.), Medicinal and Aromatic Plants: Expanding their Horizons through Omics. Elsevier Inc. http://dx.doi.org/10.1016/B978-0-12-819590-1.00014-8

"Real saffron can cost you over $10,000 per kilo. Here's what makes it so expensive" Insider. 2020. https://www.businessinsider.com/why-real-saffron-is-so-expensive-2020-6 Retrieved 28 January 2025.

Rahbani, M., Mohajeri, D., Rezaie, A. and Nazeri, M. (2012). Protective effect of ethanolic extract of saffron (dried stigmas of Crocus sativus L.) on hepatic tissue injury in streptozotocin-induced diabetic Rats. *J. Anim. Vet. Adv. 11*: 1985–1994.

Rangarajan, P., Subramaniam, D., Paul, S., Kwatra, D., Palaniyandi, K. et al. (2015). Crocetinic acid inhibits hedgehog signaling to inhibit pancreatic cancer stem cells. *Oncotarget. 6*(29): 27661–27673.

Rashed-Mohassel, M.H. (2006). Saffron botany. In: Kafi, M., Koocheki, A., Rashed-Mohassel, M.H., Nassiri, M. (Eds.), *Saffron, Production and Processing.* Science Publishers, New Hampshire, USA, pp. 13–38.

Rechinger, K. (1975). Flora Iranica Iridaceae, vol. 112. Academische Druck. U. Verganstalt, Granz.

Renau-Morata, B., Nebauer, S.G., Sánchez, M. and Molina, R.V. (2012). Effect of corm size, water stress and cultivation conditions on photosynthesis and biomass partitioning during the vegetative growth of saffron (Crocus sativus L.). *Ind. Crop Prod. 39*: 40–46

Roshanravan, N. and Ghaffari, S. (2022). The therapeutic potential of Crocus sativus Linn.: A comprehensive narrative review of clin ical trials. *Phytotherapy Research. 36*: 98–111.

Rubio-Moraga, A., Gómez-Gómez, L., Trapero, A., Castro-Díaz, N., Ahrazem, O. et al. (2013). Saffron corm as a natural source of fungicides: the role of saponins in the underground. *Ind. Crop. Prod. 49*: 915–921.

"Saffron (Crocus sativus L.)" Gernot Katzer's Spice Pages. 2010. http://gernot-katzers-spice-pages.com/engl/Croc_sat.html. Retrieved 27 January 2025.

Samarghandian, S. and Borji, A. (2014). Anticarcinogenic effect of saffron (Crocus sativus L.) and its ingredients. *Pharmacognosy Research. 6*: 99–107.

Samarghandian, S., Borji, A., Farahmand, S.K., Afshari, R. and Davoodi, S. et al. (2013). Crocus sativus L. (safron) stigma aqueous extract induces apoptosis in alveolar human lung cancer cells through caspase-dependent pathways activation (pp. 1–12). *Biomed Research International.*

Sanaie, S., Nikanfar, S., Kalekhane, Z.Y., Azizi-Zeinalhajlou, A., Sadigh-Eteghad, S. et al. (2023). Saffron as a promising therapy for diabetes and Alzheimer's disease: Mechanistic insights. *Metabolic Brain Disease. 38*(1): 137–162.

Schmidt, T., Heitkam, T., Liedtke, S., Schubert, V. and Menzel, G. et al. (2019). Adding color to a century-old enigma: multi-color chromosome identification unravels the autotriploid nature of saffron (Crocus sativus) as a hybrid of wild Crocus cartwrightianus cytotypes. *The New Phytologist. 222*(4): 1965–1980. https://doi.org/10.1111/nph.15715

Sepahi, S., Mohajeri, S.A., Hosseini, S.M., Khodaverdi, E., Shoeibi, N. et al. (2018). Effects of crocin on diabetic maculopathy: A placebo-controlled randomized clinical trial. *American Journal of Ophthalmology. 190*: 89–98.

Serrano-Díaz, J., Sánchez, A.M., Martínez-Tomé, M., Winterhalter, P. and Alonso, G.L. et al. (2013). A contribution to nutritional studies on Crocus sativus flowers and their value as food. *Journal of Food Composition and Analysis. 31*(1): 101–108.

Shahi, T., Assadpour, E. and Jafari, S.M. (2016). Main chemical compounds and pharmacological activities of stigmas and tepals of 'red gold'; saffron. *Trends Food Science Technology. 58*: 69–78.

Shokrpour, M. (2019). Saffron (Crocus sativus L.). Breeding: opportunities and challenges. In: Al-Khayri, J., Jain, S., Johnson, D. (Eds.), Advances in Plant Breeding Strategies: Industrial and Food Crops. *Springer, Cham*, pp. 675–706.

Tahiri, A., Mazri, M.A., Karra, Y., Ait Aabd, N., Bouharroud, R. et al. (2023). Propagation of saffron (Crocus sativus L.) through tissue culture: A review. *The Journal of Horticultural Science and Biotechnology. 98*(1): 10–30.

Tanra, M.A., Dar, B.A. and Sing, S. (2017). Economic analysis of Production and Marketing of saffron in Jammu and Kashmir. *J. Social Relevance Concern. 5*(10): 12–19

The observatory of economic complexity, https://oec.world/en/ (accessed 06 March 2023).

Velasco-Negueruela, A. (2001). Saffron. In *Handbook of herbs and spices*, edited by K.V. Peter, CRC Press, pp 276–285.

Wallis, T.E. (2005). *Textbook of pharmacognosy*. CBS.

Wang, Y., Sun, J., Liu, C. and Fang, C. (2014). Protective effects of crocetin pretreatment on myocardial injury in an ischemia/reperfusion rat model. *European Jornal of Pharmacology. 741*: 290–296.

Wendelbo, P. (1977). Tulips and Irises of Iran and Their Relatives. Botanical Institute of Iran, *Botanical Garden*, Tehran.

Winterhalter, P. and Straubinger, M. (2000). Saffron—renewed interest in an ancient spice. *Food Reviews International. 16*(1): 39–59.

Yildirim, M.U., Asil, H., Hajyzadeh, M., Sarihan, E.O. and Khawar, K.M. et al. (2017). Effect of changes in planting depths of saffron (Crocus sativus L.) corms and determining their agronomic characteristics under warm and temperate (Csa) climatic conditions of Turkish province of Hatay. *Acta Horticulturae. 1184*: 47–53.

Zeinali, M., Zirak, M.R., Rezaee, S.A., Karimi, G. and Hosseinzadeh, H. et al. (2019). Immunoregulatory and anti-inflammatory properties of Crocus sativus (saffron) and its main active constituents: A review. *Iranian Journal of Basic Medical Sciences. 22*(4): 334–344.

Zhang, A., Shen, Y., Cen, M., Hong, X., Shao, Q. et al. (2019). Polysaccharide and crocin contents, and antioxidant activity of saffron from different origins. *Indian Crops Production. 133*: 111–117.

Zheng, Y.Q., Liu, J.X., Wang, J.N. and Xu, L. (2007). Effects of crocin on reperfusion-induced oxidative/nitrative injury to cerebral microvessels after global cerebral ischemia. *Brain Research. 1138*: 86–94.

2 Cultivation and Ecology of Saffron

Iqra Siddique[1] and Alvina Gul[2*]

Saffron (*Crocus sativus* L.) is a perennial herb from the Iridaceae family, cultivated globally for its prized stigmas, which serve as a valuable spice known for its medicinal, aromatic, and coloring properties. It thrives in dry climates with mild winters and hot summers, primarily in the Mediterranean, Iran, and India, with Iran leading global production. Saffron farming depends on ecological factors such as temperature, light, water, and soil, requiring precise management for optimal yield. Key cultivation practices include selecting large corms, applying cow manure at time at planting, ensuring proper irrigation, intercropping, timely flower harvesting, and effective post-harvest processing. This chapter explores saffron's ecology, geographical distribution, and cultivation requirements.

1. Introduction

Saffron (*Crocus sativus* L.) is a sterile, autumn-blooming geophyte from the Iridaceae family, cultivated for its prized crimson stigmas used as a spice. It reproduces vegetatively via corms, with each corm producing one to four purple flowers. The flower contains three golden-yellow stamens and a single red pistil with three stigmas, which are dried to obtain saffron (Dar et al., 2017). Thriving in loose, well-drained clayey calcareous soils, saffron cultivation depends on climate, soil conditions, planting depth, corm size, crop density, nutrient management, and post-harvest processing (Kumar et al., 2008). It is the world's most expensive spice, known as "red gold" or "golden spice," with a commercial price of $40–60 per gram (Amanpour et al., 2019; Garavand et al., 2019). Iran leads global production, supplying 80% of the world's saffron, with Khorasan province accounting for 92% of this output (Menia et al., 2018). Saffron is cultivated in Iran, China, Spain, India, Azerbaijan, Italy, Afghanistan, UAE, Israel, Turkey, France, Egypt, Switzerland, Greece, Japan, Iraq, and Australia, with an estimated annual production of 418 Mg (Cardone et al., 2020). It holds significant value in international markets for its culinary, dyeing, religious, and medicinal applications, exhibiting aphrodisiac, antimicrobial, antifungal, antibacterial, anti-inflammatory, and antiseptic properties (Terrado, 2007; Bukhari et al., 2018; Mzabri et al., 2019; Siddique et al., 2020).

[1,2] Department of Plant Biotechnology, Atta-Ur-Rahman School of Applied Biosciences, National University of Sciences and Technology, Islamabad, Pakistan.

* Corresponding author: alvina_gul@asab.nust.edu.pk

DOI: 10.1201/9781003403760-2

Saffron's aroma,taste and color stem from three key compounds: picrocrocin (bitterness), crocin (coloring), and safranal (fragrance) (Cagliani et al., 2015). Its growth and production vary across regions due to topographic and environmental differences, which play a crucial role in cultivation (Amirnia et al., 2014; Kamyabi et al., 2014). Additional factors such as planting material quality, innovative cultivation techniques, and corm production methods significantly impact saffron yield (Gupta, 2007). Economically, saffron holds high market value and requires intensive labor—harvesting 150,000–200,000 flowers and 400 hours of work to produce 1 kg (Mzabri et al., 2019). While global saffron production has declined over the past 30–40 years, Iran remains an exception, showing growth (Khan et al., 2011). Due to its high cost, saffron is prone to adulteration, complicating trade (Babaei et al., 2014). Advancing cultivation techniques globally is essential to improve yield and expand production areas. This chapter explores saffron's cultivation requirements and ecological factors.

2. Origin and History of Saffron Cultivation

The name "saffron" likely derives from the Persian "sahafaran" (Kafi et al., 2006), while some link it to the Arabic "Zafaran," meaning yellow (Winterhalter and Straubinger, 2000). Its cultivation origin remains debated: Vavilov proposed the Middle East, others suggest Asia Minor or Greek islands (Tammaro, 1990). Saffron was cultivated in the Middle East and Egypt around 3,500 years ago, with claims of Anatolia as its birthplace. Turks carried it from Middle Asia to Anatolia, where it thrived in Iran's Horasan region and India. Its presence in Crete during the Late Bronze Age (1500–1100 BC) supports a Mediterranean origin theory (Negbi, 1999; Gresta et al., 2009; Coşkun et al., 2017). The earliest fabric-dye evidence dates to Castilla-Mancha, Spain (~2400 BC), with culinary use documented in Hammurabi's reign (Dar et al., 2017). Genetic studies trace saffron to *Crocus cartwrightianus* in Attica, Greece (Nemati et al., 2019). Medicinal use in Sicily dates to the Graeco-Roman period (8th–3rd century BC) (Manganaro, 2001). Some scholars cite Iran's Alvand and Zagros mountains as its origin, linking cultivated saffron to wild species "Kerkomise" or "Gouishi" (Kafi et al., 2006). Its use in Persia dates to the Achaemenid dynasty, and Persians later introduced it to Kashmir (Manganaro, 2001). By the Medieval Age, saffron spread across Europe, reaching Britain in the 14th century (Fernández, 2004). Turkey's Safranbolu hosts an annual saffron festival,while Spain's Consuegra celebrates the "Saffron Rose Festival" each October (Halvorson, 2008).

3. Saffron Types Obtained from Different Countries

Saffron quality varies due to different cultivation and drying methods across regions. Various cultivars produce distinct strands unique to specific locations. Boutique saffron types are available in the U.S., France, New Zealand, Switzerland, and England, including Pennsylvania Dutch saffron, known for its earthy flavor

(Willard, 2002). Spain produces saffron marketed as "Creme" and "Spanish Superior", which have milder color, taste, and aroma, evaluated under government standards (Carmona et al., 2006; Lopez-Corcoles et al., 2015). Italian saffron is more potent, with Krokos (Greece) enjoying PDO protection for its intense flavor and hue. Zafferano dell'Aquila, cultivated in Navelli Valley, is rich in crocin and safranal, giving it a strong fragrance (Abruzzo, Italy). Iranian saffron is among the most intense globally. The highly sought-after "Mongra" (Lacha) from Kashmir, recognized as one of the darkest varieties, is distinguished by its deep maroon-purple color, indicating strong taste, aroma, and potency (Vlahova, 2022).

4. Saffron Ecology and Cultivation Demands

4.1 Habitat and Geographic Distribution

Saffron is cultivated across diverse geographic regions, spanning 10°W-80°E longitude and 30°-50° N latitude (Yildirim et al., 2016). It thrives at altitudes from sea level to 2,000 m, with optimal growth in 600–1,700 m mountainous valleys and plateaus (Kumar et al., 2008). Its resilience allows cultivation in drought-prone summer regions (Yasmin & Nehvi, 2013). Major saffron-producing countries include Iran, Pakistan, Turkey, India, Greece, Spain, Morocco, Italy, Switzerland, France, China, Australia and Japan.In Iran, key cultivation areas include Sahr-Kord (2,066 m) and Southern Khorasan (1,000 m) (Ehsanzadeh et al., 2003). Kashmir's Pampore tehsil (34° 1' N, 74° 56' E) is a primary saffron hub, with villages like Zewan, Chandhara, Lethpora, and Khrew leading production (Nehvi & Salwee, 2011).

Azerbaijan, Greece, Italy, Turkey, France and Morocco collectively contribute 2% of global saffron output. Krokos Kozani is Greece's main saffron region (Mitsopoulou & Tsimidou, 2004). Baku region cultivates saffron despite low precipitation (223 mm) (Milyaeva, 1999). Italy's saffron thrives in Cerdena, Emilia-Romagna, San Gimignano and, Aquil (650–1,100 m) (Galigani & Pegna, 1999). Turkey's saffron is grown in Hivan, Hapier, Saffronbolu and Urfa Viran. Morocco's Taliouine (1,200–1,400 m) near the Atlas mountains is a key cultivation site (Ait-Oubahou and El-Otmani 1999, Yasmin and Nehvi 2013). Spain's saffron grows at 610 m, while Italy's cultivation ranges 650–1,100 m (Dhar, 1990; Kumar et al., 2008).

4.2 Soil Requirements for Saffron Cultivation

Saffron crocus thrives in fertile, well-drained soils, though it is often cultivated in dry, nutrient-poor conditions (Gresta et al., 2008). Sandy and loamy soils are ideal, while clay calcareous soil aids root penetration and development (Menia et al., 2018). Increasing organic matter significantly boosts yield, with 20–30 Mg/ha of farmyard manure recommended (Yarami & Sepaskhah, 2015; Golmohammadi, 2014).

Optimal soil conditions include:

- pH range: 6.2–8.3 (average 7.5)
- Electrical conductivity: 0.09–0.30 dS/m (average 0.17 dS/m) (Nehvi, 2009)
- Saline tolerance: Saffron grows well in saline soils, but calcium carbonate deficiency can hinder growth (Dar et al., 2017).
- Clay preference: Clay is considered ideal (Fernández, 2004).
- Loam and sandy-loam suitability: Neutral to sub-alkaline pH, low electrical conductivity, organic matter (5.46–8.67 g/kg), and active lime (21.25–26.25 g/kg) support saffron growth, stigma production, and daughter corm development (Cardone et al., 2020).

4.3 Climatic Conditions

Saffron cultivation is highly influenced by temperature and precipitation, with temperature being a key factor in flowering regulation (Kamyabi et al., 2014; Bazoobandi et al., 2020). It thrives in temperate semi-arid or arid regions (1,500–2,800 m) with rainy autumns, mild winters and summers yielding the best results (Rahimi et al., 2017). The crop endures dry summers, winter rains, and snowfall down to −10°C (Dar et al., 2017). Saffron tolerates −22 to 40°C, with optimal vegetative growth at 27°C and flowering at 17°C (Rezvani-Moghaddam et al., 2020; Koocheki et al., 2011). Early growth benefits from 23–25°C, while below 16°C favors daughter corm production (Zahmati et al., 2018). In Kashmir, anthesis occurs when air temperature is ≤ 20°C, with 474 mm precipitation during vegetative growth (Yasmin & Nehvi, 2018). 1,100 chilling hours are crucial for vernalization. Flower size is affected by temperature shifts, with peak flowering in October–November at 15–20°C (day) and 6–8°C (night). Autumn rains boost flower yield, while spring rains aid corm multiplication (Menia et al., 2018). Temperature rise during flowering has minimal impact, slightly reducing sprouts per corm (Molina et al., 2010). Cold winters and hot summers are ideal, with saffron surviving extreme frost (−100 to −180°C) (Seyyedi et al., 2018). Studies confirm 23–27°C as optimal for growth (Jami et al., 2020), while 12–28°C suits cultivation in New England (Gheshm & Brown, 2021). 15–20°C is best for flower production (Salas et al., 2020). Climate conditions vary by region: Italy sees 700 mm annual precipitation (Kumar et al., 2008), while Greece's corms grow in March–April, flowering in September under slightly different environmental conditions (Golmohammadi, 2014).

4.4 Preparation of Land for Corms Plantation

Saffron crocus requires well-prepared soil for optimal growth. The land should be ploughed 3–5 times between May and July to achieve a loose, crumbly texture (30 cm depth) for almost 3 to 5 times.10 Mg/ha of decomposed farmyard manure supports corm multiplication (Menia et al., 2018). A top dressing of 10 Mg/ha manure, 60 kg/ha K_2O, 60 kg/ha P_2O_5, and 22.5 kg/ha nitrogen is recommended every 12–15 days. subsequent years, hoeing, leveling, and ploughing ensure proper nutrient distribution (Tanra et al., 2017). Before planting, soil moisture should be checked, and irrigation applied if necessary. Saffron reproduces vegetatively via underground corms, which persist for years. Hand-planted corms develop aerial shoots that generate daughter corms (Tanra et al., 2017). The corm requirement per hectare depends on crop duration, corm size, and spacing, typically 500,000 corms or 2.5–3 metric tons with a mean diameter of 2.5 cm (Kumar et al., 2008).

Planting Time and Density

Saffron corms are best planted from late August to mid-September, though June–July is preferred in Iran and August in Italy (Menia et al., 2018; Bayat et al., 2016; Gresta et al., 2016). Spring and autumn rains enhance corm multiplication and flower yield. Delayed planting reduces growth and flowering, with peak production observed in June and October (Koocheki et al., 2016).Planting density significantly affects stigma yield. High-density planting produces more flowers and larger stigmas, while low-density corms yield heavier daughter corms with higher phosphorus and nitrogen levels (Cardone et al., 2020; Seyyedi et al., 2018). Yield of saffron is effected more by flowers number than weight of stigma (Andabjadid et al., 2015). Studies show 300–200 corms/m^2 optimize flower production, with 4–6 g mother corms yielding the most flowers (Koocheki et al., 2014). A 10–20 cm planting pattern results in 9.2 g average corm weight, 1.5 cm diameter, 170 kg/ha flower yield, and 12 kg/ha dry stigma output (Mohammad Abadi et al., 2011). In Kashmir, farmers use rectangular strips (2–3 m long, 1.5–2 m wide) with 20 cm deep, 30 cm wide drainage lines, planting corms 12–15 cm deep at 10 × 20 cm spacing (Husaini et al., 2010). Spain follows a two-row system in 20 cm deep ditches, with 12–15 cm row spacing and 30–35 cm soil coverage (Kafi et al., 2018). Conventional methods involve hill planting with 25 cm spacing, sometimes randomly sowing up to 15 corms per hill. Flatbed seeding is preferred over groove planting (Kafi & Showket, 2007).

Irrigation of Saffron

Saffron requires minimal irrigation due to its five-month dormancy, during which water is unnecessary. Semi-arid regions often provide sufficient rainfall (Kafi et al., 2018). Water availability impacts saffron yield based on climate, soil type, and management. Applying 50% of water requirements had no effect in the first year but reduced flower number and dry stigma yield in the second year compared to 75% or 100% irrigation (Koocheki et al., 2014). Mid-June irrigation reduced yield by 17%, while late August irrigation increased flower yield by 20%. Summer irrigation raises fungal disease risks. Supplemental basin irrigation is recommended, with 100 mm irrigation before flowering in dry autumns and 50 mm post-anthesis irrigation in areas with 600 mm seasonal precipitation (Mosaferi, 2001). Regions with 400–200 mm rainfall require irrigation every 24–15 days, applying 50–75%

potential crop evapotranspiration (Sepaskhah & Kamgar, 2009). Short irrigation intervals (~3000 m^3/ha) are optimal for saffron cultivation (Shajari et al., 2020).

Saffron Plant Cycle

Saffron cultivation follows a two-stage yearly cycle: active (Aug–Sep to Apr–May) and dormant. During the active phase, the plant resumes leaf, root, flower, and shoot growth, while corms remain unchanged in dormancy. The transitional period allows mitotic division and differentiation, gradually decreasing over time. After two years, flowering peaks, and by the sixth year, corms are uprooted for future planting. Saffron corms can sustain cycles for 10–15 years, producing new corms annually while older ones decompose (Koocheki & Rezvani Moghaddam, 2008; Ghorbani & Koocheki, 2017). Cultivation varies by region: Italy practices an annual cycle, while Spain (3–4 years), India & Greece (6–8 years), and Morocco (up to 12 years) follow perennial cycles. As fields age, spice yield declines due to competition for water, nutrients, and fungal issues (Gresta et al., 2009).

4.5 Fertilization Requirements

Saffron has low nutrient requirements, but excessive nitrogen boosts vegetative growth while reducing yield (Kafi et al., 2018). Organic manure (20–30 Mg/ha) is commonly used (Koocheki, 2004). Combining 20–30 Mg/ha farmyard manure with 23 kg/ha chemical nitrogen improves soil fertility. Applying 20 Mg/ha cow manure +50 kg/ha urea resulted in 0.45 g/m^2 yield, 0.89 g fresh flower weight, and 29 mm stigma length (Mohammad et al., 2012). After planting, no fertilization is practiced (Dar et al., 2017). Organic and inorganic fertilizers enhance saffron production (Omidi et al., 2009). Vermicompost (24 Mg/ha) + mycorrhiza (400 kg/ha) improved flower count by 28.08% (Li et al., 2017). Increasing vermicompost (4–8 Mg/ha) raised yield by 10.2% (Alidadi et al., 2013). Iranian farmers apply 10–80 Mg/ha farmyard manure, with 100 kg ammonium phosphate + 100 kg urea/ha added after first weeding (Behnia et al., 1999; Ghorbani & Koocheki, 2017). Urea foliar fertilizer (Jan–Mar) increased flower production (Hosseini et al., 2004).

4.6 Weeds and Pests' Management

Effective weed management is crucial for saffron cultivation, but challenges arise due to long weed persistence and narrow saffron leaves, making herbicide application difficult. Manual weeding is common but labor-intensive, as no registered herbicide exists for saffron (Porselvi, 2021). Common weeds in Iran include mouse barley, hoary cress, yarrow, and wild gold barley (Zare Hosseini et al., 2014), while Jammu and Kashmir fields host Galium tricorne, Lepidium virginicum, Chenopodium album, and others. In India, weed growth is controlled during October–April, but dormancy (May–September) allows weeds to thrive (Husaini et al., 2010). In Italy, manual weeding is used for annual crops, while Atrazine (Gesaprim 50) and Simazine (Gesatop 50) (1.0 kg/ha) manage weeds in perennial saffron fields (Dar et al., 2017). Tribenuron methyl (18.75 g ai/ha) and Ioxynil (750 g ai/ha) applied at the 6–8 leaf stage effectively control weeds. Metribuzin (560 g ai/ha) is used in spring/autumn without harming saffron (Menia et al., 2018). Iranian saffron fields use Sencor (Metribuzin) and Gallant (Haloxy fopetoxy-ethyl) post-harvest, while Buster (2,4-D, 2,4-DP) and Roundup (Glyphosate) are applied during summer dormancy. Indirect weed control methods include wood chips and sawdust mulch (Galigani & Pegna, 1999). Torsion and finger weeders face challenges due to rocky soil and emerging corms (Cirujeda et al., 2014). Pathogens such as Penicillium cyclopium and F. oxysporum f. sp. gladioli infect saffron corms in Italy (Fiori et al., 2011), while Pythium spp., Rhizoctonia spp., Phoma spp., and others cause corm dry rot (Hassan & Devi, 2003). 294 fungal endophytes have been identified in Jammu and Kashmir saffron corms (Wani et al., 2016). Brassicol (1.5–3.0 g ai/m^2) applied via root drenching helps manage rot symptoms (Ghorbani & Koocheki, 2017). However, eco-friendly pest control is preferable. Standardizing cultivation practices reduces pathogen attacks, while grading and sorting corms before sowing improves yield. Crop rotation enhances productivity and quality by preventing weed and pathogen buildup (Dita et al., 2018).

4.7 Crop Rotation

Crop rotation and intercropping enhance soil fertility and pest management in saffron fields. In Kashmir, saffron fields are rotated every 15 years, left fallow or planted with maize, oat, or linseed for 2–3 years to prevent disease buildup and restore nutrients (Nehvi et al., 2008; Kafi et al., 2018). 46% of soil in long-term saffron fields is contaminated by fungi, increasing infection risks (Sameer et al., 2012). In Italy, saffron is rotated with legumes and wheat, though data on its impact is limited. A 3–8 year gap before replanting saffron is recommended to prevent stigma yield reduction and weed proliferation (Tammaro, 1999; Dar et al., 2017). Faba beans before saffron improve stigma and corm production, with 45 corms/m^2 yielding better results than lower densities (Gresta et al., 2016). In Iran, saffron is often intercropped as a natural divider between grapes, almonds, and barberry (Khosravi, 2005).

4.8 Harvesting of Saffron

Saffron harvesting begins with flower plucking, followed by manual stigma separation and air-drying at 35°C (Kafi et al., 2018; Gresta et al., 2008). Flowers should be harvested immediately upon blooming, and stigmas separated before wilting (Molafilabi et al., 2006). In Spain, saffron harvest is celebrated as the "day of mantel", while Italy's flowering occurs mid-October to November (Dar et al., 2017). Flower timing depends on temperature and soil moisture (Molafilabi, 2004). In Khorasan, Iran, harvesting spans October–November, varying by region, climate, and irrigation (Kafi & Showket, 2007). In Kashmir, flowers are picked early morning, then shade-dried for 2–3 days in sunny conditions, longer if cloudy (Husaini et al., 2010). Limited mechanization in saffron harvesting is due to irregular flower emergence, delicate stigmas, and contamination risks (Gresta et al., 2008).

4.9 Post Harvest Processing

Drying

Saffron quality depends on proper stigma drying, as poor execution can compromise its characteristics (Carmona et al., 2005). Drying methods vary by region: sun drying in Iran and India, dark-room drying (30–35°C) in Greece, and toasting over hot ashes in Spain (Carmona et al., 2005; Gregory et al., 2005). Hot air drying (70°C for six minutes) yields high-quality saffron, while Italy uses sun or air drying, and Navelli employs charcoal fire dehydration (Tammaro, 1999). Studies comparing drying methods found solar drying superior to traditional techniques (Feili et al., 2012). Freeze drying (–40°C for four hours) produced higher safranal and crocin levels than sun drying (18°C) (Acar et al., 2011). Various drying methods—freezer, vacuum, microwave, infrared, and oven drying—are used globally. Oven drying yields the highest crocin I, while infrared drying results in the lowest. Freeze drying produces the highest crocin II (Chen et al., 2020). Despite their advantages, freeze and oven drying may be less accessible to small-town farmers.

Preservation and Storage

Saffron must be stored in a dark, controlled environment to preserve its color, aroma, and taste, as it is sensitive to light, temperature, and oxygen. Dark glass pots at low temperatures are ideal for long-term storage (Alonso et al., 1990). Proper corm storage prevents unwanted sprouting (Kumar et al., 2008). In Kashmir, corms are kept in polybags or earthen pots without monitoring water levels (Mir et al., 2008). 25°C storage enhances production, while 30°C storage boosts daughter corm and stigma yield (Hajyzadeh et al., 2016; Siracusa et al., 2009). Extended storage at 8°C reduces yield attributes (Çavuşoğlu, 2010). Saffron with 8–10% moisture content can be stored at 10°C for six months in airtight containers (Menia et al., 2018).

Saffron Production

Saffron yield varies across countries due to biological, environmental, and agronomic factors, including corm size, storage conditions, climate, sowing time, and crop management (Molina et al., 2004; Lombardo et al., 2005; Gresta et al., 2009). Larger corms (>45 g) yield 12 flowers per corm, while 20–30 g corms produce six daughter corms (Dar et al., 2017). Three flowers per plant yield commercially valuable stigmas (Emadi, 2009). In Khorasan, Iran, 78.5 kg fresh flowers (~170,000 flowers) are needed for 1 kg of dry saffron (Husaini et al., 2010).

Typical saffron yield per hectare is 10–15 kg dried stigmas. Annual production includes 418+ tons in Iran (108,000 ha), 1,000 ha in Greece, 150 ha in Spain, 7,557 ha in Afghanistan, 3,674 ha in India, 70 ha in Italy, 37 ha in France, and 850 ha in Morocco (Cardone et al., 2020). Iran leads global production (88.52%), followed by India (5.76%), Azerbaijan (1.37%), Greece (2.03%), Spain (1.22%), Italy (0.06%), and Morocco (1.02%) (Tanra et al., 2017). In 2015, Iran produced 238 MT, India 11.2 MT, Greece 4.3 MT, Spain 5.7 MT, Azerbaijan 3.7 MT, Morocco 1 MT, Afghanistan 3 MT, Turkey 0.4 MT, and Italy 0.24 MT (Ali & Hakim, 2017).

4.10 Mechanization

Saffron cultivation remains mostly manual, with limited mechanization due to the delicate nature of corms and flowers, which are prone to damage by machines. Hoeing machines help manage weeds, while modified potato diggers assist in corm picking. General soil preparation machines can be used, but no specialized equipment exists for flower harvesting and stigma separation (Diaz-Marta et al., 2006).

Efforts to mechanize include:

- Self-moving machines that allow workers to harvest flowers without physical strain.
- "Saffron All in One", aiding corm planting, weed removal, and harvesting for four workers at a time.
- Automated flower-cutting machine concept, where a computer program analyzes flower images to position a cutting system, successfully reducing flower cutting time in prototype tests (Gracia et al., 2009).

Mechanization remains a challenge, but innovations aim to improve efficiency while preserving saffron quality.

4.11 Adulteration

Saffron fraud involves mixing cheap materials like silk fibers, pomegranate, and marigold with genuine saffron to increase profit (Heidarbeigi et al., 2015). Adulterants fall into three categories: look-alike materials, dyed low-quality saffron, and weight-enhancing substitutes (Wallis, 2005). Common adulterants include onion skin, safflower, poppy, turmeric, maize stigmas, and vegetable oil coatings. Cape saffron, though similar in odor and color, is falsely marketed as genuine saffron. European markets have reported chemical adulteration using borates, salts, and artificial colorants, making saffron unsuitable for pharmaceutical use (Dar et al., 2017).

Detection methods include:

- Water test: Genuine saffron dissolves, releasing its distinct color and aroma.
- Sulfuric acid test: A blue-to-purple-to-reddish-brown shift confirms purity (Dar et al., 2017).

- Spectroscopic techniques: ATR-FTIR, LIBS, and Raman spectroscopy detect adulterants (Varliklioz Er et al., 2017).
- 1 H NMR metabolite fingerprinting identifies plant-based adulteration (Petrakis et al., 2015).
- Diffuse reflectance infrared Fourier transform spectroscopy efficiently detects powdered saffron adulteration (Petrakis & Polissiou, 2017)

Saffron Marketing

Marketing routes ensure product availability, but saffron market channels face efficiency issues (Qadri, 2018; Hamid et al., 2017). Key challenges include noncompliance with international standards, packaging concerns, weak branding, market mismatches, bulk exports, and inadequate marketing. Establishing export centers and strategic marketing can improve saffron trade (Mohammadi & Reed, 2020). Effective pricing, product positioning, promotion, and distribution enhance saffron's value. Growers lacking market knowledge, transportation, and storage face financial risks (Ganie & Nusrath, 2016). Marketing channels directly impact farmer profits, making market awareness crucial to avoid losses.

4.12 Challenges and Possible Solutions to Improve Saffron Cultivation

Saffron production faces challenges such as limited cultivation land, labor-intensive harvesting, and manual management, increasing market costs (Gresta et al., 2009). Synthetic pesticides, preservatives, and artificial coloring raise consumer health concerns, making organic certification more relevant. Organic La-Mancha saffron is now available in Spain (Ghorbani & Koocheki, 2017). Farmers have neglected saffron cultivation, often relegating it to poor soil and climate conditions. Lack of innovation hinders progress, requiring revised agronomic practices that integrate traditional farming knowledge with scientific research (Dar et al., 2017). Traditional agroecosystems emphasize long-term sustainability, using local resources, soil protection, and risk reduction (Koocheki & Ghorbani, 2005). Saffron thrives in native regions, making expansion difficult. Abiotic and biotic stresses, including pathogens, weeds, and rodents, reduce yield. Mycorrhizal technology and soil microorganisms like rhizobia, arbuscular mycorrhizal fungi, and plant growth-promoting rhizobacteria enhance resilience (Yin et al., 2010).

4.13 Prospects of Saffron Cultivation

Climate changes have impacted saffron cultivation, increasing global demand and necessitating expanded cultivation areas. Most floral parts beyond the stigma remain unused, presenting an opportunity for future research (Kothari et al., 2021). Hydroponic saffron cultivation is gaining traction, with IoT-integrated systems showing promising results. A smart hydroponic system using nutrient film technique and renewable energy aims to enhance yield and quality cost-effectively (Kour et al., 2022). To improve propagation, biotechnology and breeding techniques such as clonal selection, mutagenesis, molecular markers, and micro-propagation offer alternatives to saffron's low natural propagation rate. Establishing germplasm banks of elite clonal selections aids crop improvement, while mutation techniques have shown promising results in genetic enhancement (Mir et al., 2015).

Conclusion

Saffron crocus, primarily grown in Iran, is a high-value spice used in food, dye, and medicine. Due to labor-intensive cultivation, it remains the costliest spice, yet demand continues to rise. Expanding cultivation areas and exploring new growing regions is crucial to meeting global demand. Beyond stigmas, other flower parts hold potential for various applications, presenting opportunities for further research. With innovative technologies and strategic measures, saffron cultivation can boost exports and economic growth in producing countries.

References

Acar, B., H. Sadikoglu and M. Ozkaymak (2011). "Freeze drying of saffron (Crocus sativus L.)." *Drying Technology 29*(14): 1622–1627.

Agayev, Y.M.-o., J.-A. Fernandez and E. Zarifi (2009). "Clonal selection of saffron (Crocus sativus L.): the first optimistic experimental results." *Euphytica* 169: 81–99.

Ait-Oubahou, A. and M. El-Otmani (1999). Saffron cultivation in Morocco". En: Saffron En: Saffron Crocus sativus L. Medicinal and Aromatic Plants. *Industrial Profiles*. Negbi M. Ed, Harwood Academic Publishers Amesterdam, Holanda.

Ali, A. and I.A. Hakim (2017). "An overview of the production practices and trade mechanism of saffron in Kashmir Valley (India): Issues and challenges." *Pac. Bus. Rev. Int 10*: 97–106.

Alidadi, H., A.R. Saffari and R. Peiravi (2013). "Effects of biofertilizers effects of compost, vermicompost and sulfur compost on yield of saffron." *World Appl Sci* J *21*(9): 1386–1390.

Alonso, G., R. Varon, R. Gomez, F. Navarro and M. Salinas et al. (1990). "Auto-oxidation in Saffron at 40C and 75% Relative Humidity." *Journal of Food Science 55*(2): 595–596.

Amanpour, A., H. Kelebek and S. Selli (2019). GLC/HPLC Methods for Saffron (L.). *Bioactive Molecules in Food, Springer:* 1987–2035.

Amirnia, R., M. Bayat and M. Tajbakhsh (2014). "Effects of nano fertilizer application and maternal corm weight on flowering at some saffron (Crocus sativus L.) ecotypes." *Turkish Journal of Field Crops 19*(2): 158–168.

Andabjadid, S.S., B.P. Eslam, A.R.S. Bakhtavari and H. Mohammadi (2015). "Effects of corm size and plant density on Saffron (Crocus sativus L.) yield and its components." *International Journal of Agronomy and Agricultural Research 6*(3): 20–26.

Babaei, S., M. Talebi and M. Bahar (2014). "Developing an SCAR and ITS reliable multiplex PCR-based assay for safflower adulterant detection in saffron samples." *Food Control 35*(1): 323–328.

Bayat, M., M. Rahimi and M. Ramezani (2016). "Determining the most effective traits to improve saffron (Crocus sativus L.) yield." *Physiology and Molecular Biology of Plants 22*: 153–161.

Bazoobandi, M., H. Rahimi and M.-R. Karimi-Shahri (2020). *Saffron Crop Protection*: 169–185.

Behnia, M., A. Estilai and B. Ehdaie (1999). "Application of fertilizers for increased saffron yield." *Journal of Agronomy and Crop Science 182*(1): 9–15.

Bukhari, S.I., M. Manzoor and M. Dhar (2018). "A comprehensive review of the pharmacological potential of Crocus sativus and its bioactive apocarotenoids." *Biomedicine & Pharmacotherapy 98*: 733–745.

Cagliani, L.R., N. Culeddu, M. Chessa and R. Consonni (2015). "NMR investigations for a quality assessment of Italian PDO saffron (Crocus sativus L.)." *Food Control 50*: 342–348.

Cardone, L., V. Candido, D. Castronuovo, M. Perniola and N. Cicco et al. (2021). "Comparing annual and biennial crop cycle on the growth, yield and quality of saffron using three corm dimensions." *Scientia Horticulturae 288*: 110393.

Cardone, L., D. Castronuovo, M. Perniola, N. Cicco and V. Candido et al. (2020). "Saffron (Crocus sativus L.), the king of spices: An overview." *Scientia Horticulturae 272*: 109560.

Cardone, L., D. Castronuovo, M. Perniola, L. Scrano, N. Cicco et al. (2020). "The influence of soil physical and chemical properties on saffron (Crocus sativus L.) growth, yield and quality." *Agronomy 10*(8): 1154.

Carmona, M., A. Zalacain, J.E. Pardo, E. López, A. Alvarruiz et al. (2005). "Influence of different drying and aging conditions on saffron constituents." *Journal of Agricultural and Food Chemistry 53*(10): 3974–3979.

Carmona, M., A. Zalacain, M.R. Salinas and G.L. Alonso (2006). "Generation of saffron volatiles by thermal carotenoid degradation." *Journal of agricultural and food chemistry 54*(18): 6825–6834.

Çavuşoğlu, A. (2010). "The effects of cold storage of saffron (Crocus sativus L.) corms on morphology, stigma and corm yield." *African Journal of Agricultural Research 5*(14): 1812–1820.

Chen, D., B. Xing, H. Yi, Y. Li, B. Zheng et al. (2020). "Effects of different drying methods on appearance, microstructure, bioactive compounds and aroma compounds of saffron (Crocus sativus L.)." *LWT 120*: 108913.

Cirujeda, A., A.I. Marí, J. Aibar, S. Fenández-Cavada, G. Pardo et al. (2014). "Experiments on mechanical weed control in saffron crops in Spain." *Journal of Plant Diseases and Protection 121*: 223–228.

Coşkun, M., M. Gök and S. Coşkun (2017). "Climate characteristics of Safranbolu (Karabük) and saffron cultivation." *International Journal of Geography and Geology 6*(3): 58–69.

Dar, M.H., Groach, R., Razvi, S.M. and Singh, N. (2017). "Saffron crop (golden crop) in modern sustainable agricultural systems." *Intl. J. Res. Appl. Sci. Eng. Technol. 5*: 247–259.

Dar, M., R. Groach, S. Razvi and N. Singh (2017). "Saffron crop (golden crop) in modern sustainable agricultural systems." *Intl. J. Res. Appl. Sci. Eng. Technol 5*: 247–259.

Dhar, A. (1990). "Saffron breeding and agrotechnology-A status report." *PAFAI J 12*: 18–22.

Diaz-Marta, A., A.A. GL, K. Astraka, T. Betza, E. Camba et al. (2006). "Saffron in Europe: Problems and strategies for improving the quality and strengthen competitiveness." SAFFRON Project, European Union INTERREG IIIC South Programme, Valencia, Spain.

Dita, M., M. Barquero, D. Heck, E.S. Mizubuti and C.P. Staver et al. (2018). "Fusarium wilt of banana: current knowledge on epidemiology and research needs toward sustainable disease management." *Frontiers in Plant Science 9*: 1468.

Ehsanzadeh, P., A.A. Yadollahi and A.M. Maibodi (2003). Productivity, growth and quality attributes of 10 Iranian saffron accessions under climatic conditions of Chahar-Mahal Bakhtiari, Central Iran. I *International Symposium on Saffron Biology and Biotechnology* 650.

Emadi, B. (2009). "Separating saffron flower parts using vertical air column." *World Academy of Science, Engineering and Technology 49*: 25–28.

Feili, H.R., E.M. Aghaee and A. Taslimi (2012). "Study of chemical and microbiological properties of saffron dehydrated by using solar drying system." *International Journal of Renewable Energy Research 2*(4): 627–630.

Fernández, J.-A. (2004). "Biology, biotechnology and biomedicine of saffron." *Recent Research Developments in Plant Science*. Vol. *2*: 127–159.

Fiori, M., V. Ligios and A. Schiaffino (2011). "Identification and characterization of Burkholderia isolates obtained from bacterial rot of saffron (Crocus sativus L.) grown in Italy." *Phytopathologia Mediterranea 50*(3): 450–461.

Galigani, P.F. and F.G. Pegna (1999). Mechanized saffron cultivation, including harvesting, Harwood Academic Publishers Australia.

Ganie, M.A. and A. Nusrath (2016). "Marketing and Trade Mechanism of Saffron (Crocus sativus)." *IOSR Journal of Economics and Finance* 7(5): 55–62.

Garavand, F., S. Rahaee, N. Vahedikia and S.M. Jafari (2019). "Different techniques for extraction and micro/nanoencapsulation of saffron bioactive ingredients." *Trends in Food Science & Technology 89*: 26–44.

Gheshm, R. and R.N. Brown (2021). "Growing Saffron Crocus in the Northeastern United States: Effects of Winter Protection and Planting Density." *HortTechnology 31*(4): 524–531.

Ghorbani, R. and A. Koocheki (2017). Sustainable Cultivation of Saffron in Iran. Sustainable Agriculture Reviews. E. Lichtfouse. Cham, Springer International Publishing: 169–203.

Ghorbani, R. and A. Koocheki (2017). "Sustainable cultivation of saffron in Iran." *Sustainable Agriculture Reviews*: 169–203.

Giorgi, A., D. Pentimalli, L. Giupponi and S. Panseri (2017). "Quality traits of saffron (Crocus sativus L.) produced in the Italian Alps." *Open Agriculture 2*(1): 52–57.

Golmohammadi, F. (2014). "Saffron and its farming, economic importance, export, medicinal characteristics and various uses in South Khorasan Province-East of Iran." *International Journal of Farming and Allied Sciences 3*(5): 566–596.

Gracia, L., C. Perez-Vidal and C. Gracia-López (2009). "Automated cutting system to obtain the stigmas of the saffron flower." *Biosystems Engineering 104*(1): 8–17.

Gregory, M.J., R.C. Menary and N.W. Davies (2005). "Effect of drying temperature and air flow on the production and retention of secondary metabolites in saffron." *Journal of Agricultural and Food Chemistry 53*(15): 5969–5975.

Gresta, F., Lombardo, G.M., Siracusa, L. and Ruberto, G. (2008). "Saffron, an alternative crop for sustainable agricultural systems: A review." *Agron. Sustain. Dev. 28*: 95–112.

Gresta, F., G. Avola, G. Lombardo, L. Siracusa and G. Ruberto et al. (2009). "Analysis of flowering, stigmas yield and qualitative traits of saffron (Crocus sativus L.) as affected by environmental conditions." *Scientia Horticulturae 119*(3): 320–324.

Gresta, F., G. Lombardo, L. Siracusa and G. Ruberto (2009). "Saffron, an alternative crop for sustainable agricultural systems: a review." *Sustainable Agriculture*: 355–376.

Gresta, F., G.M. Lombardo, L. Siracusa and G. Ruberto (2009). Saffron, An Alternative Crop for Sustainable Agricultural Systems: A Review. Sustainable Agriculture. E. Lichtfouse, M. Navarrete, P. Debaeke, S. Véronique and C. Alberola. *Dordrecht*, Springer Netherlands: 355–376.

Gresta, F., C. Santonoceto and G. Avola (2016). "Crop rotation as an effective strategy for saffron (Crocus sativus L.) cultivation." *Scientia Horticulturae 211*: 34–39.

Gupta, R. (2007). "Saffron status and cultivation in north western Himalayas." *Vegetos 20*(1): 1–7.

Hajyzadeh, M., H. Asil, M. Yildirim, E. Sarihan, F. Ayanoglu et al. (2016). Evaluating effects of corm circumference and storage temperatures on yield and yield components of saffron at different elevations. V International Symposium on Saffron Biology and Technology: Advances in Biology, *Technologies, Uses and Market* 1184.

Halvorson, S. (2008). "Saffron cultivation and culture in central Spain." FOCUS on Geography *51*(1): 17.

Hamid, N., J. Kachroo, A. Bhat and Q.J.A. Peer (2017). "An economic analysis of marketing and price spread of saffron in J&K State." *Journal of Pharmacognosy and Phytochemistry 6*(5): 1231–1239.

Hassan, M. and L.S. Devi (2003). "Corm rot diseases of saffron in Kashmir valley." *Indian Phytopathology 56*(1).

Heidarbeigi, K., S.S. Mohtasebi, A. Foroughirad, M. Ghasemi-Varnamkhasti, S. Rafiee et al. (2015). "Detection of adulteration in saffron samples using electronic nose." *International Journal of Food Properties 18*(7): 1391–1401.

Hosseini, M., B. Sadeghiand and S. Aghamiri (2004). "Influence of foliar fertilization on yield of saffron (Crocus sativus L.)." *Acta horticulturae.*

Husaini, A.M., B. Hassan, M.Y. Ghani, J. Teixeira da Silva and N.A. Kirmani et al. (2010). "Saffron (Crocus sativus Kashmirianus) cultivation in Kashmir: Practices and problems." *Functional Plant Science and Biotechnology 4*(2): 108–115.

Jami, N., A. Rahimi, M. Naghizadeh and E. Sedaghati (2020). "Investigating the use of different levels of Mycorrhiza and Vermicompost on quantitative and qualitative yield of saffron (Crocus sativus L.)." *Scientia Horticulturae 262*: 109027.

Kafi, M., A.N. Kamili, A.M. Husaini, M. Ozturk and V. Altay et al. (2018). "An expensive spice saffron (Crocus sativus L.): a case study from Kashmir, Iran, and Turkey." *Global Perspectives on Underutilized Crops*: 109–149.

Kafi, M., A. Koocheki and M. Rashed (2006). Saffron (Crocus sativus): production and processing, Science Publishers.

Kafi, M. and T. Showket (2007). "Comparative study of saffron agronomy and production systems of Khorasan (Iran) and Kashmir (India)." Acta horticulturae.

Kamyabi, S., M. Habibi Nokhandan and A. Rouhi (2014). "Effect of climatic factors affecting saffron using analytic hierarchy process (AHP);(Case Study Roshtkhar Region, Iran)." *Saffron Agronomy and Technology 2*(1): 75–90.

Karaoglu, C., S. Cocu, A. Ipek, S. Uranbey, I. Parmaksiz, E. Sarihan and N. Arslan (2007). "In vitro micropropagation of saffron."

Khan, M.A., S. Naseer, S. Nagoo and F. Nehvi (2011). "Behaviour of saffron (Crocus sativus L.) corms for daughter corm production." *Journal of Phytology 3*(7).

Khosravi, M. (2005). Agroecological and economic prospects for intercropping cumin with saffron and annual crops, MSc. Thesis, Ferdowsi University of Mashhad.

Koocheki, A., Seyyedi, S.M., Koocheki, A. and Khajeh-Hosseini, M. (2020). Saffron "seed", the corm, Woodhead Publishing.

Koocheki, A. (2004). "Indigenous knowledge in agriculture with particular reference to saffron production in Iran." *Acta Horticulturae.*

Koocheki, A., A. Alizadeh and A. Ganjali (2011). "The effect of increased temperature on flowering behaviour of saffron (Crocus sativus L.)." *Iranian Agricultural Research.*

Koocheki, A. and R. Ghorbani (2005). "Traditional agriculture in Iran and development challenges for organic agriculture." *The International Journal of Biodiversity Science and Management 1*(1): 52–57.

Koocheki, A., P.R. Moghaddam and H. Fallahi (2016). "Effects of planting dates, irrigation management and cover crops on growth and yield of saffron (Crocus sativus L.)." *Agroecology 8*(3): 435–451.

Koocheki, A. and P. Rezvani Moghaddam (2008). "AgroEcological Zoning and potential yield of Saffron in Khorasan-Iran." *Journal of Biological Science 8.*

Koocheki, A., S.M. Seyyedi and M.J. Eyni (2014). "Irrigation levels and dense planting affect flower yield and phosphorus concentration of saffron corms under semi-arid region of Mashhad, Northeast Iran." *Scientia Horticulturae 180*: 147–155.

Kothari, D., R. Thakur and R. Kumar (2021). "Saffron (Crocus sativus L.): gold of the spices—a comprehensive review." *Horticulture, Environment, and Biotechnology 62*(5): 661–677.

Kour, K., D. Gupta, K. Gupta, G. Dhiman, S. Juneja et al. (2022). "Smart-hydroponic-based framework for saffron cultivation: a precision smart agriculture perspective." *Sustainability 14*(3): 1120.

Kumar, R., V. Singh, K. Devi, M. Sharma and M. Singh et al. (2008). "State of art of saffron (Crocus sativus L.) agronomy: A comprehensive review." *Food Reviews International 25*(1): 44–85.

Kumar, R., V. Singh, K. Devi, M. Sharma, M. Singh et al. (2008). "State of art of saffron (Crocus sativus L.) agronomy: A comprehensive review." *Food Reviews International 25*(1): 44–85.

Kumar, R., V. Singh, K. Devi, M. Sharma, M.K. Singh et al. (2008). "State of Art of Saffron (Crocus sativus L.) Agronomy: A Comprehensive Review." *Food Reviews International 25*(1): 44–85.

Li, N., F. Chen, F. Cui, W. Sun, J. Zhang, L. et al. (2017). "Improved postharvest quality and respiratory activity of straw mushroom (Volvariella volvacea) with ultrasound treatment and controlled relative humidity." *Scientia Horticulturae 225*: 56–64.

Lombardo, G., F. Gresta, G. La Malfa and A. Scoto (2005). Primi risultati sulla coltivazione dello zafferano nella collina interna siciliana. Proceedings of the first congress on "Piante mediterranee", Agrigento.

Lopez-Corcoles, H., A. Brasa-Ramos, F. Montero-Garcia, M. Romero-Valverde and F. Montero-Riquelme et al. (2015). "Phenological growth stages of saffron plant (Crocus sativus L.) according to the BBCH Scale." *Spanish Journal of Agricultural Research 13*(3): e09SC01–e09SC01.

Manganaro, G. (2001). "Tra archeologia ed epigrafia: due note." *Zeitschrift für Papyrologie und Epigraphik*: 189–195.

Menia, M., S. Iqbal, R. Zahida, S. Tahir, R. Kanth et al. (2018). "Production technology of saffron for enhancing productivity." *Journal of Pharmacognosy and Phytochemistry 7*(1): 1033–1039.

Milyaeva, N.S.A.L. (1999). Saffron Cultivation in Azerbaijan. *Saffron*, CRC Press: 69–77.

Mir, J., N. Ahmed, D. Singh, M. Khan, S. Zaffer et al. (2015). "Breeding and biotechnological opportunities in saffron crop improvement." *African Journal of Agricultural Research 10*(9): 970–974.

Mir, M., F. Nehvi and S. Agarwal (2008). "Post harvest processing and value addition." *Saffron Production in Jammu and Kashmir, Directorate of Extension Education. SKUAST-K*, India: 242–255.

Mitsopoulou, T. and M. Tsimidou (2004). "Morphological characteristics of Greek saffrom stigmas from Kozani Region." *Acta horticulturae*.

Mohammad Abadi, A., P. Rezvani Moghaddam and J. Fallahi (2011). "Effects of planting pattern and the first irrigation date on growth and yield of saffron (Crocus sativus L.)." *Journal of Agroecology 3*(1): 84–93.

Mohammad, M., M.E. Amiri and Y. Sharghi (2012). "Respond of saffron (Crocus sativus L.) to animal manure application." *Journal of Medicinal Plants Research 6*(7): 1323–1326.

Mohammadi, H. and M. Reed (2020). Chapter 22 - *Saffron marketing: challenges and opportunities. Saffron*. A. Koocheki and M. Khajeh-Hosseini, Woodhead Publishing: 357–365.

Molafilabi, A. (2004). "An experimental finding of production and ecophysiological aspect of saffron (Crocus sativus L.)." *Acta Hort. 650*: 195–200.

Molafilabi, A., Kafi, M., Koocheki, A. and Rashed-Mohasel, M.H. (2006). *Saffron Production Technology*, Science Publishing.

Molina, R., B. Renau-Morata, S. Nebauer, A. García-Luis and J. Guardiola et al. (2010). "Greenhouse saffron culture-temperature effects on flower emergence and vegetative growth of the plants." *Acta Horticulturae* (850): 91–94.

Molina, R., M. Valero, Y. Navarro, A. Garcıa-Luis and J. Guardiola et al. (2004). "The effect of time of corm lifting and duration of incubation at inductive temperature on flowering in the saffron plant (Crocus sativus L.)." *Scientia Horticulturae 103*(1): 79–91.

Mosaferi, Z. (2001). Effects of different irrigation regimes on saffron yield, *M. Sc. Thesis*, Ferdowsi University of Mashhad, Iran. (In Persian).

Mzabri, I., M. Addi and A. Berrichi (2019). "Traditional and Modern Uses of Saffron (Crocus Sativus)." *Cosmetics 6*(4): 63.

Negbi, M. (1999). Saffron cultivation: past, present and future prospects. *Saffron*, CRC Press: 14–28.

Nehvi, F. (2009). Forthcoming challenges for improving saffron farming systems in Kashmir. III International Symposium on Saffron: Forthcoming Challenges in Cultivation, *Research and Economics* 850.

Nehvi, F., M. Ghani, S. Dar and B. Allaie (2008). "Saffron production technology." *Saffron Production in Jammu and Kashmir, Directorate of Extension Education. SKUAST-K*, India: 114–141.

Nehvi, F. and Y. Salwee (2011). Saffron Heritage of Kashmir in India. Proceedings of International Symposium on Globally Important Agric. *Heritage Syst. (GIAHS)*.

Nemati, Z., D. Harpke, A. Gemicioglu, H. Kerndorff and F.R. Blattner et al. (2019). "Saffron (Crocus sativus) is an autotriploid that evolved in Attica (Greece) from wild Crocus cartwrightianus." *Molecular Phylogenetics and Evolution 136*: 14–20.

Omidi, H., H.N. Badi, A. Golzad, H. Torabi and M. Footoukian et al. (2009). "The effect of chemical and bio-fertilizer source of nitrogen on qualitative and quantitative yield of saffron (Crocus sativus L.)." *Journal of Medicinal Plants 8*(30): 98–163.

Petrakis, E.A., L.R. Cagliani, M.G. Polissiou and R. Consonni (2015). "Evaluation of saffron (Crocus sativus L.) adulteration with plant adulterants by 1H NMR metabolite fingerprinting." *Food Chemistry 173*: 890–896.

Petrakis, E.A. and M.G. Pollissiou et al. (2017). "Assessing saffron (Crocus sativus L.) adulteration with plant-derived adulterants by diffuse reflectance infrared Fourier transform spectroscopy coupled with chemometrics." *Talanta 162*: 558–566.

Porselvi, T. (2021). "Automatic Control and Monitoring of Greenhouse System Using Iot." *Turkish Journal of Computer and Mathematics Education (TURCOMAT) 12*(11): 2708–2715.

Qadri, B. (2018). "Examining saffron income and choice of marketing channel connection." *Asian Journal of Managerial Science 7*(1): 25–30.

Rahimi, H., M. Shokrpour, L. Tabrizi Raeini and E. Esfandiari (2017). "A study on the effects of environmental factors on vegetative characteristics and corm yield of saffron (Crocus sativus)." *Iran J. Hortic. Sci 48*: 45–52.

Rezvani-Moghaddam, P., Koocheki, A. and Khajeh-Hosseini, M. (2020). *Saffron Ecophysiology,* Woodhead Publishing.

Salas, M.d.C., J.L. Montero, J.G. Diaz, F. Berti, M.F. Quintero et al (2020). "Defining optimal strength of the nutrient solution for soilless cultivation of saffron in the Mediterranean." *Agronomy 10*(9): 1311.

Sameer, S., S. Bashir, F. Nehvi, A. Iqbal, S. Naseer et al. (2012). Effect of biofertilizers, biological control agents and soil amendments on the control of saffron corm rot (Crocus sativus L.). *IV International Symposium on Saffron Biology and Technology* 1200.

Schmidt, T., T. Heitkam, S. Liedtke, V. Schubert, and G. Menzel, et al. (2019). "Adding color to a century-old enigma: Multi-color chromosome identification unravels the autotriploid nature of saffron (Crocus sativus) as a hybrid of wild Crocus cartwrightianus cytotypes." *New Phytol. 222*: 1965–1980.

Sepaskhah, A.R. and H.A. Kamgar (2009). "Saffron irrigation regime."

Seyyedi, S.M., E. Ebrahimian and E. Rezaei-Chiyaneh (2018). "Saffron daughter corms formation, nitrogen and phosphorous uptake in response to low planting density, sampling rounds, vermicompost and mineral fertilizers." *Communications in Soil Science and Plant Analysis 49*(5): 585–603.

Shajari, M.A., P.R. Moghaddam, R. Ghorbani and A. Koocheki (2020). "The possibility of improving saffron (Crocus sativus L.) flower and corm yield through the irrigation and soil texture managements." *Scientia Horticulturae 271*: 109485.

Siddique, H.R., H. Fatma and M.A. Khan (2020). "Medicinal properties of saffron with special reference to cancer—A review of preclinical studies." *Saffron*: 233–244.

Siracusa, L., E. Napoli, G. Ruberto, F. Gresta and G. Lombardo (2009). Effect of corms storage conditions on quantitative and qualitative traits of saffron: an agro-chemical study. *III International Symposium on Saffron: Forthcoming Challenges in Cultivation, Research and Economics* 850.

Tammaro, F. (1990). Crocus sativus L.–cv. Piano di Navelli (L'Aquila saffron): environment, cultivation, morphometric characteristics, active principles, uses. *Proceedings of the international conference on saffron.*

Tammaro, F. (1999). Saffron (Crocus sativus L.) in Italy. *Saffron*, CRC Press: 61–68.

Tanra, M., B.A. Dar and S. Sing (2017). "An economic analysis of production and marketing of saffron in Jammu and Kashmir." *J. Social Relevance Concern 5*(10): 12–19.

Terrado, P.R. (2007). "El azafrán. Aspectos socioeconómicos y culturales." *Studium: Revista de Humanidades* (13): 199–228.

Varliklioz Er, S., H. Eksi-Kocak, H. Yetim and I.H. Boyaci (2017). "Novel spectroscopic method for determination and quantification of saffron adulteration." *Food analytical methods 10*: 1547–1555.

Vlahova, V. (2022). "Saffron (Crocus Sativus) As an Alternative Crop in Sustainable Agricultural Systems. A Review." Scientific Papers: Management, Economic Engineering in Agriculture & Rural Development *22*(2).

Wallis, T.E. (2005). *Textbook of Pharmacognosy*, CBS.

Wani, Z.A., D.N. Mirza, P. Arora and S. Riyaz-Ul-Hassan (2016). "Molecular phylogeny, diversity, community structure, and plant growth promoting properties of fungal endophytes associated with the corms of saffron plant: An insight into the microbiome of Crocus sativus Linn." *Fungal Biology 120*(12): 1509–1524.

Willard, P. (2002). Secrets of saffron: the vagabond life of the world's most seductive spice, Beacon press.

Winterhalter, P. and M. Straubinger (2000). "Saffron—renewed interest in an ancient spice." *Food Reviews International 16*(1): 39–59.

Yarami, N. and Sepaskhah, A. (2015). "Saffron response to irrigation water salinity, cow manure, and planting method." Agr. Water Mgt. *150*: 57–66.

Yasmin, S. and F. Nehvi (2013). "Saffron as a valuable spice: A comprehensive review." *African Journal of Agricultural Research 8*(3): 234–242.

Yasmin, S. and F. Nehvi (2018). "Phenological growth stages of Saffron (Crocus sativus L.) under temperate conditions of Jammu and Kashmir-India." *International Journal of Current Microbiology and Applied Sciences 7*(4): 3797–3814.

Yildirim, M., H. Asil, M. Hajyzadeh, E. Sarihan and K. Khawar et al. (2016). Effect of changes in planting depths of saffron (Crocus sativus L.) corms and determining their agronomic characteristics under warm and temperate (Csa) climatic conditions of Turkish province of Hatay. *V International Symposium on Saffron Biology and Technology: Advances in Biology, Technologies, Uses and Market* 1184.

Yin, B., Y. Wang, P. Liu, J. Hu and W. Zhen (2010). "Effects of vesicular-arbuscular mycorrhiza on the protective system in strawberry leaves under drought stress." *Frontiers of Agriculture in China 4*: 165–169.

Zahmati, R., H.A. Shekari and M.H. Fotokian (2018). "Growth and development of saffron (Crocus sativus L.) in response to temperature pre-treatment and environmental conditions." *Journal of BioScience and Biotechnology 7*(1): 47–50.

Zare Hosseini, H., R. Ghorbani, M.H. Rashed Mohassel and H. Rahimi (2014). "Effects of weed management strategies on weed density and biomass and saffron (Crocus sativus) yield." *Saffron Agronomy and Technology 2*(1): 45–58.

3 Genetics and Cytogenetics of Saffron

Shruti Chattaraj[1*], Vijaykumar Shivaji Gadekar[2],
Abdel Rahman Al-Tawaha[3], Sayanti Mandal[4] and
Devendra Kumar Pandey[5*]

Saffron (*Crocus sativus* L.) is the world's most expensive spice valued for its aroma, flavor, and medicinal properties. It is male sterile and autotriploid (2n=3x=24), with a haploid size of 3.45 Gbp. Its triploidy causes unique genomic traits, with eight triplet chromosomes of metacentric, submetacentric and subacrocentric types. During meiosis, trivalents, bivalents and univalents form, with trivalents appearing in ring-rod Y, V, chain and triangular shapes at metaphase I, leading to imbalanced disjunction at anaphase. Laggard chromosomes, chromatid bridges and fragmentation further contribute to sterile gamete formation. The autotriploidy and irregular meiotic behavior explain saffron's sterility. However, its exact origin—whether autotriploid or allotriploid—remains unresolved.

1. Introduction

Crocus sativus L., or saffron, belongs to the Iridaceae family and propagates vegetatively via corms, thriving in tropical and subtropical regions (Fallahi et al., 2021). The name "saffron" originates from the Arabic "zafran," meaning yellow. It is the world's most expensive spice, derived from dried styles and golden-red stigmas, valued for aroma, flavor, and food coloring. Additionally, it is used in dyeing, pharmaceuticals, and perfumery (Cardon et al., 2020; Husaini et al., 2021). Saffron contains bioactive compounds with medicinal properties, aiding in stress, depression, and post-COVID-19 recovery (Husaini et al., 2021). Cultivation dates back to 1500–2500 B.C., with origins traced to Iran, Asia Minor, or Greece (Grilli Caiola et al., 2004). Iran produces 90% of global saffron, while India (Kashmir), Spain, Italy, Greece, Turkey, and Afghanistan contribute the remainder (Shahnoushi et al., 2020).

[1] Department of Botany, Ananda Mohan College, Kolkata, India.
[2] Zoology Department, Sangola College (Affiliated to Punyashlok Ahilyadevi Holkar Solapur University), Solapur, India.
[3] Department of Biological Sciences, Al-Hussein Bin Talal University, Maan, Jordan
[4] Department of Chemistry & Biochemistry, School of Basic Sciences & Research, Sharda University, Greater Noida, India
[5] Department of Biotechnology, Lovely Professional University, Punjab, India.
* Corresponding author: shrutichattaraj@gmail.com; dkpandey1974@gmail.com

DOI: 10.1201/9781003403760-3

Saffron (*Crocus sativus* L.) is a natural autotriploid, self-incompatible and sterile plant, drawing research interest into its chromosome structure and meiotic behavior. Its 2n=24 chromosome count was first reported by Himmerbaur (1926), while Karasawa (1932) identified its autrotriploidy. Brighton (1977) documented the karyotype of cultivated saffron across different regions. Sterility is attributed to irregular meiosis, leading to inviable pollen production (Estilai and Aghamohamadi, 1977).

The origin of *Crocus sativus* L. was long debated, with uncertainty over its autopolyploid or allopolyploid nature. Morphological and cytological studies showed close similarity to *C. cartwrightianus*, leading Mathew (1982) to propose its autopolyploid evolution. Molecular analyses later confirmed that *C. sativus* L. shares more DNA similarity with *C. cartwrightianus* than any other Crocus species (Grilli Caiola et al., 2004). Genotyping-by-sequencing (GBS) analysis further established its autotriploid origin, revealing 99.3% allele similarity between the two species (Nemati et al., 2019). This chapter explores cytology, mitotic division, and meiotic behavior of saffron's triplet chromosomes, emphasizing its natural autotriploid status.

2. Cytogenetics of Saffron

Saffron (*Crocus sativus* L.) is a male sterile, triploid plant with eight triplet chromosomes (2n = 3x = 24) and a haploid genome size of 3.45 Gbp (Pandita, 2021). Cytological studies confirmed its autotriploidy, with eight trivalents forming during meiosis (Karasawa 1932, 1933, 1943). Karyotype analysis across France, Turkey, Iran, England and Majorca identified four metacentric triplets and four acrocentric triplets, including two large acrocentric chromosomes with satellites and two small acrocentric chromosomes (Brighton, 1977). Sterility is linked to autotriploidy, non-disjunction, laggard chromosomes at anaphase I, and rare inversions (Ghaffari, 1986). Despite its sterility, saffron maintains its triploid genotype through vegetative propagation via corms.

2.1 Karyotype of Saffron

Karyological studies on *Crocus sativus* L. across France, Turkey, Iran, England, and Majorca revealed a genome with four metacentric triplet chromosomes, two large acrocentric triplets with satellites, and two small acrocentric triplets. The Majorca sample showed asymmetry, with one large satellite-bearing acrocentric chromosome and an elongated metacentric chromosome (Brighton, 1977). In L'Aquila, Italy, 24 somatic chromosomes were arranged in seven triplets, one pair, and one single chromosome. These included three metacentric, two submetacentric, and two subtelocentric triplets, with one triplet carrying a small satellite (Chichiriccò, 1984). Agayev (2002) analyzed saffron karyotypes from Azerbaijan and Iran, identifying eight triplets, with 1st and 2nd triplets as subacrocentric, and the 1st triplet being the largest. Ghaffari (1986) reported variation in satellite size, with two acrocentric chromosomes having large satellites and the third with a smaller satellite. The 3rd, 4th, and 8th triplets were metacentric, while the 6th and 7th were submetacentric. The 5th triplet was heteromorphic, containing one metacentric and two subacrocentric chromosomes. Chromosome lengths ranged from 11.58±0.13 to 4.57±0.05 μm, with type 2 chromosomes having the highest arm ratio (r = 6.47) and type 4 the lowest (r = 1.11).

C-banding revealed sharp heterochromatic segments on type 3 (short arm telomere) and type 1 (long arm proximal part). Telomeric heterochromatin was prominent in type 1, 4, and 5(1) short arms, while type 2 satellites also showed heterochromatin. Centromeres of all 24 chromosomes contained weak heterochromatin, but in 5(1) chromosomes, centromeric heterochromatin was stronger than telomeric heterochromatin, unlike type 4 chromosomes (Agayev, 2002; Agayev et al., 2010). These studies aided karyotype and ideogram development, while comparative Fluorescence *in situ* hybridisation (FISH) provided deeper insights into chromosome structure (Schmidt et al., 2019).

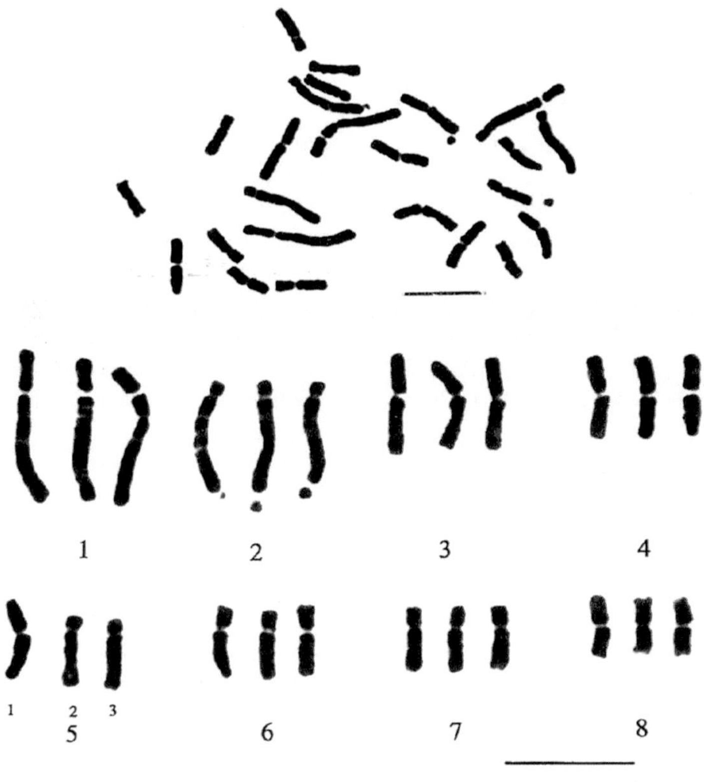

Fig. 1 *C. sativus* L. chromosomes in metaphase plate and the Karyogram (Agayev, 2002)

2.2 *Meiotic behaviour of C. sativus L. chromosomes*

The mitotic and meiotic division of saffron (*Crocus sativus* L.) has been widely studied due to its autotriploidy and sterility (Chichiriccò, 1984). Ghaffari (1986) examined pollen mother cells, confirming the 24-chromosome count reported earlier (Karasawa, 1933; Brighton, 1977; Chichiriccò, 1984). During diakinesis and metaphase I, eight trivalent chromosomes with two or three chiasmata were observed, along with bivalents and univalents in some cells. Trivalents were more frequent than other forms and appeared in Y, V, chain, ring-rod (frying pan), and triangle shapes (Figure 2A). The ring-rod type was most common (3–8 per cell), while triangle-shaped trivalents were rare (Ghaffari, 2022).

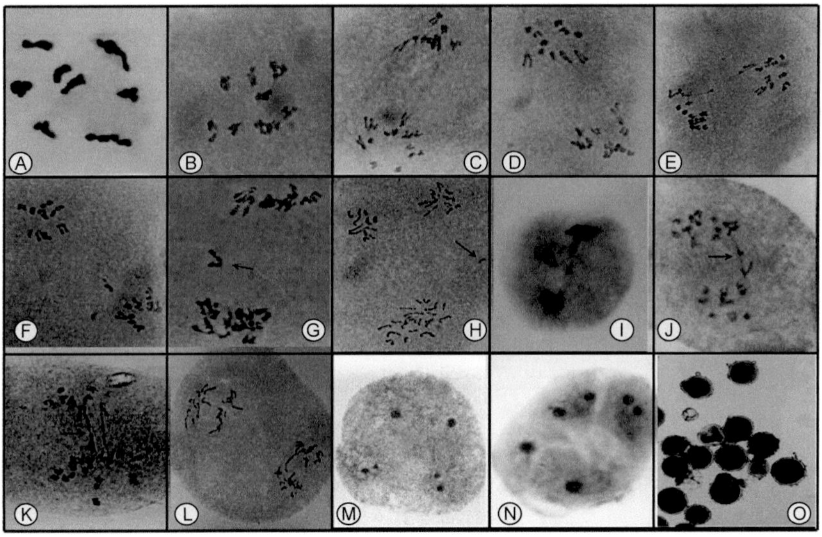

Fig. 2 Different stages of meiosis in *C. sativus* L.: A. meiotic metaphase I showing eight trivalents; B. metaphase I, showing 7 trivalents, one bivalent and one univalent; C-F. Anaphase I showing chromosome disjunction in 10-14, 11-13, 12-12 and 15-9 numbers repectively; G-I. Laggard chromosome shown by arrow; J-K. chromatid bridge in anaphase I; L. chromatid bridge in anaphase II; M. binucleated cell in telophase II; N. two binucleated microspores in tetrad stage; O. pollen grains. (Ghaffari, 2022) (licensed under a Creative Commons Attribution 4.0 International License).

Gametes exhibited chromosome numbers ranging from 8 to 16, with segregation ratios of 9-15, 10-14, 11-13, and 12-12 at anaphase I. The 11-13 disjunction was most frequent, while 9-15 was rare. Laggard chromosomes were observed at anaphase I, telophase I, and anaphase II, resulting from failure to migrate to either pole, leading to micronuclei formation (Ghaffari, 2022).

Despite 83.49 percent pollen stainability, fertility is not reflected in stainablility. In autotriploid plants, low pollen stainability is common, with most pollen grains non-functional due to irregular meiosis, reinforcing sterility and autotriploidy

(Chen et al., 2007; Diaom et al., 2009). Mitotic chromosomal studies in saffron root tip meristematic cells revealed discrete chromosomes at prometaphase, with two chromatids attached to the centromere. The metaphase plate displayed 24 chromosomes in eight triplets (Figure 3). Sister chromatids separated normally during early anaphase (Ghaffari, 1986, 2022). These findings further confirmed saffron's autotriploidy, as previously reported (Karasawa, 1933; Brighton, 1977; Chichiricco, 1984).

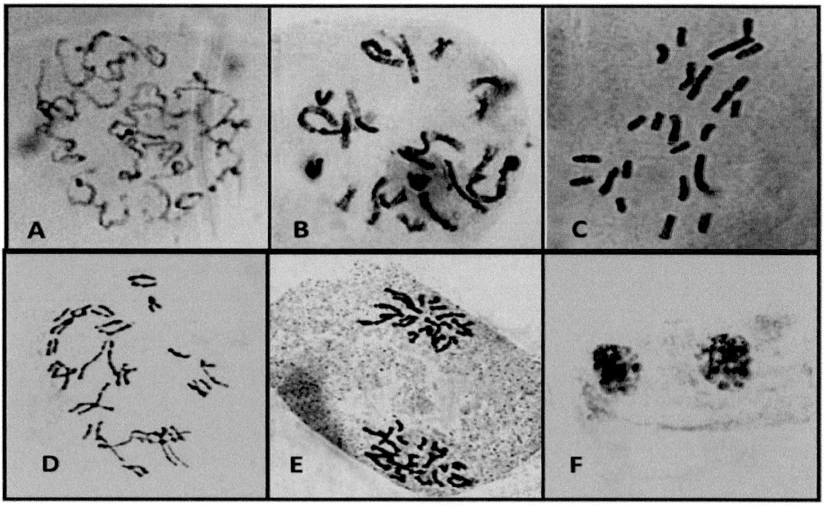

Fig. 3 *Crocus sativus* L. mitotic phases. A. prophase; B. Prometaphase; C. Metaphase; D. Chromatids start to separate for anaphase; E. Anaphase; F. Telophase. (Ghaffari, 2022) (licensed under a Creative Commons Attribution 4.0 International License).

3. Origin of Autotriploid *C. Sativus* L.

The origin of *Crocus sativus* L. remains debated, with some researchers proposing an allotriploid origin from crosses between *C. cartwrightianus, C. serotinus*, or *C. oreocreticus* (Agayev et al., 2010). Nuclear gene diversity and molecular markers (RAPD, AFLP, SSR) indicate genetic similarity to diploid Crocus species, particularly *C. cartwrightianus*, *C. hadriaticus*, and *C. oreocreticus* (Grilli Caiola et al., 2004; Larsen et al., 2015). Hybrid traits like heterosis and amphiplasty further support this hypothesis. Agayey (2002) suggested *C. sativus* L. originated from the fusion of one n and one $2n$ gametes of closely related Crocus species. Its chromosome morphology closely matches *C. cartwrightianus* (2n = 16), indicating it as a likely parent. Phylogenetic analysis, nuclear loci, and genome-wide DNA polymorphism confirm genetic closeness (Nemati et al., 2019).

Conversely, other studies support autotriploidy, considering *C. cartwrightianus* (2n = 16) as the sole parent (Karasawa, 1933; Brighton, 1977; Mathew, 1983; Chichiricco, 1984; Ghaffari, 1986; Nemati et al., 2018). Morphological, cytological, and genotyping-by-sequencing (GBS) analyses reinforce this view. Comparative Fluorescence *in situ* hybridisation (FISH) and genome sequencing suggest fusion of two *C. cartwrightianus* genomes with minor chromosomal variations, leading to autotriploid saffron (Schmidt et al., 2019).

Conclusion

Saffron (*Crocus sativus* L.) is the world's most expensive spice, derived from golden-red dried stigmas. Cytogenetic and genetic studies confirm its sterility, attributed to triploidy and irregular meiotic behavior, leading to non-viable pollen and failed ovule penetration (Chichiriccò & Caiola, 1986). Megaspore chromosome irregularities further contribute to infertility and genetic imbalance (Caiola & Canini, 2010). Saffron reproduces only through vegetative propagation via corms, maintaining a consistent genotype across generations. Phenotypic variations arise from epigenetic variability. While studies provide insights into saffron's genome and autotriploid origin, questions remain, including the unexplained heteromorphic chromosomes in the 5th triplet and the debate over its auto- or allotriploid nature. Future genomic investigations into saffron epigenetics could clarify its genetic composition and parental lineage, aiding breeding advancements for higher-quality saffron production.

References

Agayev, Y.M. 2002. New Features in Karyotype Structure and Origin of Saffron, *Crocus sativus* L. *Cytologia. 67*(3): 245–252. doi:10.1508/cytologia.67.245

Agayev, Y.M., Zarifi, E. and Fernández, J.A. 2010. A study of karyotypes in the *Crocus sativus* L. Aggregate and origin of cultivated saffron. *Acta Horticulturae. 850*: 47–54. doi:10.17660/actahortic.2010.850

Brighton, C.A. 1977. Cytology of *Crocus sativus* and its allies (Iridaceae). *Plant Systematics and Evolution. 128*: 137–157.

Cadone, L., Castronuovo, D., Perniola, M., Cicco, N. and Candido, V et al. 2020. Saffron (*Crocus sativus* L.), the king of species: An overview. *Scientia Horticulturae. 272*: 109560.

Caiola, M.G. and Canini, A. 2010. "Looking for saffron's (*Crocus sativus* L.) parents," in *Saffron*. ed. A. M. Husaini (UK: Global Science Books), 1–14.

Chen, Q.F., Hsam Sai, L.K. and Zeller, F.J. 2007. Cytogenetic studies on diploid and autotetraploid common buckwheat and their autotriploid and trisomics. *Crop Science. 47*: 2340–2345.

Chichiriccò, G. 1984. Karyotype and Meiotic Behaviour of the Triploid *Crocus sativus* L. *Caryologia. 37*(3): 233–239. doi:10.1080/00087114.1984.10797702.

Chichiriccò, G. and Caiola, M.G. 1986. *Crocus sativus* pollen germination and pollen tube growth in vitro and after intraspecific and interspecific pollination. *Can. J. Bot.* 64: 2774–2777. doi: 10.1139/b86-369.

Diaom, W.P., Bao, S.Y., Jiang, B., Cui, L., Qian, C.T. et al., 2009. Cytogenetic studies on microsporogenesis and male gametophyte development in autotriploid cucumber (*Cucumis sativus* L.): implication for fertility and production of trisomics. *Plant Systematics and Evolution. 279*: 87–92.

Estilai, A. and Aghamohammadi, Z. 1977. Pollen stainability and pollen germination in relation sterility of saffron (*Crocus sativus*). *Bulletin Faculty of Science Tehran University. 9*: 10–15.

Fallahi, H.R., Abbasi Avval Bohlooli, S., Pahlavan, Z., Hosseini, S.M., Hosseini, S.A.H. et al., 2021. Saffron vegetative growth as affected by transplanting and direct corm planting under field conditions. *Journal of Horticulture and Postharvest Research. 4*: 1–10.

Ghaffari, S.M. 1986. Cytogenetic studies of cultivated Crocus sativus (Iridaceae). *Plant Systematics and Evolution. 153*(3-4): 199–204. doi:10.1007/bf00983687

Ghaffari, S.M. 2022. Cytogenetics of saffron (*Crocus sativus*), a natural autotriploid species; an investigation on the reasons for its sterility. *Iran. J. bot. 28*(1): 57–63.

Grilli Caiola, M., Caputo, P. and Zaier, R. 2004. RAPD analysis in *Crocus sativus* L. accession and related Crocus species. *Biolgia Plantarum. 48*: 375–380.

Himmerbaur, W. 1926: *Zur Entwicklungsgeschichte von Crocus sativus* L. *Festschrift, Tschirch Taunitz, Leipzig*. pp. 335–350.

Husaini, A.M., Nadiya Jan, K. and Wani, G.A. 2021. Saffron: A potential drug-supplement for severe acute respiratory syndrome coronavirus (COVID) management. *Heliyon* 7:e07068

Karasawa, K. 1932. Karyological studies of some flowering Crocus. *Bot Mag. Tokyo. 46*: 800–802.

Karasawa, K. 1933. On the triploidy of *Crocus sativus* L., and its high sterility. *Jap. J. Genet. 9*: 6–8.

Karasawa, K. 1943. Karyological studies in Crocus. III. *Journal of Japanese Botany. 12*: 475–503.

Larsen, B., Orabi, J., Pedersen, C. and Ørgaard, M. 2015. Large intraspecific genetic variation within the Saffron-Crocus group (Crocus L., series Crocus; Iridaceae). *Plant Systematics and Evolution 301*: 425–437.

Mathew B. 1983. The Crocus: a revision of the genus Crocus (Iridaceae). Portland, OR, USA: Timber Press Inc.

Mathew, B. 1982: The Crocus. A revision of the genus Crocus (Iridaceae). Batsford, B.T. Ltd., London.

Nemati, Z., Blattner, F.R., Kerndorff, H., Erol, O. and Harpke, D. et al., 2018. Phylogeny of the saffron-crocus species group, Crocus series Crocus (Iridaceae). *Molecular Phylogenetics and Evolution. 127*: 891–897.

Nemati, Z., Harpke, D., Gemicioglu, A., Kerndorff, H. and Blattner, F.R. 2019: Saffron (*Crocus sativus*) is an autotriploid that evolved in Attica (Greece) from wild *Crocus cartwrightianus. Molecular Phylogenetics and Evolution. 136*: 14–20.

Pandita, D. 2021. Saffron (*Crocus sativus* L.): phytochemistry, therapeutic significance and omics-based biology. In: Pandita D (ed) Medicinal and aromatic plants. *Elsevier, Amsterdam*, pp. 325–396.

Schmidt, T., Heitkam, T., Liedtke, S., Schubert, V. and Menzel, G. et al., 2019. Adding color to a century-old enigma: multi-color chromosome identification unravels the autotriploid nature of saffron (*Crocus sativus*) as a hybrid of wild Crocus cartwrightianus cytotypes. *New Phytologist.* doi:10.1111/nph.15715

Shahnooshi, N., Abolhassani, L., Kavakebi, V., Reed, M. and Saghaian, S. et al., 2020. Economic analysis of saffron production. 10.1016/B978-0-12-818638-1.00021-6.

4 Saffron: Nature's Golden Elixir – The Emerging Role of Saffron Bioactives in Nutrition and Medicine

Tanushri Kaul[1*], Puja Chakraborty[1], Isha Gupta[1], Jyotsna Bharti[1], Rachana Verma[1], Arulprakash Thangaraj[1], Rashmi Kaul[1], R. Shubhra[1], Sugyan Preet[1] and Deepak Bhardwaj[2]

Saffron, extracted from *Crocus sativus* stigmas, is among the world's most expensive spices, valued for its culinary, cosmetic, and medicinal applications. Its bioactive compounds—crocins, crocetin, safranal, picrocrocin, and glycosides—offer anti-inflammatory, antioxidant, cardioprotective, and antidepressant benefits. Multi-omics and genome engineering help unravel saffron's biosynthesis, enhancing its biochemical quality for therapeutic and nutritional use. Research highlights its potential in mental health, neurodegenerative diseases, and aging-related conditions. Advancing omics-driven strategies can optimize saffron's bioactive profile, maximizing its health benefits and commercial value.

1. Introduction

Crocus sativus L., a triploid, male sterile crop, holds historical and cultural significance. Its dried red stigmas, known as saffron or "red gold," are the costliest condiment (Kazemi-Shahandashti et al., 2020). Cultivated for millennia across Europe and Asia, saffron originated in Persia (modern-day Iran) and spread to India and China, becoming integral to culinary and medicinal traditions. In the 20th century, synthetic medications reduced saffron's medical applications, but recent studies highlight its therapeutic potential. Saffron's bioactive compounds exhibit anticancer, antioxidant, antidepressant, and cardioprotective properties, aiding in insomnia, anxiety, bronchitis, asthma, colds, diabetes, and metabolic syndrome (José et al., 2017; Xu et al., 2023). This chapter explores saffron's chemical properties, biological effects, and medicinal uses, expanding research on its nutritional and pharmacological benefits.

[1] Nutritional Improvement of Crops Group, Plant Biology and Biotechnology Division, ICGEB, Aruna Asaf Ali Marg, New Delhi, India.

[2] Botany Department, Central University, Jammu, Rahya Suchani Samba District Bagla, J&K.

* Corresponding author: tanushri@icgeb.res.in; kaultanushri3@gmail.com

DOI: 10.1201/9781003403760-4

2. Saffron's Biochemical Landscape: Elucidating the Presence and Biochemistry of Key Metabolites

Saffron's dried stigmas contain ~150 bioactive compounds, including anthocyanins, flavonoids, proteins, amino acids and vitamins. Key secondary metabolites—crocin, crocetin, picrocrocin, and safranal—define its color, flavor, and fragrance, making it essential in global cuisine and perfumery. Beyond these, phenol, delphinidin, kaempferol,and flavonoids contribute antioxidant properties. Tepals and perianth are rich in flavanols, tricin, and anthocyanins, while quercetin, luteolin, acacetin, apigenin, and scutellarein are found in leaves and pollen. Delphinidin, petunidin, and malvidin glycosides account for 30% of total phenolic content, giving tepals their vibrant hue (Maqbool et al., 2022; Elfardi et al., 2024).

Flavonoids, secondary polyphenol metabolites , are abundant in *Crocus sativus* and exhibit antiproliferative, antioxidant, cardioprotective, and anti-inflammatory properties. They play a vital role in UV protection, flower coloration, defense mechanisms, and plant-microbe interactions. Kaempferol glucosides, along with vitexin, orientin, naringenin, quercetin, and rhamnetin, contribute to saffron's medicinal and nutraceutical value. Saffron's chemical profile is influenced by geography, climate, cultivation, and processing. ISO standards (3632-1:2011, 3632-2:2010) assess its crocin, crocetin, picrocrocin, and safranal content for quality (García-Rodríguez et al., 2017; Criado-Navarro et al., 2024). Beyond stigmas, edible saffron flowers are gaining popularity. Rich in carotenoids, saffron offers anti-inflammatory, immunomodulatory, and antioxidant benefits. It contains alkaloids, amino acids, and phenolic acids like caffeic acid, chlorogenic acid, and gallic acid, enhancing flavor and pharmacological properties. Phytosterols, including stigmasterol and β-sitosterol, support cholesterol reduction and overall health (Hosseinzadeh et al., 2013; Mounira et al., 2015).

The carotenoid and apocarotenoid pathways share common steps, beginning with isopentenyl diphosphate (IPP) to zeaxanthin, originating from the plastid-localized MEP pathway. Key enzymes regulate this process, leading to the formation of geranyl geranyl diphosphate (GGPP), a precursor for carotenoid and apocarotenoid biosynthesis (Frusciante et al., 2014; Dhar et al., 2017; Stelluti et al., 2024). The pathway branches at lycopene, producing α-carotene or β-carotene, which undergo hydroxylation to form lutein and zeaxanthin. Zeaxanthin cleavage dioxygenase (ZCD)/carotenoid cleavage dioxygenase 2 (CCD2) catalyzes zeaxanthin cleavage, yielding crocetin dialdehyde, which converts to crocetin and is glycosylated into crocin. During saffron development, the stigma color transitions from white to scarlet (Fig. 2) Understanding gene expression of in saffron's apocarotenoid pathway is crucial for metabolic and genome engineering to enhance bioactive compound production. CCD enzymes, classified into CCD1, CCD4, CCD7, CCD8, and NCEDs, influence pigmentation and bioactive compound synthesis. Crocin, a yellowish-red polyene ester, serves as a natural food colorant but is unstable under environmental stress, requiring proper drying and storage. Crocetin, derived from crocin hydrolysis, offers antioxidant, antidepressant, and anti-inflammatory benefits, widely used in pharmaceuticals and food industries.

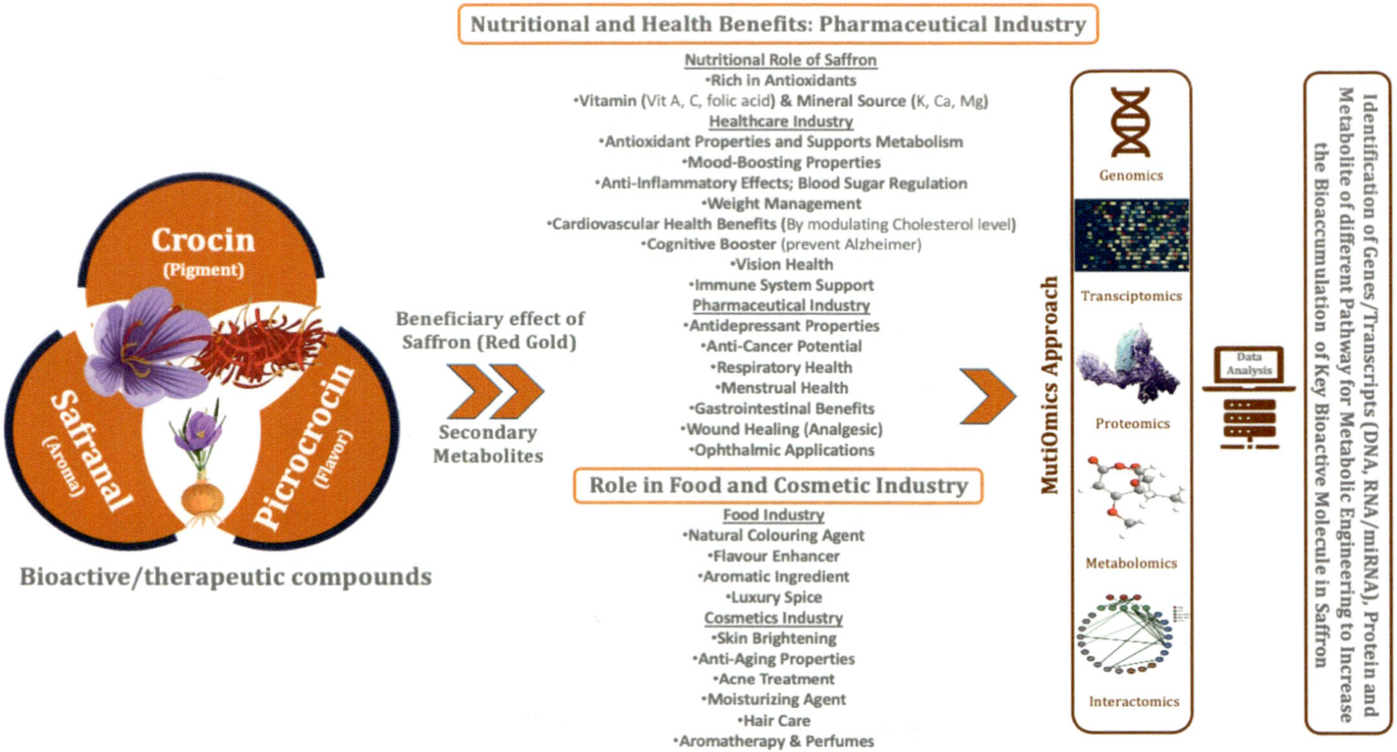

Fig. 1 Saffron, known as red gold, contains secondary metabolites with nutritional and health benefits, playing a vital role in the food and cosmetic industries. These three bioactive compounds, exclusive to saffron, serve as key molecules. A multi-omics approach can help identify genes and regulators, enabling metabolic engineering for enhanced bioaccumulation of target compounds.

Monoterpenoids like picrocrocin and safranal, synthesized from zeaxanthin degradation, contribute to saffron's bitter taste and aroma. Safranal, a volatile monoterpene, exhibits memory-enhancing, antidepressant, and neuroprotective effects (Tong et al., 2015). Optimizing saffron biosynthesis can improve quality and yield.

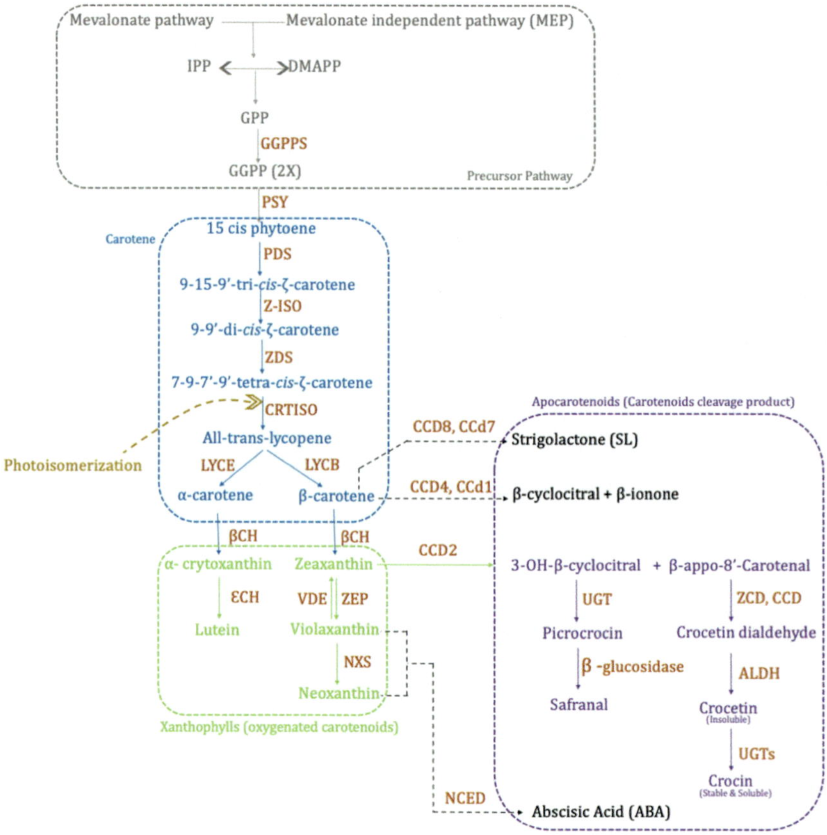

Fig. 2 Biosynthetic pathway of of crocin, picrocrocin, and safranal, where grey represents precursor pathway molecules, blue denotes carotenoids, green indicates xanthophylls, and purple signifies apocarotenoids. Black marks downstream pathways beyond this chapter's scope.

Key abbreviations:

- IPP: Isopentenyl pyrophosphate
- DMAPP: Dimethylallyl pyrophosphate
- GPP: Geranyl pyrophosphate
- GGPP: Geranylgeranyl pyrophosphate
- GGPPS: Geranylgeranyl diphosphate synthase
- PSY: Phytoene synthase
- PDS: Phytoene desaturase

- ZDS: ζ-carotene desaturase
- Z-ISO: 15-cis-ζ-carotene isomerase
- CRTISO: Carotenoid isomerase
- LCYB: Lycopene β-cyclase
- LYCE: Lycopene Ɛ-cyclase
- βCH: β-ring carotene hydroxylase
- ƐCH: Ɛ-ring carotene hydroxylase
- VDE: Violaxanthin de-epoxidase
- ZEP: Zeaxanthin epoxidase
- NXS: Neoxanthin synthase
- CCD: Carotenoid cleavage dioxygenases
- ZCD: Zeaxanthin cleavage dioxygenase
- UGT: Uridine diphosphate glycosyltransferase
- ALDH: Aldehyde dehydrogenase
- NCED: 9-cis-epoxycarotenoid dioxygenase

3. Saffron's Bioactive Compounds: Unlocking its Role in Health and Nutrition

The World Health Organization (WHO) is increasingly emphasizing traditional and herbal medicines due to their minimal side effects. Researchers are exploring natural therapeutic agents for disease management, with saffron bioactive compounds showing anti-inflammatory, antioxidant, cardioprotective, antidepressant, memory-enhancing, and anticancer properties. Crocin, a major saffron constituent, exhibits anticancer activity, inhibiting colitis-associated colon carcinogenesis. Saffron has been used medicinally since the 10th century, with Persian physician Avicenna documenting its traditional applications. Current studies highlight saffron's role in mental health, neurodegenerative diseases, and age-related conditions like cardiovascular diseases, ocular disorders, and type 2 diabetes (Salim et al., 2023). Saffron's safranal and crocins act as antidepressants, modulating serotonergic, noradrenergic, and dopaminergic pathways. Safranal also exhibits anti-anxiety effects, comparable to diazepam, likely interacting with the GABA_A receptor. Saffron's antioxidant metabolites help reduce oxidative stress and regulate blood sugar, supporting diabetes management. Studies show saffron extract (3% crocin) protects against glaucoma-related retinal damage (Fernández et al., 2019). It also preserves retinal antioxidant reserves, reducing lipid peroxidation in diabetic retinopathy. Crocins inhibit microglia activation, controlling inflammation linked to diabetic retinal damage (Fernández et al., 2019). Saffron extract antagonizes the 5-HT2C receptor, mitigating obsessive-compulsive behaviours in animal models. Preclinical studies suggest crocin may alleviate neuropsychiatric disorders like depression, anxiety, and schizophrenia. Clinical trials confirm saffron's cognitive benefits in Alzheimer's patients. Saffron's bioactive compounds aid Parkinson's disease management, muscle preservation, and mitochondrial protection. Despite limited systemic bioavailability, oral crocin administration achieves 56-81 times higher serum crocetin concentrations than direct crocetin intake. Crocins cross the blood-brain barrier (BBB), accumulating in the brain 90 minutes post-administration. Saffron exhibits anti-inflammatory, anticonvulsant, antihistamine, anti-asthmatic, and gastroprotective effects. Traditionally used for menstrual regulation, it also enhances oxygen diffusion and blood circulation, benefiting diabetic patients. Clinical trials report retinal protection at 20 mg daily doses. Saffron's anticancer potential stems from apoptosis induction, immune modulation, and oxidative stress reduction. Crocin inhibits colorectal cancer cells, while safranal and crocin help prevent cancer and cardiovascular diseases by eliminating reactive oxygen species (ROS). Kaempferol, crocetin, crocins, and quercetin suppress pro-inflammatory cytokines, mitigating chronic inflammation. Emerging evidence supports saffron's efficacy against gastric, colorectal, pancreatic, bladder, breast, lung, and liver cancers. Crocins and crocetin regulate gene expression and immune responses, promoting cancer cell apoptosis while sparing healthy cells. Picrocrocin inhibits adenocarcinoma and hepatocarcinoma cell proliferation. Overall, saffron's bioactive compounds offer neuroprotective, psychiatric, and neuromuscular health benefits. Further research is needed to standardize dosages and confirm long-term efficacy across populations (Tashakori et al., 2023; Bandegi et al., 2023; Bao et al., 2023; Ghofrani et al., 2022; Abdulkareem et al., 2021; Fotoohi et al., 2021; Salama et al., 2020; Rafieipour et al., 2019; Cao et al., 2017; Shafei et al., 2017; Sadeghnia et al., 2017; Christodoulou et al., 2017; Kyriakoudi et al., 2015; Razavi et al., 2015., Nam et al., 2010). The role and health benefits of Saffron's secondary metabolites are mentioned in Table 1.

Table 1 Role and health benefits of different Secondary phytochemical constituents from different parts of saffron

Compound	*Biological Activities*	*Present in plant part*	*References*
	Metabolites		
Crocin	Anti-schizophrenia, Antifatigue, Anti-Alzheimer's, Neuroprotective, Anti-depressant, Anti-diabetic, Therapeutics for multiple sclerosis, Antioxidative, Anti-tumor, Therapeutics for neuro-retinal diseases, Anti-inflammatory, Anti-apoptotic, Mitigates blood pressure and heart rate	Stigmas	Mir et al., 2023 Salma et al., 2020 Nam et al., 2010 Mizuma et al., 2009 Finley et al., 2017 Ochiai et al., 2007 Cao et al., 2017 Ghofrani et al., 2022 Tashakori et al., 2023 Bandegi et al., 2014 Bao et al., 2023 Heydari et al., 2023 Abdulkareem et al., 2021 Shafei et al., 2017

Contd.

Table 1 *Contd.*

Compound	*Biological Activities*	*Present in plant part*	*References*
Safranal	Anti-convulsant Anti-Alzheimer's Anti-Huntington disease Therapeutics for retinal degeneration Antioxidative Prevention of respiratory distress	Stigmas	Tiribuzi et al., 2017 Sadeghnia et al., 2008 Rafieipour et al., 2019 Fotoohi et al., 2021 Sadeghnia et al., 2017
Picrocrocin	Antiproliferative Anti-cancer activity	Stigmas	Kyriakoudi et al., 2015 Hoshyar et al., 2008
Crocusatin D, Crocusatin F, Crocusatin G, Crocusatin H, Crocusatin E, Crocusatin I	Withdrawal syndrome, depression, spatial memory	Stigmas	Christodoulou et al., 2015
Isophorone	Hyperglycemia–glucose uptake/ metabolism Parkinson's disease	Stigmas	Christodoulou et al., 2015
Lycopene	Reduces effects of CVD	Stigmas	Razavi et al., 2015
Crocetin	Anti-Parkinson's disease Protection of heart diseases Anti-apoptotic	Stigmas	Ahmad et al., 2005 Wang et al., 2014
	Phenolic Acids		
Chlorogenic acid	Prevents weight gain, reduces development of liver steatosis, and reduces lipid accumulation in liver and enhances insulin sensitivity	Stigmas	Santana-Gálvez et al., 2017
Caffeic acid	Anti-viral especially against herpes and HIV virus, and treatment for cancer	Stigmas	Gismondi et al., 2018
Methylparaben	Antibacterial; antifungal	Stigmas	Gismondi et al., 2018 Karimi et al., 2010
Gallic acid	Antioxidative, antineoplastic, and anti-inflammatory; useful for cardiovascular disorders	Stigmas	Gismondi et al., 2018 Karimi et al., 2010
Pyrogallol	Antibacterial; antifungal; and anti-malarial	Petals	Karimi et al., 2010 Alfaqih et al., 2010

Contd.

Table 1 *Contd.*

Compound	*Biological Activities*	*Present in plant part*	*References*
	Phytosterols		
β-Sitosterol	Reduces cholesterol levels; treatment for heart disease and rheumatoid arthritis	Flowers (stigma, petals, and pollens)	Feizy et al., 2016
Stigmasterol	Used as a drug for cancer therapy by inducing intracellular signaling pathways in various types of cancers. It affects the pathways of gastric and ovarian cancers; these pathways include Akt/mTOR and JAK/STAT	Flower (Stigmas, Stamen, Perianth, corms)	Feizy et al., 2016 Bakrim et al., 2022
Fucosterol	Anti-cancerous, anti-diabetic, antifungal, anti-osteoporotic, and anti-cholinergic; reduces blood cholesterol level	Flowers (petals)	Feizy et al., 2016 Abdul et al., 2016
	Other Compounds		
Vitexin	Antioxidative, anti-cancerous, anti-inflammatory, anti-hyperalgesic, neuroprotective	Leaves	He et al., 2016; Sanchez-Vioque et al., 2016
Orientin	Anti-viral, antibacterial, anti-depressant, cardioprotective, anti-aging, and antioxidant	Tepals	Lam et al., 2016 Sanchez-Vioque et al., 2016
Kaempferol	Reduces the risk of cancer and metastasis Anti-depressive Inflammation Antioxidant	Tepals	Hosseinzadeh et al., 2007 Chen et al., 2013
Isoorientin	Prevents liver damage; antioxidative, anti-inflammatory, and anti-nociceptive properties	Tepals	Sanchez-Vioque et al., 2016 Yaun et al., 2016
Naringenin	Antimicrobial, anti-mutagenic, and anti-cancer; reduces cardiovascular and gastrointestinal diseases	Tepals	Salehi et al., 2019

Contd.

Table 1 *Contd.*

Compound	*Biological Activities*	*Present in plant part*	*References*
Astragalin	Anti-obesity, anti-diabetic, anti-ulcer, anti-osteoporotic	Stigmas	Riaz et al., 2018
Dihydrokaempferol	Used as a treatment for rheumatoid arthritis (RA)	Petals	Zhang et al., 2018 Montoro et al., 2012
Myricetin	Plays role in the scavenging of ROS; anti-allergic, anti-platelet aggregation, and anti-hypertensive	Stigmas	Gismondi et al., 2018
Quercetin	Reduces blood sugar level and reactive oxygen species and prevents heart diseases	Flowers	Sanchez-Vioque et al., 2016
Rhamnetin	Plays role in attenuation of melanogenesis by reducing oxidative stress and pro-inflammatory mediators	Stamens and Petals	Kim et al., 2013
Populin	Anti-diabetic, reduces the activity of aldose reductase enzyme, and antioxidant	Stigmas	Hosseinzadeh et al., 2007

Plants provide food, shelter, fiber, and medicinal resources, especially in underdeveloped regions. The shift toward natural food consumption is increasing as an alternative to processed foods. Saffron contains proteins, amino acids, minerals, lipids, fibers, sugars, and vitamins (B1, B2), along with macro- and micronutrients like Ca, K, Mg, P, and Fe. Omics and metabolic engineering are crucial for exploring saffron's nutrient profile and enhancing bioactive compound production (Mir et al., 2024).

4. Omics and Genome Engineering/Metabolic Engineering to Improve Bioactive Compounds

Saffron has yet to undergo comprehensive multiomics and metabolic engineering studies for bioactive compound synthesis, despite its economic significance and rising demand. Its sterility and lack of a complete genome sequence hinder traditional genetic research, while biotic and abiotic stress threatens cultivation, necessitating new strategies for sustainable production. Omics-based approaches—genomics, transcriptomics, proteomics, and metabolomics—help decode saffron's functional properties, offering insights into biochemical pathways, floral development, stigma quality, and genome assembly (Fig. 1). Early research focused on microRNAs (miRNAs) regulating growth, development, and stress responses. In silico EST analysis predicted three novel miRNAs (csa-miR1, csa-miR2, csa-miR3) (Guleria et al., 2012). Additionally, miR414 and miR837-5p, identified via EST analysis, were linked to protein kinases and transcription factors, playing a key role in saffron's apocarotenoid pathway (Zinati et al., 2016). The draft genome of saffron (*Crocus sativus* L.), assembled via Illumina sequencing, spans 3.01 Gb, covering ~84.24% of the genome. Annotation identified 53,546 functional genes, 5,726 transcription factors, 862,275 repetitive regions, and 964,231 SSR markers. This analysis uncovered key genetic components in the apocarotenoid biosynthetic pathway, crucial for crocin, crocetin, picrocrocin, and safranal production, offering insights into saffron's biochemical properties (Ambardar et al., 2022). Site-directed mutations at CsCCD2 identified key amino acids (I143, L146, R161, E181, T259, and S292) involved in zeaxanthin cleavage, aiding research on biosynthetic gene duplication and metabolite evolution (Xu et al., 2024). Xuan et al. (2019) uncovered regulatory genes in saffron's flowering pathway, using NGS and SMRT sequencing to analyze flowering vs. non-flowering transcripts. They identified 62 genes across five pathways—vernalization, gibberellin, photoperiod, autonomous, and age-related regulation. Two novel genes, PB.20221.2 and PB.38952.1, were found to be tissue-specific, showing higher expression in floral organs during flowering compared to non-flowering plants. These findings enhance understanding of saffron's flowering mechanisms and contribute to future Iridaceae research (Qian et al., 2019). Saffron cultivation is significantly affected *Fusarium oxysporum*, with RNA-seq analysis identifying 2169 differentially expressed genes, including 337 upregulated defense-related genes such as chitinase, serine/threonine kinase, and WRKY transcription factor (Darvishian et al., 2025). These findings aid breeding programs for pathogen resistance. Apical bud samples from small and large corms analyzed via RNA sequencing (Illumina Novaseq 6000) revealed the regulatory role of starch and sucrose metabolism, along with auxin and ABA involvement in floral development (Chaudhary et al., 2024). Hu et al. (2020) identified a three-stage floral transition, with sugar, GA, IAA, and zeatin pathways upregulated, while ABA and starch were downregulated. 165 flowering-related unigenes were characterized (Hu et al., 2020). Jain et al. (2016) conducted transcriptomic sequencing of *Crocus sativus* tissues, identifying 105,269 unique transcripts with 54% functional annotation. Key transcription factors such as MYB, WRKY, and bHLH, regulating secondary compound metabolism, were highlighted (Jain et al., 2016). These studies enhance understanding of saffron's genetic and metabolic pathways. PacBio SMRT sequencing of *Crocus sativus* identified 31,755 protein-coding genes, with 50.1% forming paralogous gene pairs, suggesting two whole-genome duplication events. These duplications significantly influenced secondary metabolite genes, enhancing crocin, picrocrocin, and safranal accumulation (Junyang et al., 2020). RNA sequencing of apical buds from small and large corms revealed size-dependent flowering regulation, with phenylpropanoid biosynthesis, hormone signaling, and carbon metabolism playing key roles. Genes involved in starch and sucrose metabolism, auxin, ABA regulation, and flowering development were upregulated in flowering buds. Network analysis identified enolase as a central regulator, with pathway analysis validating gene expression data (Chaudhary et al., 2024). iTRAQ-based proteomic analysis of crocus buds under cold stress identified 5,624 proteins, with 201 differentially abundant proteins (DAPs). Upregulated proteins were linked to sucrose metabolism, lipid transport, and glutathione metabolism, while downregulated proteins affected starch biosynthesis and oxidative stress response. Additionally, three novel flower-related proteins and eight key genes were validated in floral development and sucrose/starch biosynthesis pathways. Cold stress disrupts floral development, leading to non-flowering buds (Chen et al., 2021). Proteomic analysis of saffron stigmas identified two antioxidant peptides (DGGSDYLGK and VDPYFNK) with high antioxidant activity. These peptides enhanced endogenous antioxidant enzyme activity and reduced malondialdehyde content, demonstrating their potential as health-promoting food components (Long et al., 2024). Metabolomic profiling using GC-MS and LC-MS compared Iranian and Chinese saffron, identifying 325 metabolites, including lipids, alkaloids, amino acids, terpenoids, organic acids, and flavonoids. Significant differences were observed in flavonoid and terpenoid composition between the two varieties (Li et al., 2024). Additionally, yellow crocus tepals contain crocins and flavonoids similar to saffron, suggesting their potential as an alternative source to meet global saffron demand (Nemati et al., 2024). Non-targeted metabolomics of *Crocus sativus* detected over 800 metabolites, with petals containing flavonoids, alkaloids, amino acids, and vitamins. 339 differential metabolites were identified—55 upregulated (including rutin, delphinidin-3-O-glucoside, and quercetin) and 284 downregulated. These findings highlight saffron petals' potential in medicine, food, cosmetics, and health products (Zhou et al., 2022). Near-infrared spectroscopy (NIR) studies effectively assess saffron quality, accurately predicting crocin and kaempferol-3-O-β-sophoroside levels (Escobar-Talavera et al., 2024). Researchers analyzed saffron stigma development, identifying phytoene synthase and carotenoid cleavage dioxygenase genes, with CsAG, CsAP3, and CsNAP playing key roles in stigma formation and apocarotenoid biosynthesis. A metabolite fingerprinting study using 1H NMR spectroscopy and chemometric analysis differentiated Italian and Iranian saffron, validating adulteration detection and confirming essential attributes retained for up to four years. UPLC-HR-MS/MS analysis of five saffron samples identified 62 metabolites, detecting auramine-O and Gardenia jasminoides Ellis as adulterants. High adulteration rates in commercial saffron reduce effectiveness and pose health risks (Inas et al., 2023). GC-MS analysis of saffron corm development revealed metabolite variations across phases from dormancy to active metabolism, with a sharp increase in carbohydrates and energy-related metabolites (Bagri et al., 2017). The impact of blue, red, and white LED light on saffron flowering and growth was examined, highlighting photoperiod effects on stigma metabolites through transcriptome and metabolomic studies, offering insights for indoor saffron farming (Li et al., 2024). GC-MS and UPLC-MS/MS analysis of saffron floral organs identified 248 metabolites, including amino acids, flavonoids, and secondary compounds. The study highlights floral regulatory networks and the therapeutic potential of saffron's bioactive compounds (Bagri et al., 2024).

Synthetic biology and metabolic engineering enable targeted metabolite synthesis, with CRISPR/Cas technology enhancing precise gene editing for bioactive molecule production. Modifying apocarotenoid biosynthesis genes significantly increased crocin and picrocrocin yields (Zhang et al., 2020). Metabolic engineering in Nicotiana species successfully produced crocin and related apocarotenoids, with *N. glauca* yielding up to 400 μg/g dry weight, surpassing *N. tabacum.* Co-expression of CsCCD2L with CrtZ and Or genes further boosted crocin production (Ahrazem et al., 2022). A virus-driven method in *Nicotiana benthamiana* achieved higher crocin and picrocrocin yields compared to agroinfiltration, demonstrating efficient bioactive compound production Martí et al. (2022). Researchers engineered Tomaffron, a tomato variety expressing CsCCD2, to accumulate high levels of saffron apocarotenoids. Combining Tomaffron with Xantomato, a zeaxanthin- and β-carotene-rich tomato, resulted in a 15-fold increase in crocin and picrocrocin production. Coupling hP3 and BETA mutations maintained fruit yield stability while enhancing lycopene and β-carotene, making it an optimal combination for apocarotenoid production (Lobato et al., 2024). The A126L mutation in CstLcyB2a (stigma-specific lycopene-β-cyclase) enhances apocarotenoid biosynthesis by redirecting metabolic flux toward β-carotene formation, significantly increasing apocarotenoid production. This advancement could boost saffron quality, resilience, and productivity, paving the way for next-generation crop development (Mir JA et al., 2024). Over the past decade, CRISPR-Cas genome editing has enabled precise gene modifications, enhancing biosynthetic pathways for targeted metabolite production. Successfully applied to rice and tomatoes, it has increased carotenoid and lycopene accumulation. In saffron, CRISPR-Cas technology holds promise for boosting secondary metabolite yields and quality. The Nutritional Improvement Crop research group has optimized an in vitro regeneration and genome editing protocol using CRISPR/Cas9 (Chib et al., 2020). This study paves the way for gene editing and metabolite enhancement in saffron. Additionally, our research group conducted untargeted GC/MS metabolic profiling across different geographical origins, analyzing sugar, organic acids, amino acids, and vitamins to assess climate-driven quality variations. Findings indicate that Kashmir (India) saffron ranks highest in quality among tested varieties (unpublished data).

Conclusion

In conclusion, saffron's bioactive compounds offer significant health and nutritional benefits. Continued research will unravel its biochemical complexities, solidifying its role in human well-being. Future studies should focus on integrating omics technologies and genome engineering to enhance saffron's bioactive profile, paving the way for innovative health-promoting products.

References

Abdul, Q.A., Choi, R.J., Jung, H.A. and Choi, J.S. Health benefit of fucosterol from marine algae: A review. 2016, *J. Sci. Food Agric. 96*, 1856–1866.

Abdulkareem Aljumaily, S.A., Demir, M., Elbe, H., Yigitturk, G., Bicer, Y. et al. Antioxidant, anti-inflammatory, and anti-apoptotic effects of crocin against doxorubicin-induced myocardial toxicity in rats. Environ. *Sci. Pollut. Res.* 2021, 28, 65802–65813.

Abdulkareem Aljumaily, S.A., Demir, M., Elbe, H., Yigitturk, G., Bicer, Y. et al. Antioxidant, anti-inflammatory, and anti-apoptotic effects of crocin against doxorubicin-induced myocardial toxicity in rats. *Environ. Sci. Pollut. Res.* 2021, *28*, 65802–65813.

Ahmad, A.S., Ansari, M.A., Ahmad, M., Saleem, S., Yousuf, S. et al. Neuroprotection by crocetin in a hemi-parkinsonian rat model. *Pharmacol. Biochem. Behav.* 2005, *81*, 805–813.

Ahrazem, O., Zhu, C., Huang, X., Rubio-Moraga, A., Capell, T. et al. Metabolic Engineering of Crocin Biosynthesis in Nicotiana Species. *Front Plant Sci.* 2022 Mar 8; 13:861140. doi: 10.3389/fpls.2022.861140. PMID: 35350302; PMCID: PMC8957871.

Alfaqih, H. and Abu-Bakar, N. The potential of pyrogallol as a possible antimalarial drug candidate (**2010**). *Acad. J. Microbiol. Immun.*, *1*, 1–4.

Ali Darvishian, Farhad Nazarian-Firouzabadi, Ahmad Ismaili and Mostafa Darvishnia, Saffron corms transcriptome analysis reveals genes associated with corm rot disease, *Plant Gene*, Volume 41, 2025, 100488, ISSN 2352-4073, https://doi.org/10.1016/j.plgene.2024.100488.

Ambardar, S., Vakhlu, J. and Sowdhamini, R.. Insights from the analysis of draft genome sequence of Crocus sativus L. *Bioinformation* (2022). 18(1): 1–13. doi: 10.6026/97320630018001.

Bagri, J., Yadav, A. and Anwar, K. Metabolic shift in sugars and amino acids regulates sprouting in Saffron corm (2017). Sci Rep 7, 11904. https://doi.org/10.1038/s41598-017-10528-2.

Bakrim, S., Benkhaira, N., Bourais, I., Benali, T. and Lee, L.H. et al. Health Benefits and Pharmacological Properties of Stigmasterol. *Antioxidants* 2022, *11*, 1912.

Bandegi, A.R., Rashidy-Pour, A., Vafaei, A.A. and Ghadrdoost, B. Protective effects of Crocus sativus L. extract and crocin against chronic-stress induced oxidative damage of brain, liver and kidneys in rats. *Adv. Pharm. Bull.* 2014, 4, 493–499.

Bandegi, A.R., Rashidy-Pour, A., Vafaei, A.A. and Ghadrdoost, B. Protective effects of *Crocus sativus* L. extract and crocin against chronic-stress induced oxidative damage of brain, liver and kidneys in rats. *Adv. Pharm. Bull.* 2014, *4*, 493–499.

Bao, X., Hu, J., Zhao, Y., Jia, R., Zhang, H. et al. Advances on the anti-tumor mechanisms of the carotenoid Crocin. PeerJ 2023, 11, e15535.

Bao, X., Hu, J., Zhao, Y., Jia, R., Zhang, H. et al. Advances on the anti-tumor mechanisms of the carotenoid Crocin. *PeerJ* 2023, *11*, e15535.

Cao, W., Cui, J., Li, S., Zhang, D., Guo, Y. et al. Crocetin restores diabetic endothelial progenitor cell dysfunction by enhancing NO bioavailability via regulation of PI3K/AKTeNOS and ROS pathways. *Life Sci.* 2017, 181, 9–16.

Cao, W., Cui, J., Li, S., Zhang, D., Guo, Y. et al. Crocetin restores diabetic endothelial progenitor cell dysfunction by enhancing NO bioavailability via regulation of PI3K/AKTeNOS and ROS pathways. *Life Sci.* 2017, *181*, 9–16.

Chaudhary, A. and Singh, K. Transcriptome analysis of apical meristem enriched bud samples for size dependent flowering commitment in Crocus sativus reveal role of sugar and auxin signalling. *Mol Biol Rep.* 2024 May 3;51(1):605. doi: 10.1007/s11033-024-09574-7. PMID: 38700570.

Chen, J., Zhou, G., Dong, Y., Qian, X., Li, J. et al. Screening of Key Proteins Affecting Floral Initiation of Saffron Under Cold Stress Using iTRAQ-Based Proteomics. *Front Plant Sci.* 2021 May 11; 12:644934. doi: 10.3389/fpls.2021.644934. PMID: 34046047; PMCID: PMC8144468.

Chen, A.Y. and Chen, Y.C. (2013). A review of the dietary flavonoid, kaempferol on human health and cancer chemoprevention. *Food chemistry*, *138*(4), 2099–2107.

Chib, S., Thangaraj, A. and Kaul, S. Development of a system for efficient callus production, somatic embryogenesis and gene editing using CRISPR/Cas9 in Saffron (Crocus sativus L.). *Plant Methods* 16, 47 (2020). https://doi.org/10.1186/s13007-020-00589-2.

Christodoulou, E., Kadoglou, N.P.E., Kostomitsopoulos, N. and Valsami, G. Saffron: A natural product with potential pharmaceutical applications. *J. Pharm. Pharmacol.* 2015, 67, 1634–1649.

Christodoulou, E., Kadoglou, N.P.E., Kostomitsopoulos, N. and Valsami, G., Saffron: A natural product with potential pharmaceutical applications. *J. Pharm. Pharmacol.* 2015, *67*, 1634–1649.

Criado-Navarro, I., Ledesma-Escobar, C.A., Pérez-Juan, P. and Priego-Capote, F. Distribution of Main Bioactive Compounds from Saffron Species as a Function of Infusion Temperature and Time in an Oil/Water System. Molecules 2024, 29, 3080. https://doi.org/10.3390/ molecules29133080

Dhar, M.K., Sharma, M., Bhat, A., Chrungoo, N.K. and Kaul, S. (2017). Functional genomics of apocarotenoids in saffron: insights from chemistry, molecular biology and therapeutic applications. *Briefings in functional genomics*, 16(6), 336–347. https://doi.org/10.1093/bfgp/elx003

El Midaoui, A., Ghzaiel, I., Vervandier-Fasseur, D., Ksila, M., Zarrouk, A. et al. Saffron (Crocus sativus L.): A source of nutrients for health and for the treatment of neuropsychiatric and age-related diseases. *Nutrients.* 2022; 14:597. doi: 10.3390/nu14030597.

Elfardi, Y.R., El Boukhari, R., Fatimi, A. and Bouissane, L. The Multifaceted Therapeutic Potential of Saffron: An Overview Based on Research and Patents. Drugs Drug Candidates 2024, 3, 437–454. https://doi.org/10.3390/ddc3030026.

Escobar-Talavera, J.F., Martínez-Navarro, M.E., Alonso, G.L. and Sánchez-Gómez, R. Determination of Saffron Flower Metabolites by Near-Infrared Spectroscopy for Quality Control. *Horticulturae* 2024, 10, 593. https://doi.org/10.3390/ horticulturae10060593.

Feizy, J. and Reyhani, N. Gas chromatographic determination of phytosterols and fatty acids profile in saffron petals. *Can. Chem. Trans.* 2016, *4*, 389–397.

Fernández-Albarral, J.A., Ramírez, A.I., de Hoz, R., López-Villarín, N. and Salobrar-García, E. et al. Neuroprotective and anti-inflammatory effects of a hydrophilic saffron extract in a model of glaucoma. *Int. J. Mol. Sci.* 2019, 20, 4110.

Finley, J.W. and Gao, S. A perspective on Crocus sativus L.(saffron) constituent crocin: A potent water-soluble antioxidant and potential therapy for Alzheimer's disease. J. Agric. *Food Chem.* 2017, 65, 1005–1020.

Finley, J.W. and Gao, S. A perspective on *Crocus sativus* L.(saffron) constituent crocin: A potent water-soluble antioxidant and potential therapy for Alzheimer's disease. *J. Agric. Food Chem.* 2017, *65*, 1005–1020.

Fotoohi, A., Moloudi, M.R., Hosseini, S., Hassanzadeh, K., Feligioni, M. and Izadpanah, E. A novel pharmacological protective role for safranal in an animal model of Huntington's disease. Neurochem. Res. 2021, 46, 1372–1379.

Fotoohi, A., Moloudi, M.R., Hosseini, S., Hassanzadeh, K. and Feligioni, M. et al. A novel pharmacological protective role for safranal in an animal model of Huntington's disease. *Neurochem. Res.* 2021, *46*, 1372–1379.

Frusciante, Sarah, Diretto, Gianfranco, Bruno, Mark, Ferrante, Paola, Pietrella, Marco. et al. (2014). Novel carotenoid cleavage dioxygenase catalyzes the first dedicated step in saffron crocin biosynthesis. *Proceedings of the National Academy of Sciences.* 5. 10.1073/pnas.1404629111.

García-Rodríguez, M.V., López-Córcoles, H., Alonso, G.L., Pappas, C.S., Polissiou, M.G. et al. Comparative evaluation of an ISO 3632 method and an HPLC-DAD method for safranal quantity determination in saffron. *Food Chem.* 2017, 221, 838–843.

Ghofrani, S., Joghataei, M.-T., Afshin-Majd, S., Baluchnejadmojarad, T. and Roghani, M. et al. Crocin, a bioactive constituent of Crocus sativus, alleviates trimethyltin-induced cognitive deficits through down-regulation of hippocampal apoptosis and oxidative stress. *J. Basic. Clin. Pathophysiol.* 2022, 10, 38–44.

Ghofrani, S., Joghataei, M.-T., Afshin-Majd, S., Baluchnejadmojarad, T., Roghani, M. et al. Crocin, a bioactive constituent of *Crocus sativus*, alleviates trimethyltin-induced cognitive deficits through down-regulation of hippocampal apoptosis and oxidative stress. *J. Basic. Clin. Pathophysiol.* 2022, *10*, 38–44.

Gismondi, A., Serio, M., Canuti, L. and Canini, A. Biochemical, antioxidant and antineoplastic properties of Italian saffron (*Crocus sativus* L.). *Am. J. Plant Sci.* 2012, *3*, 1573–1580.

Guleria, P., Goswami, D., Yadav, K.S. (2012). Computational identification of miRNAs and their targets from Crocus sativus L. *Arch Biol Sci 64*(1): 65–70.

He, M., Min, J.W., Kong, W.L., He, X.H., Li, J.X. et al. (2016). A review on the pharmacological effects of vitexin and isovitexin. *Fitoterapia*, *115*, 74–85.

Heydari, M., Zare, M., Badie, M.R., Watson, R.R., Talebnejad, M.R. et al. Crocin as a vision supplement. *Clin. Exp. Optom.* 2023, 106, 249–256.

Heydari, M., Zare, M., Badie, M.R., Watson, R.R., Talebnejad, M.R. et al. Crocin as a vision supplement. *Clin. Exp. Optom.* 2023, *106*, 249–256.

Hoshyar, R., Bathaie, S.Z. and Ashrafi, M. Interaction of safranal and picrocrocin with ctDNA and their preferential mechanisms of binding to GC- and AT-rich oligonucleotides. *DNA Cell Biol.* 2008, *27*, 665–673.

Hosseinzadeh, H., Motamedshariaty, V. and Hadizadeh, F. (2007). Antidepressant effect of kaempferol, a constituent of saffron (Crocus sativus) petal, in mice and rats. *Pharmacologyonline*, *2*, 367–370.

Hosseinzadeh, H., Shakib, S.S., Sameni, A.K. and Taghiabadi, E. Acute and subacute toxicity of safranal, a constituent of saffron, in mice and rats. *Iran. J. Pharm. Res. IJPR* 2013, 12, 93–99.

Hosseinzadeh, H. and Talebzadeh, F. Anticonvulsant evaluation of safranal and crocin from Crocus sativus in mice. *Fitoterapia* 2005, 76, 722–724.

Hu, J., Liu, Y. and Tang, X. Transcriptome profiling of the flowering transition in saffron (Crocus sativus L.). *Sci Rep* 10, 9680 (2020). https://doi.org/10.1038/s41598-020-66675-6.

Inas, Y., Younis, Engy Mohsen, Rana M. Ibrahim, Alisdair R. Fernie, Saleh Alseekh et al., Non-targeted metabolomics and chemometrics for saffron (Crocus sativus L.) authentication and adulteration detection in relation to its anticholinesterase activity, *Food Chemistry Advances*, Volume 2, 2023, 100217, ISSN 2772-753X, https://doi.org/10.1016/j.focha.2023.100217.

Jain, M., Srivastava, P., Verma, M. De novo transcriptome assembly and comprehensive expression profiling in Crocus sativus to gain insights into apocarotenoid biosynthesis. *Sci Rep* 6, 22456 (2016). https://doi.org/10.1038/srep22456.

Jayram Bagri, Vikash Kumar Singh, Khushboo Gupta, Jeremy Dkhar, Aijaz Ahmad Wan et al. *Physiologia plantaram*. https://doi.org/10.1111/ppl.14598.

José Bagur M, Alonso Salinas GL, Jiménez-Monreal AM, Chaouqi S, Llorens S et al. Saffron: An Old Medicinal Plant and a Potential Novel Functional Food. *Molecules*. 2017 Dec 23; *23*(1):30. doi: 10.3390/molecules23010030.

Junyang Yue, Ran Wang, Xiaojing Ma, Jiayi Liu, Xiaohui Lu et al., Full-length transcriptome sequencing provides insights into the evolution of apocarotenoid biosynthesis in Crocus sativus, *Computational and Structural Biotechnology Journal*, Volume 18, 2020, Pages 774–783, ISSN 2001-0370, https://doi.org/10.1016/j.csbj.2020.03.022.

Karimi, E., Oskoueian, E., Hendra, R., Jaafar, H.Z. Evaluation of *Crocus sativus* L. stigma phenolic and flavonoid compounds and its antioxidant activity. *Molecules* 2010, *15*, 6244–6256.

Kazemi-Shahandashti, S-S., Mann, L., El-nagish, A., Harpke, D., Nemati, Z. et al. (2022) Ancient Artworks and Crocus Genetics Both Support Saffron's Origin in Early Greece. Front. *Plant Sci.* 13:834416. doi: 10.3389/fpls.2022.834416.

Kim, Y.J. Rhamnetin attenuates melanogenesis by suppressing oxidative stress and pro-inflammatory mediators. *Biol. Pharm. Bull.* 2013, *36*, 1341–1347.

Kyriakoudi, A., O'Callaghan, Y.C., Galvin, K., Tsimidou, M.Z., O'Brien, N.M. et al. Cellular transport and bioactivity of a major saffron apocarotenoid, picrocrocin (4-(β-D-glucopyranosyloxy)-2, 6, 6-trimethyl-1-cyclohexene-1-carboxaldehyde). J. Agric. *Food Chem.* 2015, 63, 8662–8668.

Kyriakoudi, A., O'Callaghan, Y.C., Galvin, K., Tsimidou, M.Z., O'Brien, N.M. et al. Cellular transport and bioactivity of a major saffron apocarotenoid, picrocrocin (4-(β-D-glucopyranosyloxy)-2, 6, 6-trimethyl-1-cyclohexene-1-carboxaldehyde). *J. Agric. Food Chem.* 2015, *63*, 8662–8668.

Lam, K.Y., Ling, A.P.K., Koh, R.Y., Wong, Y.P. and Say, Y.H. (2016). A review on medicinal properties of orientin. *Advances in Pharmacological and Pharmaceutical Sciences*, *2016*(1), 4104595.

Li, D., Zuo, W., Ma, S. Metabolomics characterization of two saffron from Iran and China using GC–MS and LC–MS methods. *J Anal Sci Technol* 15, 7 (2024). https://doi.org/10.1186/s40543-024-00421-9.

Li, L., Zhou, Y., Wang, J. Effects of supplementary light treatment on saffron: integrated physiological, metabolomic, and transcriptome analyses. *BMC Plant Biol* 24, 1247 (2024). https://doi.org/10.1186/s12870-024-05944-2.

Lobato-Gómez, M., Drapal, M., Fernández-Muñoz, R., Presa, S., Espinosa, A. et al. Maximizing saffron apocarotenoid production in varied tomato fruit carotenoid contexts. *Plant J.* 2024 Nov;120(3):966-983. doi: 10.1111/tpj.17030. Epub 2024 Sep 18. PMID: 39292868.

Long, Yiyang and Tao, Han and Wang, Shiyu and Xing, Bingcong and Wang, Zhineng et al. Identification and Functional Validation of Two Novel Antioxidant Peptides in Saffron: An Example of the Exploitation and Utilization of Scarce Plants.

Maqbool, Z., Arshad, M.S., Ali, A., Aziz, A., Khalid, W., Afzal, M.F. et al. Potential Role of Phytochemical Extract from Saffron in Development of Functional Foods and Protection of Brain-Related Disorders. *Oxid Med Cell Longev.* 2022 Sep 22; 2022:6480590. doi: 10.1155/2022/6480590.

Maricarmen Martí, Gianfranco Diretto, Verónica Aragonés, Sarah Frusciante, Oussama Ahrazem et al., Efficient production of saffron crocins and picrocrocin in Nicotiana benthamiana using a virus-driven system,Metabolic Engineering, Volume 61, 2020, Pages 238–250, ISSN 1096-7176, https://doi.org/10.1016/j.ymben.2020.06.009.

Mir, J.A., Yadav, A.K., Singh, D., Ashraf, N. A novel mutation in non-constitutive lycopene beta cyclase (CstLcyB2a) from Crocus sativus modulates carotenoid/apocarotenoid content, biomass and stress tolerance in plants. Planta. 2024 Aug 27; 260(4):80. doi: 10.1007/s00425-024-04515-x.

Mir, R.A., Tyagi, A., Hussain, S.J., Almalki, M.A., Zeyad, M.T. et al. (2023). Saffron, a Potential Bridge between Nutrition and Disease Therapeutics: Global Health Challenges and Therapeutic Opportunities. Plants, 13(11), 1467. https://doi.org/10.3390/plants13111467

Mir, R.A., Tyagi, A., Hussain, S.J., Almalki, M.A., Zeyad, M.T. et al. Saffron, a Potential Bridge between Nutrition and Disease Therapeutics: Global Health Challenges and Therapeutic Opportunities. *Plants* 2024, 13, 1467. https://doi.org/10.3390/ plants13111467.

Mizuma, H., Tanaka, M., Nozaki, S., Mizuno, K., Tahara, T. et al. Daily oral administration of crocetin attenuates physical fatigue in human subjects. *Nutr. Res.* 2009, *29*, 145–150.

Montoro, P., Maldini, M., Luciani, L., Tuberoso, C.I., Congiu, F. et al. Radical scavenging activity and LC-MS metabolic profiling of petals, stamens, and flowers of *Crocus sativus* L. *J. Food Sci.* 2012, *77*, C893–C900.

Moratalla-López, N., Bagur, M.J., Lorenzo, C., Salinas, MEMR, Alonso, G.L. et al. Bioactivity and Bioavailability of the Major Metabolites of Crocus sativus L. *Flower. Molecules.* 2019 Aug 2; 24(15): 2827. doi: 10.3390/molecules24152827.

Mounira, L., Bernardo, M., Luigi, C.P., Guido, F., Fatima, G. et al. Phytochemical composition of Moroccan saffron accessions by headspacesolid-phase-microextraction. *Am. J. Essent. Oils Nat. Prod.* 2015, 2, 1–7.

Nam, K.N., Park, Y.-M., Jung, H.-J., Lee, J.Y., Min, B.D. et al. Anti-inflammatory effects of crocin and crocetin in rat brain microglial cells. *Eur. J. Pharmacol.* 2010, 648, 110–116.

Nam, K.N., Park, Y.-M., Jung, H.-J., Lee, J.Y., Min, B.D. et al. Anti-inflammatory effects of crocin and crocetin in rat brain microglial cells. *Eur. J. Pharmacol.* 2010, *648*, 110–116.

Nemati, Z., Kazemi-Shahandashti, S.S., Garibay-Hernández, A. Metabolomic and transcriptomic analyses of yellow-flowered crocuses to infer alternative sources of saffron metabolites. *BMC Plant Biol* 24, 369 (2024). https://doi.org/10.1186/s12870-024-05036-1.

Ochiai, T., Shimeno, H., Mishima, K.I., Iwasaki, K., Fujiwara, M. et al. Protective effects of carotenoids from saffron on neuronal injury *in vitro* and *in vivo*. *Biochim. Biophys. Acta (BBA)—Gen. Subj.* 2007, *1770*, 578–584.

Qian, X., Sun, Y., Zhou, G. Single-molecule real-time transcript sequencing identified flowering regulatory genes in Crocus sativus. *BMC Genomics* 20, 857 (2019). https://doi.org/10.1186/s12864-019-6200-5

Rafieipour, F., Hadipour, E., Emami, S.A., Asili, J., Tayarani-Najaran, Z. et al. Safranal protects against beta-amyloid peptide-induced cell toxicity in PC12 cells via MAPK and PI3 K pathways. *Metab. Brain Dis.* 2019, 34, 165–172.

Rafieipour, F., Hadipour, E., Emami, S.A., Asili, J., Tayarani-Najaran, Z. et al. Safranal protects against beta-amyloid peptide-induced cell toxicity in PC12 cells via MAPK and PI3 K pathways. *Metab. Brain Dis.* 2019, *34*, 165–172.

Razavi, B.M., Hosseinzadeh, H. Saffron as an antidote or a protective agent against natural or chemical toxicities. DARU J. Fac. *Pharm. Tehran Univ. Med. Sci.* 2015, *23*, 31.

Razavi, B.M., Hosseinzadeh, H. Saffron as an antidote or a protective agent against natural or chemical toxicities. *DARU J. Fac. Pharm. Tehran Univ. Med. Sci.* 2015, *23*, 31.

Riaz, A., Rasul, A., Hussain, G., Zahoor, M.K., Jabeen, F. et al. A bioactive phytochemical with potential therapeutic activities. *Adv. Pharmacol. Sci.* 2018, *2*, 9794625.

Sadeghnia, H., Cortez, M., Liu, D., Hosseinzadeh, H., Snead, O.C. et al. Antiabsence effects of safranal in acute experimental seizure models: EEG and autoradiography. *J. Pharm. Pharm. Sci.* 2008, *11*, 1–14

Sadeghnia, H.R., Shaterzadeh, H., Forouzanfar, F., Hosseinzadeh, H. Neuroprotective effect of safranal, an active ingredient of Crocus sativus, in a rat model of transient cerebral ischemia. *Folia Neuropathol.* 2017, *55*, 206–213.

Salama, R.M., Abdel-Latif, G.A., Abbas, S.S., El Magdoub, H.M., Schaalan, M.F. et al. Neuroprotective effect of crocin against rotenone-induced Parkinson's disease in rats: Interplay between PI3K/Akt/mTOR signaling pathway and enhanced expression of miRNA-7 and miRNA-221. Neuropharmacology 2020, *164*, 107900.

Salehi, B., Fokou, P.V.T., Sharifi-Rad, M., Zucca, P., Pezzani, R. et al. The therapeutic potential of naringenin: A review of clinical trials. *Pharmaceuticals* 2019, *12*, 11.

Salim, R., Nehvi, I.B., Mir, R.A., Tyagi, A., Ali, S. et al. A review on anti-nutritional factors: Unraveling the natural gateways to human health. *Front. Nutr.* 2023, *31*, 1215873.

Sánchez-Vioque, R., Santana-Méridas, O., Polissiou, M., Vioque, J., Astraka, K. et al. (2016). Polyphenol composition and *in vitro* antiproliferative effect of corm, tepal and leaf from Crocus sativus L. on human colon adenocarcinoma cells (Caco-2). *Journal of Functional Foods*, *24*, 18-25.

Santana-Gálvez, J., Cisneros-Zevallos, L., Jacobo-Velázquez, D.A. Chlorogenic acid: Recent advances on its dual role as a food additive and a nutraceutical against metabolic syndrome. *Molecules* 2017, *22*, 358.

Serrano-Díaz, J., Sánchez, A.M., Alvarruiz, A., Alonso, G.L. Preservation of saffron floral bio-residues by hot air convection. *Food Chem.* 2013; 141:1536–1543. doi: 10.1016/j.foodchem.2013.04.029.

Shafei, M.N., Faramarzi, A., Rad, A.K., Anaeigoudari, A. Crocin prevents acute angiotensin II-induced hypertension in anesthetized rats. *Avicenna J. Phytomedicine* 2017, *7*, 345–352.

Shafei, M.N., Faramarzi, A., Rad, A.K., Anaeigoudari, A. Crocin prevents acute angiotensin II-induced hypertension in anesthetized rats. *Avicenna J. Phytomedicine* 2017, *7*, 345–352.

Singla, R.K., Bhat, G. Crocin: An overview. *Indo Glob. J. Pharm. Sci.* 2011, *1*, 281–286.

Stelluti, S., Grasso, G., Nebauer, S.G., Alonso, G.L., Renau-Morata, B. et al. (2024). Arbuscular mycorrhizal symbiosis modulates the apocarotenoid biosynthetic pathway in saffron. *Scientia Horticulturae*.

Tashakori, A., Hassanpour, S., Vazir, B. Protective effect of crocin on the cuprizone-induced model of multiple sclerosis in mice. Naunyn-Schmiedeberg's Arch. *Pharmacol.* 2023, *396*, 1713–1725.

Tashakori, A., Hassanpour, S., Vazir, B. Protective effect of crocin on the cuprizone-induced model of multiple sclerosis in mice. *Naunyn-Schmiedeberg's Arch. Pharmacol.* 2023, *396*, 1713–1725.

Tiribuzi, R., Crispoltoni, L., Chiurchiù, V., Casella, A., Montecchiani, C. et al. Trans-crocetin improves amyloid-β degradation in monocytes from Alzheimer's disease patients. *J. Neurol. Sci.* 2017, *372*, 408–412.

Tong, Y., Zhu, X., Yan, Y., Liu, R., Gong, F. et al. The influence of different drying methods on constituents and antioxidant activity of saffron from China. *Int. J. Anal. Chem.* 2015, 2015, 953164.

Wang, Y., Sun, J., Liu, C., Fang, C. Protective effects of crocetin pretreatment on myocardial injury in an ischemia/reperfusion rat model. *Eur. J. Pharmacol.* 2014, *741*, 290–296.

Xu, Z., Chen, S., Wang, Y., Tian, Y., Wang, X. et al. Song J. Crocus genome reveals the evolutionary origin of crocin biosynthesis. *Acta Pharm Sin B.* 2024 Apr;14(4): 1878–1891. doi: 10.1016/j.apsb.2023.12.013

Xu, Zhichao & Chen, Shanshan & Wang, Yalin & Tian, Ya & Wang et al. (2023). Crocus genome reveals the evolutionary origin of crocin biosynthesis. *Acta Pharmaceutica Sinica B.* 14. 10.1016/j.apsb.2023.12.013.

Yuan, L., Wang, J., Wu, W., Liu, Q. and Liu, X. et al. (2016). Effect of isoorientin on intracellular antioxidant defence mechanisms in hepatoma and liver cell lines. *Biomedicine & Pharmacotherapy, 81*, 356–362.

Zhang, Y., Yan, G., Sun, C., Li, H., Fu, Y. et al. Apoptosis effects of dihydrokaempferol isolated from *Bauhinia championii* on Synoviocytes. *Evid. Based Complement. Altern. Med.* 2018, *2*, 9806160.

Zhou, L., Cai, Y., Yang, L., Zou, Z., Zhu, J. et al. Comparative Metabolomics Analysis of Stigmas and Petals in Chinese Saffron (Crocus sativus) by Widely Targeted Metabolomics. Plants (Basel). 2022 Sep 17; *11*(18): 2427. Doi: 10.3390/plants11182427.

Zinati, Z., Shamloo-Dashtpagerdi, R., Behpouri, A. (2016). In silico identification of miRNAs and their target genes and analysis of gene co-expression network in saffron (Crocus sativus L.) stigma. *Mol Biol Res Commun* 5(4):233.

5 Economic Importance of Saffron

Bushra Hafeez Kiani[1*]

The market for aromatics and therapeutic plants continues to grow, with increasing demand for naturally derived products. Saffron (*Crocus sativus* L.), a triploid monocot herb, is a globally valuable export and one of the finest spices. Its key components—safranal, crocin, and picrocrocin—define its fragrance, color, and flavor. The crimson Saffron stigmas and styles are essential in textile, pharmaceutical, cosmetic, and perfume industries. Global cultivation averages 300 tons annually, projected to grow by 12% in five years. Historically valued for fragrance, medicinal, and dyeing properties, saffron is now primarily marketed as a culinary spice. This chapter explores its economic importance, spanning therapeutic, traditional, cosmetology, perfumery, culinary, and coloring applications.

1. Introduction

Crocus sativus L., one of the most expensive aromatic crops, has wide therapeutic applications globally (Rios et al., 1996; Ferrence & Bendersky, 2004). Its three-branched crimson styles hold economic significance, particularly in culinary, pharmaceutical, and dye industries. Safranal, an apocarotenoid phytoconstituent, gives saffron its distinct fragrance. Saffron is cultivated in Pakistan, Israel, India, Iran, Italy, UAE, Afghanistan, Turkey, Azerbaijan, Egypt, France, Switzerland, Greece, Japan, Iraq, China, Spain, and Tasmania, Australia, with 418 Megagrams (Mg) produced annually. Iran's Khorasan region accounts for 92% of global production (Cardone et al., 2020). Saffron is manually grown via vegetative propagation of daughter corms. Its key compounds—safranal, crocin, and picrocrocin—define its fragrance, color, and bitterness (Cagliani et al., 2015). Due to its high cost, saffron is often adulterated with safflower stamens and maize stigmas, impacting commerce (Babaei et al., 2014).

Review studies highlight *Crocus sativus* applications, including anticancer activity (Bhandari, 2015), chemical constituents (Bathaie et al., 2014), antioxidative properties (Rahaiee et al., 2015), toxicology (Badie Bostan et al., 2017), pharmaceutical potential of petals (Hosseini et al., 2018), therapeutic value (Razak et al., 2017), and agricultural technology (Kumar et al., 2009). This chapter provides a comprehensive overview of saffron's economic significance, covering

[1] Department of Biological Sciences (Female Campus), International Islamic University, Islamabad, Pakistan.

* Corresponding author: tbushrahafeezkiani@gmail.com

DOI: 10.1201/9781003403760-5

its general, traditional, therapeutic, pharmacological, cosmetic, perfumery, dye, and commercial applications (Figure 1).

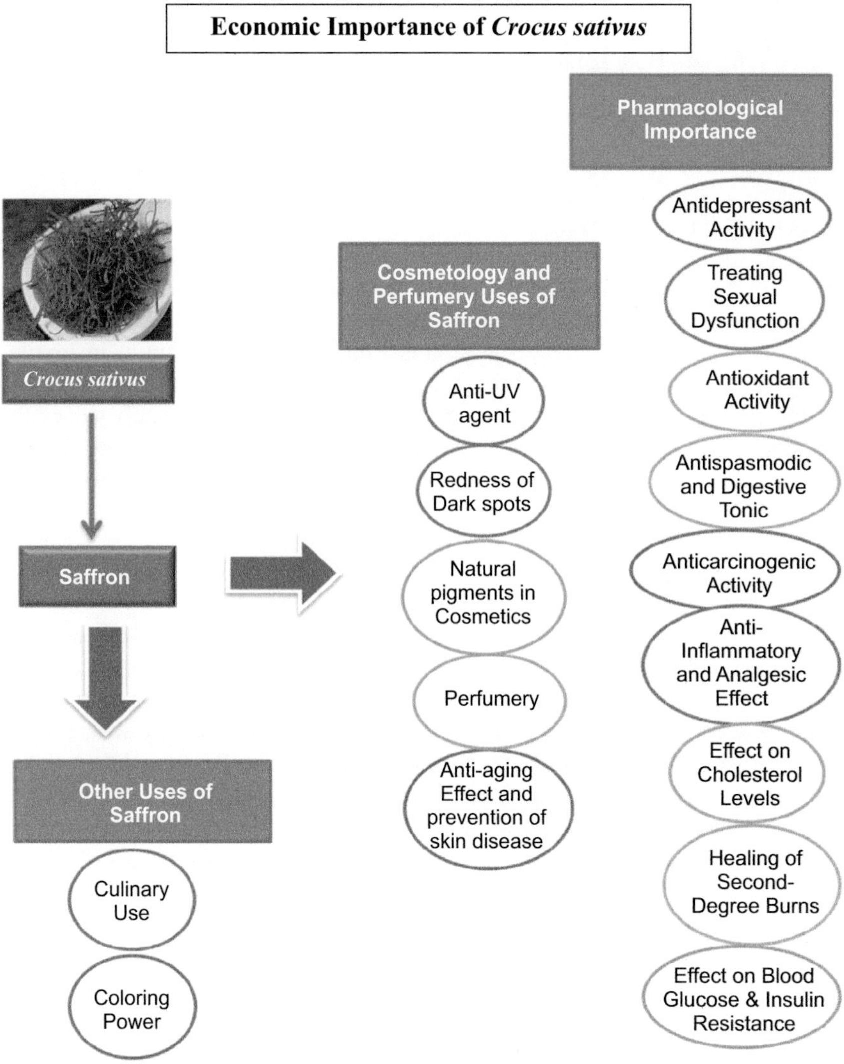

Fig. 1 Economic Importance of *Crocus sativus* (Saffron)

2. Economic Importance of Saffron

Pharmacological Study of Saffron

Saffron has been a medicinal staple for over 3,000 years, valued in Ayurveda, Arab, Egyptian, Chinese, Mongolian, and Greek medicine for its broad therapeutic properties. Key medicinal benefits include:

2.1 Antidepressant

Crocus sativus has historically been used to alleviate depression, a condition affecting 11.6 % of the global population (Suganya et al., 2016). Like conventional antidepressants, saffron may enhance serotonin uptake, with tryptophan converting into serotonin (5-hydroxytryptamine) (Hill, 2004).

While saffron styles are traditionally used for therapeutic purposes, petals—a cheaper alternative—are now being explored for their antidepressant potential. Further scientific research is needed to assess their effectiveness in treating mild depression (Murray & Lopez, 1997; Khazdair et al., 2018).

2.2 Sexual Dysfunction Treatment

Saffron has been recognized as an aphrodisiac across Rome, Greece, Egypt, and various cultures, with Phoenicians, Chinese, and Arabs historically using it for this purpose (Dwyer et al., 2011). The sex-stimulating properties of liquid extracts of *C. sativus* stigmas as well as their compounds, crocin, and safranal have been investigated. Research indicates that saffron could successfully treat various sexual problems caused by fluoxetine in females, including arousal, discomfort, or lubrication with little to no side effects (Akhondzadeh et al., 2008). Saffron has also been demonstrated to cure erectile dysfunction by increasing the longevity and frequency of erections, just following ten days of medication (Akhondzadeh et al., 2013).

2.3 Antioxidant

Crocetin and crocin, two carotenoids, act as natural antioxidants, protecting cells by neutralizing reactive oxygen species and free radicals. Crocin's antioxidative properties are enhanced when combined with dimethylcrocetin, safranal, and bioflavonoids (Hosseinzadeh et al., 2009). The brain's phospholipid-rich neurons are highly susceptible to oxidative stress, contributing to Alzheimer's disease and other neurodegenerative disorders. Studies suggest saffron may limit amyloid peptide accumulation, offering potential therapeutic benefits (Bathair et al., 2013).

2.4 Anticancer Properties

Cancer remains a major global health concern, with high-antioxidant diets linked to lower disease prevalence. Saffron carotenoids, particularly crocin and crocetin, exhibit anticancer properties, suppressing tumor growth without harming normal cells. Studies highlight their ability to induce apoptosis in various tumors. Saffron's anticancer properties have been extensively studied in laboratory and animal models (Akhondzadeh et al., 2010; Hamidreza et al., 2010). Researchers found that combining saffron extracts with disodium selenite and sodium arsenite enhances its chemopreventive effects (Abdullaev, 2010).

Additionally, saffron exhibits cytotoxicity against tumor tissues, with a five-day pre-treatment significantly reducing cell and genome damage caused by cisplatin chemotherapy (Botsoglou, 2005).

2.5 Antispasmodic and Digestive Tonic

Saffron supports digestive and reproductive health, aiding digestion, appetite control, hemorrhoid relief, and menstrual regulation (Premkumar et al., 2003). Safranal enhances gastric protection, reducing ulcer surface, balancing pH, and mitigating tissue damage from indomethacin (Hosseinzadeh et al., 2013).

2.6 Anti-Inflammatory and Analgesic Effect

Natural substances, including dietary supplements and plant therapies, have long been used to reduce pain and inflammation (Tamaddonfard et al., 2019). Saffron tinctures and extracts aid in gingivitis, abscesses,back pain, fever, sores, and teething discomfort. Aqueous and alcoholic extracts from stigmas and petals exhibit anti-inflammatory and nociceptive effects, benefiting severe and chronic pain (Eghdami et al., 2013).

2.7 Effect on Cholesterol Levels

Snacking, often linked to stress, contributes to obesity and metabolic disorders like hyperlipidemia, hypertension, heart disease, renal issues, and type 2 diabetes. Crocetin in saffron helps lower blood cholesterol, reducing the risk of myocardial infarctions and atherosclerosis (Hosseinzadeh et al., 2002). Crocin's hypolipidemic action is thought to stem from pancreatic lipase suppression, limiting cholesterol and fatty acid absorption (Akhondzadeh et al., 2011). Saffron exhibits antiobesity and anorexigenic effects, reducing caloric lipid breakdown via pancreatic lipase suppression.It regulates glucose and lipid metabolism while enhancing neurotransmitter levels to curb hunger (Melnyk et al., 2010). Crocin significantly lowers lipid accumulation and weight gain, adjusting epididymal fat proportion (Mashmoul et al., 2013).

2.8 Effect on Blood Glucose and Insulin Resistance

Crocetin (40 mg/kg) enhances insulin sensitivity, reducing compensatory hyperinsulinemia. It mitigates hypertension from fructose-rich diets and prevents hyperlipidemia by maintaining normal LDL cholesterol, fats, and free fatty acids (Eghdami et al., 2013).

2.9 Healing of Second-Degree Burns

A study compared saffron extract ointment with silver sulfadiazine in rat models for thermal burn treatment. Rats treated with saffron ointment showed smaller burn sizes and enhanced epithelial cell formation indicating accelerated healing (Mashmoul et al., 2014).

2.10 Effects on the Eyes

Saffron has been traditionally used across various cultures to treat eye conditions like corneal disease, cataracts, and infections (McGee, 2004; Khorasani et al., 2008). Ancient Egyptians used kohl pencils for eye protection against sun, wind, and sand. Recent studies suggest saffron extract enhances blood circulation and retinal function,while helping reduce cataracts, photoreceptor apoptosis, and retinal degeneration Xuan et al., 1999; Makri et al., 2013; Laabich et al., 2006; Fernandez-Sanchez et al., 2012).

3. Cosmetology and Perfumery Uses of Saffron

Saffron's cosmetic applications date back to ancient civilizations, including Cleopatra's beauty rituals and traditional Iranian and Greek medicine for skin healing and rejuvenation. Soaked in donkey milk or combined with fats, it exhibits anti-aging effects (Zargari, 1997; Mir, 2004).

Hindu women historically used saffron-based bindis, symbolizing fortune and wisdom (Li et al., 2004). Saffron petals, rich in kaempferide and crocin, serve as bioactive compounds for modern cosmetic formulations (Zeka et al., 2015; Ahrazam et al., 2018).

3.1 Anti-UV Agent

Saffron's anti-sun properties help shield skin from harmful UV rays, potentially serving as a natural UV absorber superior to homosalate-based sunscreens (Golmohammadzadeh et al., 2010; Tabrizi et al., 2003). Its antioxidant effects contribute to skin cancer prevention and solar protection (Das et al., 2004; Carmona et al., 2007).

3.2 Redness of Dark Spots

Saffron reduces melanin production, promoting skin brightening and depigmentation (Das et al., 2010). Melanocytes synthesize melanin using eumelanin (brown-black) and pheomelanin (red-yellow) (Moshiri et al., 2015). Tyrosinase,a key enzyme in melanogenesis, drives oxidative reactions (Ferrence & Bendersky, 2001). Monoterpenoids, quercetin, crocin, and kaempferide in saffron, exhibit antioxidative effects, inhibiting tyrosinase activity to reduce melanin synthesis (De et al., 2001).

3.3 Anti-Aging Effect and Diseases of the Skin

Saffron (*Crocus sativus*) is widely used in herbal cosmetic, often soaked with Ocimum basilicum leaves to treat acne scars. Combining saffron strands with unpasteurized milk and oils enhances exfoliation and blood flow. Its antioxidant properties help reduce erythema, a condition causing red, inflamed rashes, by lowering proinflammatory cytokines like Interleukins and TNF. Skin melanoma may benefit from topical application of 3% saffron extract. Studies suggest saffron aids dermal cell proliferation and irritation recovery (Xing et al., 2008).

3.4 Perfumery

Saffron's fragrance, described as sensual by Aristophanes (Clouds 51), was cherished by the Greeks (Li and Wu, 2002). Its distinct aroma comes from safranal, its primary component. Between 2000–146 BC, Greeks used saffron as a regal pigment in palaces, salons, restrooms, and theaters later making it accessible to common citizens (Giaccio, 2004). The Parthian Empire incorporated saffron into royal fragrances and rejuvenating oils (Dadkhah et al., 2003). Today, its earthy, delicate scent remains a staple in modern perfumes.

3.5 Saffron as Natural Pigments in Cosmetics

Phyto-pigments, including beetroot anthocyanins, pepper carotenoids, curcumin, saffron, and chlorophyll, are traditionally used as coloring agents in cosmetics and food industries. While synthetic dyes pose long-term risks, the cosmetic industry is shifting toward organic alternatives. Due to its high cost, saffron is used in minute concentrations for cosmetics. It serves as a natural substitute for tartrazine and turmeric, the latter being prone to sunlight fading (Colledge, 2005).

4. Other Uses of Saffron

4.1 Culinary Use

Saffron has been a culinary staple since ancient times, prized for its honey-like fragrance with metallic undertones (Basker & Negbi, 1983). It flavors rice dishes in Southeast Asia, Iran, India, and Spain, including Paella Valenciana and zarzuela stew. French cuisine incorporates saffron in bouillabaisse, while Italy's Milanese risotto and Sweden's saffron cakes highlight its versatile use across Europe. Saffron enhances Iranian cuisine, notably in chelow kabab, Iran's national dish. South Asian cuisine incorporates it into kulfi, gulab jamun, and biryani (Tsatsaroni et al., 1998). In Moroccan cuisine, saffron flavors mrouzia (dill and mutton stew) and koftas (spiced meatballs in tomato sauce). It's also a key ingredient in chermoula herb blends and saffron-infused tea (Modaghegh et al., 2008).

4.2 Coloring Power

Naturally-derived dyes, favored in regions with chemical food coloring bans, offer safer alternatives. Crocin, a water-soluble carotenoid, enables *C. sativus* to serve as a coloring agent in food, cosmetics, and textiles (Romadan et al., 2012).

It enhances pasta, dairy products (butter, margarine, cheese), and textile dyeing, imparting its distinct golden hue.

Crocus sativus treatments remain stable in acidic and basic media, due to alkyl alkanoates, dicarboxylic acids, nitrogenous compounds, and crocin pKa. Its buffer solutions help reduce cellulose oxidation, making saffron dyes ideal for silk, carpets, wool, and Buddhist monk robes. Naturally-derived dyes are safer, allergen-free, and environmentally biodegradable, offering a sustainable alternative to chemical dyes.

5. Safety and Toxicity Considerations

Saffron has been a culinary staple for decades. In mice, the LD50 is 1.6 g/kg for stigmas and 6 g/kg for petals via intraperitoneal administration (Kianbakht, 2009). Oral doses above 5 g/kg are harmful, with 20 g daily potentially fatal. In humans, saffron is safe below 1.5 g, but toxic above 5 g, causing vertigo, nausea, diarrhea, and vomiting. Severe poisoning may lead to uncontrollable bleeding, numbness, conjunctivitis, and yellowing of skin and sclera due to pigment accumulation (Schmidt et al., 2007).

Conclusions

Crocus sativus, one of the most profitable therapeutic plants, plays a key role in regional economic growth. Its dried stigmas yield saffron, a prized culinary spice with applications in pharmaceuticals, cosmetics, and textiles. Saffron's cosmetic appeal stems from its health benefits and organic composition, with recent research elevating its pharmaceutical significance in Codex listings. It aids in oxidative stress, sun damage, hyperpigmentation, and aging. However, meticulous cultivation results in low yields, driving high costs and risk of adulteration, limiting availability.

References

Abdullaev, F. Biological properties and medicinal use of saffron. *ISHS Acta Hortic.* 2006, 739, 339–345.

Abrishami, M.H. Understanding of Iranian Saffron, 1st ed., Tous: Tehran, Iran, 1987.

Ahrazem, O., Argandona, J., Fiore, A., Aguado, C., Luján, R. et al., Transcriptome analysis in tissue sectors with contrasting crocins accumulation provides novel insights into apocarotenoid biosynthesis and regulation during chromoplast biogenesis. *Sci. Rep.* 2018, 8, 2843.

Akhondzadeh, B.A., Khondzadeh, S., Ghoreishi, S., Noorbala, A., Rezazadeh, S. et al. Petal and Stigma of Crocus sativus L. in the Treatment of Depression: A Pilot Double—Blind Randomized Trial. *J. Med. Plant. Res.* 2008, 7, 29–36.

Akhondzadeh, S., Ashrafi, M., Ghaeli, P., Jamshidi, A., Kashani, L. et al., Saffron for treatment of fluoxetine-induced sexual dysfunction in women: Randomized double-blind placebo-controlled study. Hum. *Psychopharmacol.* 2013, 28, 54–60.

Akhondzadeh, S., Harirchian, M., Shaffe-Sabet, M. Saffron in the treatment of patients with mild to moderate Alzheimer's disease: A 16-week, randomized and placebo-controlled trial. *J. Clin. Pharm. Ther.* 2010, 35, 581–588.

Akhondzadeh, S., Kamalipour, M. Cardiovascular Effects of Saffron: An Evidence-Based Review. *J. Tehran Heart Cent.* 2011, 6, 59–61.

Babaei S, Talebi M, Bahar M. (2014). Developing an SCAR and ITS reliable multiplex PCR-based assay forsaf flower adulterant detection in saffron samples. *Food Control* 35: 323–328. https:// doi.org/ 10. 1016/j. foodcont. 2013. 07. 019

Badie Bostan, H., Mehri S., Hosseinzadeh, H. (2017). Toxicology effects of saffron and its constituents: a review. *Iran J Basic Med Sci 20*: 110–121. https://doi.org/10.22038/ ijbms 2017. 8230

Basker, D., Negbi, M. Uses of saffron. Econ. Bot. 1983, 37, 228–236.

Bathaie, S.Z., Farajzade, A., Hoshyar, R. (2014). A review of the chemistry and uses of crocins and crocetin, the carotenoid natural dyesin saffron, with particular emphasis on applications as colorants including their use as biological stains. *Biotech Histochem 89*: 401–411. https:// doi. org/ 10. 3109/ 10520 295. 2014. 890741

Bathaie, S., Bolhassani, A., Khavari, A. Saffron and natural carotenoids: Biochemical activities and anti-tumor effects. *Biochim. Biophys. Acta* 2013, 1845, 20–30.

Bhandari, P.R. (2015). *Crocus sativus* L. (saffron) for cancer chemoprevention: a mini review. J Tradit Complement Med *5*: 81–87. https:// doi.org/ 10. 1016/j. jtcme. 2014. 10. 009

Botsoglou, N.A., Florou, P.P., Nikolakakis, I., Giannenas, I., Dotas, V. et al. Effect of dietary saffron (*Crocus sativus* L.) on the oxidative stability of egg yolk. *Br. Poult. Sci.* 2005, 46, 701–707.

Cagliani, L.R., Culeddu, N., Chessa, M., Consonni, R. (2015). NMR investigations for a quality assessment of Italian PDO saffron (*Crocus sativus* L.). *Food Control 50*: 342–348. https:// doi. org/ 10. 1016/j. foodcont. 2014. 09. 017

Cardone, L., Castronuovo, D., Perniola, M., Cicco, N., Candido, V. (2020). Saffron (*Crocus sativus* L.), the king of spices: an overview. *Sci Hortic 272*: 109560

Carmona, M., Sanchez, A.M., Ferreres, F., Zalacain, A., Tomas-Barberan, F. et al. Identification of the flavonoid fraction in saffron spice by LC/DAD/ MS/MS: Comparative study of samples from different geographical origins. *Food Chem.* 2007, 100, 445–450.

Colledge, M.A.R. 'Parthians', Translated to Persian by M. Rajabnia, Hirmand: Tehran, Iran, 2005, pp. 1–86.

Dadkhah, M., Ehtesham, M., Fekrat, H. ëIranian Saffron an Unknown Jewelí, 1st ed., Shahr Ashub Publication: Tehran, Iran, 2003.

Das, I., Chakrabarty, R.N., Das, S. Saffron can prevent chemically induced skin carcinogenesis in Swiss albino mice. *Asian Pac. J. Cancer Prev*. 2004, 5, 70–76.

Das, I., Das, S., Saha, T. Saffron suppresses oxidative stress in DMBA-induced skin carcinoma: A histopathological study. *Acta Histochem*. 2010, 112, 317–327.

De Leeuw, S.M., Smit, N.P., Van, V.M., Pennings, E.M., Pavel, S. et al. Melanin content of cultured human melanocytes and UV-induced cytotoxicity. J. Photochem. Photobiol. 2001, 61, 106–113.

Dwyer, A., Hawrelak, J., Whitten, D. Herbal Medicines, other than St. John's Wort, in the treatment of Depression: A systematic Review. *Sci. Rev. Altern. Med.* 2011, 16, 40–49.

Eghdami, K., Erfanparast, A., Farshid, A.A., Samadi, F., Tamaddonfard, E. et al. Comparison of the effects of crocin, safranal and diclofenac on local inflammation and inflammatory pain responses induced by carrageenan in rats. *Pharmacol. Rep*. 2013, 65, 1272–1280.

Fernández-Sánchez, L., Lax, P., Esquiva, G., Martín-Nieto, J., Pinilla, I. et al. Safranal, a saffron constituent, attenuates retinal degeneration in P23H rats. PLoS ONE 2012, 7, e43074.

Ferrence, S.C. and Bendersky, G. (2004). Therapy with saffron and the goddess at Thera. *Perspective of Biological Medicine*, *47*, 199–226.

Ferrence, S.C. and Bendersky, G. Therapy with saffron and the goddess at Thera. Perspect. *Biol. Med.* 2001, 4, 199–226.

Giaccio, M. Crocetin from saffron. An active component of an ancient spice. Crit. Rev. *Food Sci. Nutr.* 2004, 44, 155–172.

Golmohammadzadeh, S., Jaafari, M.R. and Hosseinzadeh, H. Does saffron have antisolar and moisturizing effects? Iran. *J. Pharm. Res*. 2010, 9, 133–140.

Hamidreza, P., Marjan, A., Rouhollah, H., Shahriar, E., Soltan, A. et al. Effect of Crocus sativus on gentamicin induced nephrotoxicity. *Biol. Res*. 2010, 43, 83–90.

Hill, T. The Contemporary Encyclopedia of Herbs and Spices: Seasonings for the Global Kitchen, 1st ed., Wiley: Hoboken, NJ, USA, 2004.

Hosseini, A., Razavi, B.M. and Hosseinzadeh, H. (2018). Saffron (*Crocus sativus*) petal as a new pharmacological target: a review. Iran J Basic Med Sci 21:1091–1099

Hosseinzadeh, H., Molaei, M., Shakeri, M., Rajabi, O., Shamsa, A. et al. Evaluation of *Crocus sativus* L. (saffron) on male erectile dysfunction: A pilot study. *Phytomedicine* 2009, 16, 690–693.

Hosseinzadeh, H. and Nassiri-Ask, M. Avicenna's (Ibn Sina) the Canon of Medicine and Saffron (*Crocus sativus*): A review. *Phytother. Res.* 2013, 27, 475–783.

Hosseinzadeh, H. and Younesi, H.M. Antinociceptive and anti-inflammatory effects of *Crocus sativus* stigma and petal extracts in mice. *BMC Pharmacol.* 2002, 2, 7.

Khazdair, M.R., Anaeigoudari, A., Hashemzehi, M. and Mohebbati, R. Neuroprotective potency of some spice herbs, a literature review. *J. Tradit. Complement. Med.* 2018, 9, 98–105.

Khorasani, G., Hosseinimehr, S.J., Zamani, P., Ghasemi, M., Ahmadi, A. et al. The Effect of Saffron (*Crocus Sativus*) Extract for Healing of Second-degree Burn Wounds in Rats. *Keio J. Med.* 2008, 57, 190–195.

Kianbakht, S. A systematic review on pharmacology of saffron and its active constituents. *J. Med. Plants* 2009, 28, 1–23.

Kumar, R., Singh, V., Devi, K., Sharma, M., Singh, M.K. et al. (2009). State of art of saffron (*Crocus sativus* L.) agronomy: a comprehensive review. *Food Rev Int 25*(1): 44–85. https:// doi.org/ 10.1080/ 87559 12080 24585 03

Laabich, A., Vissvesvaran, G.P., Lieu, K.L., Murata, K., McGinn, T.E. et al. Protective effect of crocin against blue light- and white light-mediated photoreceptor cell death in bovine and primate retinal primary cell culture. Investig. Ophthalmol. *Vis. Sci.* 2006, 47, 3156–3163.

Li, C.Y., Lee, E.J. and Wu, T.S. Antityrosinase principals and constituents of the petals of *Crocus sativus*. *J. Nat. Prod.* 2004, 67, 437–440.

Li, C.Y. and Wu, T.S. Constituents of pollen of Crocus sativus L. and their tyrosinase inhibitory activity. *Chem. Pharm. Bull.* 2002, 50, 1305–1309.

Makri, O.E., Ferlemi, A.V., Lamari, F.N. and Georgakopoulos, C.D. Saffron administration prevents selenite-induced cataractogenesis. *Mol. Vis.* 2013, 30, 1188–1197.

Mashmoul, M., Azlan, A., Yusof, B.N., Khaza'ai, H., Mohtarrudin, N. et al. Effects of saffron extract and crocin on anthropometrical, nutritional and lipid profile parameters of rats fed a high fat diet. *J. Funct. Foods* 2014, 8, 180–187.

Mashmoul, M., Azrina, A., Huzwah, K., Barakatun, N., Mohd, Y. et al. Saffron: A Natural Potent Antioxidant as a Promising Anti-Obesity Drug. *Antioxidants* 2013, 2, 293–308.

McGee, H. On Food and Cooking: The Science and Lore of the Kitchen, Scribner: New York, NY, USA, 2004, p. 422, ISBN 0-684-80001-2.

Melnyk, J., Marcone, M. and Wang, S. Chemical and biological properties of the world's most expensive spice: Saffron. *Food Res. Int.* 2010, 43, 1981–1989.

Mir, H. Herbal Knowledge: Usage of Herbs in Prevention and Treatment of Diseases, with Latest Research around the World, 2nd ed., Daftare Nashre Farhange Eslami: Tehran, Iran, 2004.

Modaghegh, M.H., Shahabian, M. and Esmaeili, H.A. Safety evaluation of saffron (*Crocus sativus*) tablets in healthy volunteers. *Phytomedicine* 2008, 15, 1032–1037.

Moshiri, M., Vahabzadeh, M. and Hosseinzadeh, H. Clinical Applications of Saffron (*Crocus sativus*) and its Constituents: A Review. *Drug Res.* 2015, 65, 287–295.

Murray, C.J. and Lopez, A.D. Alternative projections of mortality and disability by cause 1990–2020: Global burden of disease study. *Lancet* 1997, 24, 1498–1504.

Premkumar, K., Abraham, S.K., Santhiya, S.T. and Ramesh, A. Protective effects of saffron (*Crocus sativus* L.) on genotoxinsinduced oxidative stress in Swiss albino mice. *Phytother. Res.* 2003, 17, 614–617.

Rahaiee, S., Moini, S., Hashemi M. and Shojaosadati, S.A. (2015). Evaluation of antioxidant activities of bioactive compounds & various extracts obtained from saffron (*Crocus sativus* L.): a review. *J Food Sci Technol 52*: 1881–1888. https:// doi.org/ 10. 1007/s13197-013-1238x

Ramadan, A., Soliman, G. and Mahmoud, S.S. Evaluation of the safety and antioxidant activities of *Crocus sativus* and Propolis ethanolic extracts. *J. Saudi Chem. Soc.* 2012, 16, 13–21.

Razak, S.I.A., Anwar Hamzah, M.S., Yee, F.C., Kadir, M.R.A. and Nayan, N.H.M. (2017). A review on medicinal properties of saffron toward major diseases. *J Herbs Spices Med Plants 23*: 98–116. https://doi.org/ 10. 1080/ 10496 475. 2016. 12725 22

Rios, J.L., Recio, M.C., Giner, R.M. and Manez, S. (1996). An update review of saffron and its active constituents. *Phytotherapy Research, 10*, 189–193.

Riverón, N.L. The combination of natural and synthetic agents: A new pharmacological approach in cancer chemoprevention. *Proc. West. Pharmacol. Soc.* 2002, 45, 74–75.

Schmidt, M., Betti, G. and Hensel, A. Saffron in phytotherapy: Pharmacology and clinical uses. *Wien. Med. Wochenschr.* 2007, 157, 315–319.

Suganya, K., Preethi, P.S., Suganya, M. and Nanthini, A. (2016) Natural pigments in cosmetics—Past to present. *Int. J. Pharm. Sci. Bus. Manag.*, 6, pp.7–14.

Tabrizi, S., Mortazavi, S.A. and Kamalinejad, M. An *in vitro* evaluation of various Rosa damascena flower extracts as a natural antisolar agent. *Int. J. Cosmet. Sci.* 2003, 25, 259–265.

Tamaddonfard, E., Erfanparast, A., Farshid, A.A., Imani, M. and Mirzakhani, N. et al. Safranal, a constituent of saffron, exerts gastro-protective effects against indomethacin-induced gastric ulcer. *Life Sci.* 2019, 224, 88–94.

Tsatsaroni, E.G., Liakopoulou-Kyriakides, M. and Eleftheriadis, I.C. Comparative study of dyeing properties of two yellow natural pigments. Effect of enzymes and proteins. *Dyes Pigm.* 1998, 37, 307–315.

Xing, H.G., Li, Z., Huachen, W. and Hong, D.C. Effcacy and safety of innovative cosmeceuticals. *Clin. Dermatol.* 2008, 26, 367–374.

Xuan, B., Zhou, Y.H., Li, N., Min, Z.D., Chiou, G.C. et al. Effects of crocin analogs on ocular blood flow and retinal function. *J. Ocul. Pharmacol. Ther.* 1999, 15, 143–152.

Zargari, A. Medicinal Plants, 1st ed., Tehran University Publication: Tehran, Iran, 1997.

Zeka, K., Ruparelia, K.C., Continenza, M.A., Stagos, D., Vegliò, F. et al. Petals of Crocus sativus L. as a potential source of the antioxidants crocin and kaempferol. *Fitoterapia* 2015, 107, 128–134.

6 | Beneficial Effects, Applications, Safety Evaluation and Limitations of the Medicinal Plant *Crocus sativus* (Saffron)

Mani Iyer Prasanth[1,2,3], Bhagavathi Sundaram Sivamaruthi[4,5,6], Dicson Sheeja Malar[7], James Michael Brimson[1,8] and Tewin Tencomnao[1,2*]

Saffron, derived from *Crocus sativus*, has been a traditional medicinal staple for centuries, owing its health benefits to key compounds: crocin, crocetin, picrocrocin, and safranal. These components contribute to antioxidant, antitumor, antidepressant, and anxiolytic effects. Crocin, responsible for saffron's color, offers memory-enhancing and neuroprotective properties. Safranal, a volatile terpene aldehyde, supports cardiovascular, respiratory, and metabolic health. Saffron's immunomodulatory effects make it a promising candidate for inflammatory disorders and mood regulation, including anxiety and schizophrenia. Its growing role in modern medicine highlights its versatile therapeutic applications.

1. Introduction

Saffron (*Crocus sativus*) has long been used in folk and traditional medicine for various ailments. Its key compounds—crocin, crocetin, picrocrocin, and safranal—contribute to its hypotensive, hypoglycemic, antioxidant, antitumor, anti-inflammatory, and antidepressant effects. Saffron extract combats oxidative stress, reducing lipid peroxidation and nitric oxide levels, while boosting glutathione,

[1] Center of Excellence on Natural Products for Neuroprotection and Anti-ageing (Neur-Age NatChula), Chulalongkorn University, Bangkok, Thailand.
[2] Department of Clinical Chemistry, Faculty of Allied Health Sciences, Chulalongkorn University, Bangkok, Thailand.
[3] Nitte (Deemed to be University), Nitte University Centre for Science Education and Research (NUCSER), Division of Molecular Genetics and Cancer, Mangaluru, Karnataka, India.
[4] Office of Research Administration, Chiang Mai University, Chiang Mai, Thailand.
[5] Innovation Center for Holistic Health, Nutraceuticals, and Cosmeceuticals, Faculty of Pharmacy, Chiang Mai University, Chiang Mai 50200, Thailand.
[6] Department of Microbiology, Karpagam Academy of Higher Education, Coimbatore 641021, India.
[7] Department of Biomedical Engineering, Annai Vailankanni College of Engineering, Pottakulam, Tamil Nadu, 629401, India.
[8] Innovation and International Affairs, Faculty of Allied Health Sciences, Chulalongkorn University, Bangkok, Thailand.
* Corresponding author: tewintencomnao@gmail.com; tewin.t@chula.ac.th

DOI: 10.1201/9781003403760-6

SOD, CAT, GPx, and thiol content (Rahaiee et al., 2015; Cerdá-Bernad et al., 2022). Crocin, responsible for saffron's color, offers diverse health benefits with high efficacy and minimal toxicity, phytochemical (Alavizadeh and Hosseinzadeh, 2014).

Safranal, a volatile terpene aldehyde, known for its aromatic properties and strong antioxidant effects. It offers therapeutic potential for diabetes, asthma, cancer, neurodegenerative, and cardiovascular diseases, while also exhibiting anti-photoaging benefits for the skin. Its antioxidative action stabilizes biological membranes, prevents fatty acid peroxidation, and restores antioxidant enzyme activity. The unsaturation of safranal enables efficient scavenging of singlet oxygen and peroxyl radicals, aiding in oxidative stress reduction (Nanda & Madan, 2021; Esmaealzadeh et al., 2023). Saffron is increasingly recognized in complementary, alternative, and modern medicine for its health-enhancing effects. It modulates innate and acquired immunity, regulating MAPK and NF-κB pathways to control pro-inflammatory cytokines, enzymes, adhesion molecules, and chemokines, aiding in inflammatory process management (Zeinali et al., 2019). Additionally, saffron shows promise in mood disorder treatment for children and adolescents (Saxena et al., 2023) and is a potential therapeutic agent for anxiety and schizophrenia (Pitsikas, 2016).

2. Beneficial Activities of *Crocus sativus*

2.1 Anticancer Activity

Saffron (*Crocus sativus*) exhibits anticancer properties, targeting leukemia, skin, cervical, breast, colorectal, liver, and pancreatic cancers, largely due to its carotenoid antioxidants (Samarghandian & Borji, 2014). It selectively targets cancer cells, preventing tumor formation while protecting normal cells and reducing chemotherapy toxicity (Shakeri et al., 2020). Its aqueous infusion aids in reducing papilloma formation in skin carcinogenesis, likely by modulating phase II detoxifying enzymes like GST, GPx, CAT, and SOD (Das et al., 2004). Saffron extracts (aqueous and alcoholic) and crocin effectively reduce proliferation and induce G0/G1 cell cycle arrest in malignant prostate cancer cell lines. They trigger apoptosis, downregulating Bcl-2 and upregulating Bax.

Additionally, saffron, crocin, and crocetin suppress invasion and migration in xenografted male nude mice, suggesting anti-metastatic potential (D'Alessandro et al., 2013; Festuccia et al., 2014). Safranal inhibits PC-3 cell growth, inducing apoptotic cell death (Samarghandian & Shabestari, 2013; Jiang et al., 2020)

Crocin significantly reduces pancreatic cancer cell viability, inducing apoptosis via Bcl-2 downregulation and cell cycle modulation. In murine models, it lowers tumor burden and protects against radiation-induced hepatic oxidative damage (Bakshi et al., 2020).

Crocin induces apoptosis in gastric adenocarcinoma cells, increasing the Bax/Bcl-2 ratio. It also triggers autophagy-independent programmed cell death in colon cancer cells (Hoshyar et al., 2013; Amin et al., 2015).

Saffron's anticancer effects include cell cycle arrest, apoptosis induction, immune modulation, and anti-inflammatory activity (Naeimi et al., 2019). It also inhibits viability in human alveolar basal epithelial carcinoma cells, suggesting chemotherapeutic potential for lung cancer (Samarghandian et al., 2011).

Saffron and its bioactive compounds exhibit anticancer properties, inducing apoptosis, cell cycle arrest, and modulating key signaling pathways to target various cancer types.

2.2 Anti-inflammatory Effect

Crocin lowers blood lipid levels, reducing endothelin, total cholesterol, and LDL, while increasing HDL. It inhibits lipogenesis, suppressing lipogenesis-related proteins and promoting lipid catabolism.Its anti-inflammatory effects decrease pro-inflammatory cytokines, elevate anti-inflammatory markers, and shift macrophage polarization to M2 phenotype. Additionally, crocin inhibits NF-κB p65 expression and nuclear translocation, potentially modulating coronary atherosclerosis (Li et al., 2018). *C. sativus* L. petal ethanolic extract shows anti-inflammatory effects, reducing nitric oxide and IL-6 production in RAW 264.7 murine cells. It also inhibits TRAP-positive osteoclast formation and downregulates key osteoclast-related genes, demonstrating anti-osteoclastogenic properties (Orabona et al., 2022).

Safranal activates TRPA1, inducing calcium responses in DRG neurons, leading to anti-inflammatory and analgesic effects. In non-alcoholic fatty liver disease models, saffron supplementation reduces C-reactive protein, leptin, and malondialdehyde, while boosting total antioxidant capacity (Li Puma et al., 2019; Pour et al., 2020).

Crocetin reduces liver enzyme activities, lipid levels, and oxidative stress markers in NAFLD mouse models. It enhances liver and kidney histology, modulates inflammatory cytokines, and boosts anti-inflammatory markers (Xu et al., 2021). Saffron and crocin, administered through intra-peritoneal injections, reduce adhesion band frequency, inflammatory cell infiltration, collagen composition, and oxidative stress in mice (Arjmand et al., 2021). In ulcerative colitis patients, saffron supplementation boosts SOD and GPx levels, enhancing total antioxidant capacity and reducing disease severity (Tahvilian et al., 2021). Pre-treatment in murine colitis models improves colonic mucosa histopathology, alleviates colitis, and modulates gut microbiome and short-chain fatty acids, supporting its therapeutic potential (Banskota et al., 2021)

Crocetin reverses uterine ischemia/reperfusion injury in rats, reducing inflammation and oxidative stress (Peng et al., 2023a). In hypertensive older men, saffron supplementation with resistance training lowers inflammatory biomarkers linked to cardiovascular disease (Mojtahedi et al., 2022). However, a meta-analysis finds no significant impact on C-reactive protein, TNF-α, or IL-6 levels, suggesting limited effects on inflammatory cytokines (Asbaghi et al., 2021).

2.3 Anti-obesity Effect

Saffron stigma ethanolic extract reduces rat body weight, with appetite suppression noted as a clinical side effect (Akhondzadeh et al., 2008; Noorbala et al., 2005). Crocin prevents weight gain, fat accumulation, and insulin resistance in ovariectomized mice, enhancing energy expenditure and fat oxidation (Liu et al., 2021). In a trial with 60 overweight women, saffron supplementation led to reduced snacking and greater weight loss compared to placebo, highlighting its satiety-enhancing and anti-obesity effects (Gout et al., 2010; Hasani-Ranjbar et al., 2013). It also positively impacted waist circumference and serum malondialdehyde levels (Ebrahimi et al., 2019a; Ebrahimi et al., 2019b). Saffron supplementation significantly reduces waist circumference and fasting plasma glucose (Rahmani et al., 2020). Another meta-analysis found a nonsignificant decrease in weight, BMI, and hip circumference, but a notable reduction in waist-to-hip ratio. While fasting blood sugar improved in metabolic syndrome patients, HbA1C levels remained unchanged, suggesting potential cardiometabolic benefits for overweight and obese individuals (Tahmasbi et al., 2022).

2.4 Anti-hepatotoxic Effect

Acetaminophen overdose can cause hepatotoxicity, leading to cell swelling, inflammation, and necrosis in rat livers, with elevated enzyme levels and reduced protein and albumin. However, *C. sativus* hydro-alcoholic extract (20 mg/kg) showed mild hepatocyte degeneration, lowering alanine/aspartate aminotransferase and bilirubin, while enhancing total protein and albumin, offering liver protection (Omidi et al., 2014). *C. sativus* L. petal polyphenolic-rich extract regulates liver oxidative stress and inflammatory response in rats, effectively controlling hepatic injury markers and reducing liver damage (Wali et al., 2020).

Saffron supplementation reduces hyperglycemia, hyperlipidemia, and improves liver function, lowering fasting plasma glucose, triglycerides, insulin, AST, and ALT. It also enhances depression, sleep quality, and overall well-being in diabetic patients (Tajaddini et al., 2021.

2.5 Hypoglycemic and Antidiabetic Effects

Crocin, recognized for its antidiabetic effects, mitigates insulin insensitivity induced by free fatty acids and regulates TNF-alpha mRNA expression in fructose-fed rodents (Schmidt et al., 2007; Yoshino et al., 2011). Saffron enhances glucose uptake and insulin sensitivity, linked to AMPK and MAPK phosphorylation, excluding PI 3-kinase. Combined saffron and insulin administration improves insulin sensitivity via AMPK, MAPKs, PI 3-kinase, and mTOR pathways (Winterhalter & Straubinger, 2000).

Crocin reduces malondialdehyde and xanthine oxidase activities while increasing glutathione levels in diabetic rat kidneys, leading to lower blood urea nitrogen and creatinine (Altinoz et al., 2015). Pre-treatment with crocin before renal ischemia/reperfusion provides anti-inflammatory, antioxidant, and protective effects, improving leukocyte infiltration, ICAM-1, TNF-α mRNA expression, malondialdehyde, and ferric reducing/antioxidant power levels (Yarijani et al., 2017).

Saffron supplementation (3 months) significantly reduced fasting plasma glucose, cholesterol, LDL-c, and LDL/HDL ratio, while showing no significant changes in HbA1c, HDL-C, API, and TG (Moravej Aleali et al., 2019). Saffron supplementation (8 weeks) lowers fasting blood sugar and glycosylated hemoglobin, while enhancing antioxidant activity (Karimi-Nazari et al., 2019). In type 2 diabetes patients, 100 mg/day saffron reduces fasting blood glucose and downregulates TNF-α and IL-6 gene expression (Mobasseri et al., 2020). Crocin supplementation improves antioxidant capacity in nicotinamide-streptozotocin-induced diabetic rats, regulating SOD1 expression and restoring SOD and total antioxidant capacity. It also significantly reduces ALT and AST levels.In type 2 diabetic patients, crocin lowers serum high-sensitive C-reactive protein, TNF-α, NF-κB, and Plasma IL-6, alleviating diabetes complications (Behrouz et al., 2021).

Saffron intervention lowers fasting blood sugar, lipid profile, atherogenic indices, liver enzymes, nitric oxide, and malondialdehyde, improving glycemic control, and oxidative status in diabetes (Tajaddini et al., 2023).

Systematic reviews suggest saffron may improve fasting blood glucose and HbA1c, though further validation is needed (Correia et al., 2023; Giannoulaki et al., 2020). Another review highlights saffron's positive impact on oxidative stress (Abedi et al., 2023).

2.6 Hypolipidemic Effect

Crocin, a key saffron constituent, lowers cholesterol in hyperlipidemic rats and reduces triglycerides in quails with coronary artery disease (Sheng et al., 2006; He et al., 2005). It inhibits dietary fat absorption, reducing fat hydrolysis, and promotes fecal fat excretion by blocking pancreatic lipase (Sheng et al., 2006). Additionally, saffron and crocin improve liver enzyme levels, albumin, and fatty liver histopathology (Mashmoul et al., 2016).

Crocin improves serum ALT, AST, and total bilirubin, while preventing cholesterol and triglyceride elevation. It mitigates glutathione depletion, SOD exhaustion, and malondialdehyde accumulation in liver tissue. Additionally, crocin inhibits TGF-β, alpha-smooth muscle actin, and collagen 1-α, while upregulating matrix metalloproteinase-2. It also reduces NF-κB and COX-2 expression, demonstrating protective effects against thioacetamide-induced liver fibrosis (Algandaby, 2018).

Saffron supplementation significantly reduces serum alanine aminotransferase, but has no notable effect on aspartate aminotransferase and alkaline phosphatase levels (Mousavi et al., 2022).

2.7 Antimicrobial Activity

Safranal shows antifungal efficacy against *Candida* spp., and its thiosemicarbazone derivative enhances this effect. These compounds suppress germ tube formation, a key l virulence factor in *C. albicans,* suggesting potential for novel anti-Candida agents (Carradori et al., 2016).

C. sativus stigma extracts from Morocco and Italy show greater efficacy against Gram-positive bacteria than Gram-negative strains (Zaazaa et al., 2021). Petal applications have been proposed for superficial wound treatment (Pagano et al., 2022). Additionally, complete inhibition of mycelial growth and spore germination was observed in *Rhizopus stolonifera*, *Penicillium digitatum*, and *Botrytis cinerea* (Naim et al., 2022).

2.8 Anxiety and Depression

Diazinon, an organophosphorus insecticide, irreversibly inhibits cholinesterase, reducing antioxidants activity and triggering oxidative stress. It suppresses red blood cell cholinesterase activity and elevates TNF-α, 8-iso-prostaglandin F(2α), and soluble protein-100 β. However, *C. sativus* mitigates diazinon-induced toxicity in rats (Moallem et al., 2014).

Saffron supplementation (8 weeks) reduced depression scores in healthy individuals. It also attenuated stress-induced heart rate variability decline and increased urinary crocetin levels, correlating with improved mood (Jackson et al., 2021). *C. sativus* supplementation (6 weeks) improved Hamilton Depression Rating Scale scores in depressed outpatients (Moshiri et al., 2006). Crocin enhanced depression-related inventories (Talaei et al., 2015). In postpartum depression, saffron matched fluoxetine's effects, while in major depressive disorder with anxious distress, it showed similar efficacy to citalopram (Kashani et al., 2017; Ghajar et al., 2017). Saffron regulates serotonin and dopamine secretion, reducing depression in methamphetamine users with HIV/AIDS (Jalali and Hashemi, 2018). Crocin exhibits neuroprotective effects, improving depression-related markers in stressed rats (Abbaszade-Cheragheali et al., 2022). Crocetin alleviates depression, protecting against hippocampus injury (Lin et al., 2021). Saffron pre-treatment prevents stress-induced depressive symptoms, highlighting its neuroprotective potential (Monchaux De Oliveira et al., 2021).

Systematic reviews and meta-analyses confirm saffron's efficacy in treating depression, comparable to pharmaceutical antidepressants, with fewer side effects (Dwyer et al., 2011; Modabbernia & Akhondzadeh, 2013; Hausenblas et al., 2015; Khazdair et al., 2015; Shafiee et al., 2018; Khaksarian et al., 2019). Saffron combined with other substances improves major depressive disorder (Hausenblas et al., 2015). In mice, saffron enhances time spent in the open field maze, correlating with gut microbial shifts, suggesting a mental health connection (Pontifex et al., 2022).

Saffron supplementation alleviates depression symptoms in older adults, as assessed by the Hamilton Depression Rating Scale (Ahmadpanah et al., 2019).

Circulating human metabolites from saffron intake protect neurons from oxidative stress-induced neurotoxicity, reducing depression (Wauquier et al., 2022).

In summary, saffron's bioactive compounds, particularly crocins, offer promising neuroprotective and molecular mechanisms for depressive disorder treatment.

2.9 Alzheimer's and Huntington's Disease

Saffron supplementation at (15 mg twice daily) improves cognitive function in Alzheimer's disease (AD) patients (Akhondzadeh et al., 2010). Crocin administration enhances memory deficit, autophagy, and apoptosis markers in AD-induced Wistar rats, improving spatial memory and reducing Bax/Bcl-2 ratio and cleaved Caspase-3, suggesting beta-amyloid apoptosis inhibition via antioxidant properties (Asadi et al., 2015).

Saffron supplementation regulates cognitive decline in MCI patients, improving Mini-Mental State Examination scores and positively influencing MRI, EEG, and ERP. In AD models, saffron modulates glutamate levels, reduces oxidative stress, and affects Aβ and tau aggregation, showing comparable efficacy to donepezil and memantine with a better safety profile (D'Onofrio et al., 2021).

Oral saffron intake improves cognitive assessment scores in MCI/AD patients, linked to cholinergic, dopaminergic, and glutamatergic interactions. It enhances antioxidant status, inhibits ROS, and reduces inflammatory cytokines, preventing amyloid beta-peptide fibril formation while boosting serotonin, supporting neuroprotection (Abu-Izneid et al., 2022). *C. sativus*, *Panax ginseng*, and *Ginkgo biloba* herbal mix improves memory, biochemical markers, and cognition in vascular dementia. A phase III clinical trial compared herbal formulation vs placebo over 52 weeks, followed by a 13-week follow-up, assessing cognitive function and safety (Karamacoska et al., 2023). Safranal mitigates 3-nitropropionic acid-induced alterations in Huntington's disease (HD) rat models, preventing motor dysfunction, weight loss, and locomotor deficits. It also reduces nitrite, malondialdehyde, SOD, CAT activity, and glutathione depletion, supporting neuroprotection (Fotoohi et al., 2021).

2.10 Memory and Cognition

Crocins alleviate non-spatial and spatial recognition memory deficits in rats, validated by NORT. Post-training crocin administration counteracts apomorphine-induced impairments, suggesting potential reversal of dopaminergic dysfunction-related memory deficits. Crocins also modulate memory storage and retrieval, with implications for schizophrenia-like behavioral deficits (Pitsikas and Tarantilis, 2017; Pitsikas, 2021).

Crocins reverse delay-dependent deficits in NORT, with crocin and memantine effectively counteracting natural forgetting. Their combination suggests potential for recognition memory regulation, offering an innovative approach to memory disorder treatment (Pitsikas & Tarantilis, 2018).

Crocins mitigate ketamine-induced memory deficits in rats, counteracting impairments in object recognition, object location, and habituation tests, which reflect non-spatial spatial, and associative memory (Pitsikas & Tarantilis, 2021). *C. sativus* and crocin support cognitive function, mitigating memory disorders in Alzheimer's, cerebral injuries, and schizophrenia models. Their mechanisms include long-term potentiation, anti-amyloidogenic activity, AChE inhibition, and antioxidant effects (Pitsikas, 2015).

2.11 Cerebral Ischemia

Saffron supplementation reduces stroke severity in acute ischemic stroke, lowering serum neuron-specific enolase and s100 levels while increasing Brain-derived neurotrophic factor. After three months, it maintains a higher Barthel index (Asadollahi et al., 2019). In another study, saffron (400 mg/day) enhances antioxidant enzyme activity, glutathione, and total antioxidant capacity, while reducing malondialdehyde levels and stroke severity per NIH Stroke Scale (Gudarzi et al., 2022). Oxidative stress triggers necrosis and apoptosis in ischemic tissue, leading to inflammatory responses. Saffron reduces brain natriuretic peptide levels, mitigating stroke-induced brain edema, and enhances vascular endothelial growth factor expression, aiding cerebral protection (Azami et al., 2021). Crocin modulates neuroinflammatory mitochondrial signaling, hypoxia-inducible factors, vascular endothelial growth factors, and cytokines, supporting ischemic stroke recovery (Shahbaz et al., 2022).

2.12 Other Neuroprotective Effects

Crocetin and crocin modulate vasoconstriction and hypertension with crocetin reducing phenylephrine-induced contractility in intact vessels, while crocin increases contractility in deendothelialized vessels. Crocetin's effect is abolished by NG-nitro-l-arginine methyl ester, highlighting its role in nitric oxide-mediated vasodilation (Llorens et al., 2015). Saffron protects retinal photoreceptors, preserving morphology and visual function while improving flicker sensitivity in neurodegenerative disorders like age-related macular degeneration. Its interaction with P2X7 purinergic receptors supports retinal health, highlighting its therapeutic potential (Corso et al., 2016). Additionally, saffron simple syrup (2 months) significantly reduces fatigue severity, suggesting benefits for progressive neurological disorders like multiple sclerosis (Ashtiani et al., 2020).

Crocin protects against intracerebral hemorrhage-induced brain injury, significantly reducing brain edema and neurological deficits. It also inhibits neuronal ferroptosis, altering Fe^{2+} concentration and key marker expression, suggesting its therapeutic potential (Wang et al., 2022).

Saffron effectively reduces hyperactivity, improves sleep duration, and aids sleep onset in ADHD patients (7–17 years), showing promise as an alternative to methylphenidate (Blasco-Fontecilla et al., 2022). Saffron (40 mg/kg) and endurance exercise synergistically enhance brain-derived neurotrophic factor (BDNF) and serotonin in Wistar rats. The saffron plus exercise group showed higher Neurotrophin-3 mRNA in soleus muscle, increased hippocampal 5-hydroxyindolacetic acid, and improved short-term memory, highlighting saffron's neuroprotective potential (Akbari-Fakhrabadi et al., 2021).

2.13 Improving Sleep Quality

Crocetin enhances sleep maintenance and subjective sleep quality, improving wakefulness and refreshment scores compared to placebo, though other sleep parameters remain unchanged (Umigai et al., 2018).

Saffron supplementation (28 days) significantly improved Insomnia Severity Index total score, Restorative Sleep Questionnaire, and Pittsburgh Sleep Diary sleep quality ratings in adults with sleep problems (Lopresti et al., 2020).

Saffron supplementation (14 or 28 mg, 28 days) significantly improved sleep quality, mood upon awakening, and insomnia classifications, while increasing evening melatonin levels (Lopresti et al., 2021). *C. sativus* reduces insomnia severity, enhancing sleep quality and duration compared to placebos, potentially reducing reliance on sedative-hypnotics (Munirah et al., 2022). Saffron, crocin and safranal enhance sleep duration and quality, inducing hypnotic effects (Sadat Rafiei et al., 2023).

2.14 Reproductive and Sexual-Related Problems

Saffron stigma intake alleviated premenstrual syndrome symptoms in women aged 25–40, significantly improving Total Premenstrual Daily Symptoms and Hamilton Depression Rating Scale scores (Agha-Hosseini et al., 2008; Dante and Facchinetti, 2011). Additionally, in 120 women with premenstrual dysphoric disorder, two weeks of saffron supplementation led to notable symptom improvements, recorded via daily severity assessments (Rajabi et al., 2020).

Testis irradiation (2 Gy X-rays) in pubertal mice caused severe structural and cellular damage, marked by PARP1, PCNA, SOD2, and HuR overexpression and reduced SIRT1 and CAT levels. However, 24-hour crocetin treatment, pre-and post-radiation, significantly alleviated injury and modulated DNA damage and oxidative stress responses (Rossi et al., 2021). Similarly, crocin mitigated nitrosamine-induced reproductive toxicity in male rats, improving sperm parameters, total antioxidant capacity, testosterone levels, and seminiferous tube diameter (Salahshoor et al., 2022). Saffron stigma supplementation (200 mg/day for ten days) improved nocturnal penile tumescence, base rigidity, and sexual function in individuals with erectile dysfunction (Shamsa et al., 2009). It also enhanced erectile function and intercourse satisfaction in fluoxetine-induced sexual dysfunction (Modabbernia et al., 2012). In women, saffron increased arousal,lubrication, and pain relief but had no significant effect on satisfaction, orgasm, or desire (Kashani et al., 2013).

A saffron-based topical gel significantly improved erectile dysfunction in diabetic men, validated by the International Index of Erectile Function Questionnaire (Mohammadzadeh-Moghadam et al., 2015). Saffron, combined with other substances, showed potential benefits for premenstrual syndrome, sexual dysfunction, and infertility (Hausenblas et al., 2015).

Crocetin improved bull spermatozoa quality and fertilization rates by reducing superoxide anion, hydrogen peroxide, and lipid peroxidation, while preserving motility, viability, and acrosomal integrity. DNA fragmentation remained minimal. In the fertilization medium, crocetin significantly increased acrosome-reacted spermatozoa and blastocyst production, enhancing sperm quality and its fertilizing capacity through ROS regulation and lipid peroxidation control (Sapanidou et al., 2016).

Electromagnetic field exposure negatively affects the male reproductive system, increasing HspA2 immunoreactivity, germinal epithelium apoptosis, and abnormal sperm levels while reducing sperm count, LH, and testosterone in mice. Crocin effectively countered these effects (Vafaei et al., 2020) A randomized trial with 56 women found saffron improved the Hot Flash-Related Daily Interference Scale and Hamilton Depression Rating Scale, supporting its use for treating hot flashes and depressive symptoms in post-menopausal women (Kashani et al., 2018).

Saffron improved mood and psychological symptoms in perimenopausal women, significantly reducing Greene Climacteric Scale psychological scores and PANAS negative affect compared to placebo (Lopresti and Smith, 2021). Additionally, saffron herbal lowered depression and boosted happiness in a clinical trial (Delam et al., 2023).

A randomized trial found that saffron improved sexual dysfunction in married women (18–55 years), enhancing desire, lubrication, and satisfaction by 62%, as measured by the Female Sexual Function Index (Kashani et al., 2022).

PCOS, a hormonal disorder in reproductive-age women, involves irregular cycles, elevated androgens, and metabolic disruptions. Using network pharmacology, researchers identified key therapeutic targets in *C. sativus* extract, including serine/threonine kinase 1, STAT3, MAPK3, and MAPK1. Gene enrichment analyses highlighted MAP kinase and serine-threonine pathways, while molecular docking confirmed strong binding affinities of crocetin, picrocrocin, and safranal to these targets (Tiwari et al., 2023).

2.15 Cardiovascular Activity

Saffron's cardiovascular benefits stem from its antioxidant properties, aiding cholesterol reduction and mitigating atherosclerosis risk (Kamalipour & Akhondzadeh, 2011; Verma & Bordia, 1998). Crocetin, a saffron compound, enhances antibiotic yield and reduces hypercholestcrolemia and atherosclerosis in rabbits on a high-lipid diet (Basker & Negbi, 1983; Zheng et al., 2005). Additionally, saffron (10 μg/ml) improved myocardial function in isolated rabbit hearts affected by ROS-induced cardiac dysfunction (Chahine et al., 2013).

Saffron exhibits intrinsic smooth muscle activity, with greater potency on the ileum than the aorta. Kaempferol and crocin showed selective negative inotropic effects, with kaempferol reducing KCl-induced contraction in animal models, while crocin had no effect. Under oxidative stress, both compounds decreased intracellular ROS formation and enhanced cell viability, highlighting saffron petal extract's cardiovascular potential (Zeka et al., 2020).

Crocetin enhances acetylcholine-induced vasodilation in hypertensive and normotensive rat aortas, improving endothelial nitric oxide-mediated relaxation (Mancini et al., 2014). Crocin counteracts cigarette smoke-induced oxidative stress in rats, normalizing cotinine levels, inflammatory markers, and key gene expressions (PKC, PI3K, MAPK, Nrf2, GCLc), while exhibiting antioxidant and cardioprotective effects (Dianat et al., 2018).

Crocin reduced atherosclerotic area, increased eNOS expression, and lowered HIF-1α levels in apolipoprotein E knockout mice, without affecting total, LDL, or HDL cholesterol (Makaritsis et al., 2022). In individuals with atherosclerosis, saffron improved Macnew total, physical, and social domain scores, enhancing overall quality of life (Ahmadikhatir et al., 2022). Saffron improves cardiovascular markers, including triglycerides, total cholesterol, systolic blood pressure, TNF-α, waist circumference, malondialdehyde, total antioxidant capacity, and alanine transferase (Zamani et al., 2022). Its cardioprotective effects may stem from reducing ROS, regulating apoptosis via Bax/Bcl2, inhibiting TNF-α and IL-8 through NF-κB modulation, and influencing HDL and LDL cholesterol levels (Ghaffari & Roshanravan, 2019).

2.16 Visual Functions

Crocin and crocetin show therapeutic potential for ocular diseases, including AMD, glaucoma, and diabetic maculopathy (Heitmar et al., 2019). In murine models of dry eye disease, saffron significantly reduced fluorescein scores, improved ocular surface health, and lowered inflammatory cytokines TNF-α, IL-1β, and IL-6 (Yousefi-Manesh et al., 2022). Oral saffron supplementation showed therapeutic potential for mild/moderate age-related macular degeneration, improving visual acuity and reducing multifocal electroretinogram latency compared to placebo (Broadhead et al., 2019). Clinical studies using Snellen, LogMar, and EDTRS charts consistently reported enhanced visual acuity with saffron (Falsini et al., 2010; Piccardi et al., 2012; Marangoni et al., 2013; Lashay et al., 2016; Riazi et al., 2017; Broadhead et al., 2019). Non-randomized trials further indicated sustained benefits within three months of supplementation in early AMD patients (Piccardi et al., 2012; Marangoni et al., 2013).

Daily saffron supplementation for one month significantly reduced intraocular pressure (Bonyadi et al., 2014). Crocin (15 mg/day) improved best-corrected visual acuity, lowered HbA1c levels, and reduced central macular thickness in diabetic macular edema patients compared to placebo (Sepahi et al., 2018)]. Crocin exhibits cytoprotective, antioxidative, anti-inflammatory, and vasodilatory effects in retinal tissue, benefiting glaucoma, macular dystrophies, diabetic retinopathy, and AMD. Its therapeutic impact involves PI3K/Akt activation and NF-κB inhibition, supporting its role in retinal health and vision maintenance (Heydari et al., 2023). Saffron Repron demonstrated neuroprotective effects in an

LPS-induced neuroinflammation model, preserving cognitive and visual functions. Pre-treatment mitigated neuroinflammatory damage, particularly in retinal tissue, highlighting its potential as an early marker for neurodegenerative diseases like Alzheimer's (Di Paolo et al., 2023).

2.17 Breathing Functions

A clinical trial found that saffron supplementation improved allergic asthma symptoms, reducing daytime and nighttime shortness of breath, salbutamol spray usage, and activity limitations compared to placebo. It also lowered asthma severity, systolic and diastolic blood pressure, and eosinophil and basophil concentrations (Zilaee et al., 2019).

2.18 Pancreas Protection

Crocin reduced pancreatic edema, amylase, and lipase levels while mitigating oxidative stress from cerulein challenge. It regulated inflammatory cytokines, inhibited p65-NFκB nuclear translocation, and reversed histological damage. Additionally, crocin upregulated Nrf-2 and downregulated IL-6, TNF-α, nitrotyrosine, and NFκB, suggesting its potential in preventing acute pancreatitis and supporting pancreatic health (Godugu et al., 2020).

2.19 Diuretic Activity

Saffron aqueous extract increased urine output and enhanced sodium, potassium, creatinine, and urea excretion in experimental rats, comparable to hydrochlorothiazide, a potent diuretic (Shariatifar et al., 2014).

2.20 Ergogenic Effects

Saffron supplementation (300 mg/day for ten days) significantly increased isometric and isotonic forces in nonactive male university students, improving visual and audio reaction times—key for athletic performance. Its ergogenic effects may enhance muscle mitochondrial biogenesis, motor cortex function, blood perfusion, and oxygen transport (Meamarbashi & Rajabi, 2016).

2.21 Atheroprotective Effect

Saffron aqueous extract improved glucose and triglyceride profiles without affecting body weight or total cholesterol. It also dose-dependently reduced aortic stenosis and macrophage content in apoE knockout mice (Christodoulou et al., 2018).

2.22 Arthritis Treatment

In a rheumatoid arthritis rat model, weight gain decreased while arthritis index, thymus index, and plantar temperature increased. Blood parameters worsened, with elevated enzyme levels and increased TNF-α, COX-2, and NF-κB. Safranal regulated these markers, reducing inflammatory cell infiltration and edema (Cellat et al., 2022). Saffron supplementation significantly reduced tender and swollen joints, pain intensity, and disease activity score in rheumatoid arthritis patients. It improved Physician Global Assessment and erythrocyte sedimentation rate while lowering high-sensitivity C-reactive protein, TNF-α, IFN-γ, and malondialdehyde, with an increase in total antioxidant capacity (Hamidi et al., 2020).

2.23 Anti-angiogenic Effects

Crocetin and crocin inhibited vein vessel formation in zebrafish, showing differences in effective concentrations. In the HUVEC model, both compounds suppressed cell migration and tube formation while modulating VEGFR2 phosphorylation. Crocetin exhibited stronger VEGFR2 binding than crocin, highlighting their potential as natural biomedicines for diseases involving abnormal blood vessel growth (Zhao et al., 2021).

3. Protection from Toxins

Chlorpyrifos-induced cardiotoxicity in albino rats led to severe histological changes, including fiber disorganization, hemorrhage, and increased collagen deposition. Caspase 3 immunoreactivity was elevated, while SOD and CAT levels decreased, and MDA increased. Crocin treatment significantly improved these markers, restoring cardiac tissue toward normalcy (Khalaf & El-Mansy, 2019).

Copper oxide nanoparticles disrupted oxidative balance in male albino rats, increasing serum malondialdehyde while decreasing GPX and total antioxidant capacity. They also reduced red blood cell and hemoglobin levels while raising white blood cell count. Histopathological changes in the cerebellar cortex were observed. Crocin mitigated oxidative stress and structural damage, offering protective effects (Mohamed Mowafy et al., 2021).

4. Safety Evaluation

A double-blind trial found saffron supplementation for seven days had no significant adverse effects but lowered standing systolic blood pressure, mean arterial pressure, and hematological parameters while increasing sodium, blood urea nitrogen, and creatinine levels (Modaghegh et al., 2008). Crocin supplementation showed no notable changes in blood or other parameters (Mohamadpour et al., 2013).

5. Practical Applications

Saffron petal-derived hydrogels, enriched with kaempferol and crocin,exhibit strong biocompatibility in mouse fibroblasts, making them promising for wound treatment and skincare applications (Zeka et al., 2018).

Table 1 Health benefits of Saffron.

Activity	*Extracts/ bioactive compounds*	*Model used*	*Effect observed*	*Reference*
Anti-cancer Neuroprotection	Aqueous extract	Mice	• Papilloma formation Alter the GST, GPx, CAT and SOD	Das et al., 2004
	Aqueous and alcoholic extracts and crocin	Malignant and non-malignant prostate cancer cell lines	• Malignant cell proliferation, Bcl-2 expression. • Bax expression • Caspase-9 activation	D'Alessandro et al., 2013
	Safranal	Prostate cancer cell lines	Cytotoxic to PC-3 cells	Samarghandian and Shabestari, 2013
	Safranal	Prostate cancer cell lines	• G0/G1 cell cycle regulatory proteins • Cyclin-dependent kinase inhibitors • Skp2 expression	Jiang et al., 2020
	-	C6 glioma rat cell line	• Cellular viability • LC3-II protein	Giakoumettis et al., 2019
	Crocin	Cell lines and mice	• Cell viability, Bcl-2 expression, and hepatic toxicity	Bakshi et al., 2020
	Aqueous extract (50, 100 and 200 mg/kg)	Rats	Prevents cholinesterase activity. Improves the oxidative stress and neuronal damage markers.	Moallem et al., 2014
	Crocetin	Mice	• DA and CREB, ERK1/2, CREB expressions • MKP-1, proBDNF expressions	Lin et al., 2021
	Crocin (50 mg/kg)	Rats	Anti-anxiety effects	Pitsikas and Tarantilis, 2020
	Ethanol extract of petals	Clinical trial (CT) with 40 patients	Improved the Hamilton Depression Rating Scale value	Moshiri et al., 2006

Contd.

Table 1 *Contd.*

Activity	*Extracts/ bioactive compounds*	*Model used*	*Effect observed*	*Reference*
	Crocin	CT with 40 patients	Improved the Beck depression and anxiety inventory and a general health questionnaire	Talaei et al., 2015
	Crocin and crocetin	Hypertensive rats	Improved the contractility in de-endothelizated vessels.	Llorens et al., 2015
	Saffron	CT	• Postpartum depression	Kashani et al., 2017
	Saffron	CT with 60 subjects	• Major depressive disorder and anxious distress	Ghajar et al., 2017
	Saffron	CT with 57 subjects	• Regulated the serotonin and dopamine secretions. • Depression	Jalali and Hashemi, 2018
	Saffron	CT with 50 subjects	• Depression	Ahmadpanah et al., 2019
	Saffron	CT with 80 women	Improved depression scores and quality of life.	Bahrami et al., 2023
	Crocin	Rats with unpredictable chronic mild stress	• Serum corticosterone levels, cortical malondialdehyde (MDA), TNF-αand interleukin-6 • SOD, CAT, and brain-derived neurotrophic factor (BDNF)	Abbaszade-Cheragheali et al., 2022
	Saffron	Mice	• Depressive-like behavior Regulated the players of the kynurenine pathway	Monchaux De Oliveira et al., 2021
	Saffron capsules	CT with 46 AD patients	Improved cognitive functions	Akhondzadeh et al., 2010
	Safranal	Rats	Improved activities like rotarod and locomotor. • Nitrite, MDA, glutathione, SOD and CAT.	Fotoohi et al., 2021
	Crocin	Wistar rats with AD	Improved spatial memory. • Bax/Bcl-2 ratio and cleaved Caspase-3	Asadi et al., 2015

Contd.

Table 1 *Contd.*

Activity	*Extracts/ bioactive compounds*	*Model used*	*Effect observed*	*Reference*
	Saffron	Cell culture	• Viability of mouse primary retinal cells and photoreceptor-derived 661W cells	Corso et al., 2016
	Crocin	Rats	Improved apomorphine-induced performance deficits	Pitsikas and Tarantilis, 2017
	Crocin	Rats	Improved forgetting and recognition memory	Pitsikas and Tarantilis, 2018
	Crocin	Rats	Improved anesthetic ketamine-induced performance impairments	Pitsikas and Tarantilis, 2021
	Saffron	CT in 45 adults	• Hamilton Rating Scale, hospital anxiety and depression scale	Bangratz et al., 2018
	Aqueous extract of Saffron	CT in 39 adults	• Stroke severity, serum neuron-specific enolase and s100 levels • BDNF and Barthel index	Asadollahi et al., 2019
	Saffron simple syrup	CT in 30 patients	• The mean score of fatigue	Ashtiani et al., 2020
	Saffron	CT in 40 patients	• Antioxidant enzyme activities • Stroke severity	Gudarzi et al., 2022
	Saffron	CT in 56 individuals	• Depression scores Improved heart rate and social relationships	Jackson et al., 2021
	Crocin	Mice	• Brain edema and neurological deficits • SOD and GSH-px • MDA • Nrf2 expression	Wang et al., 2022

Contd.

Table 1 *Contd.*

Activity	*Extracts/ bioactive compounds*	*Model used*	*Effect observed*	*Reference*
Neuroprotection (anti-depression and anti-anxiety)	Saffron	Clinical trial	Improved hyperactivity symptoms and sleep	Blasco-Fontecilla et al., 2022
	Saffron	Mice	• The time in the center of the Open Field maze • Akkermansia, Muribaculaceae, Christensenellacae and Alloprevotella	Pontifex et al., 2022
Cognitive health	Saffron	Clinical trial	• Mini-Mental State Examination scores	Tsolaki et al., 2016
Reproductive and sexual-related effects	Ethanol extract of stigma	CT with 50 patients	• Symptoms of premenstrual syndrome	Agha-Hosseini et al., 2008
	Saffron capsules	CT with 20 males	Improved nocturnal penile tumescence	Shamsa et al., 2009
	Crocetin	*In vitro*	• Testis injury, DNA damage and oxidative stress	Rossi et al., 2021
	Crocin	Male Rats	Improved the sperm parameters, total antioxidant capacity, testosterone level	Salahshoor et al., 2022
	Saffron capsules	CT trial with 30 males	• Erectile function and intercourse satisfaction	Modabbernia et al., 2012
	Saffron stigma	CT with 34 women	• Arousal, lubrication domains	Kashani et al., 2013
	Saffron	CT with 34 women	• Premenstrual dysphoric disorder	Rajabi et al., 2020
	Saffron gel	CT with 50 diabetic males	• International Index of Erectile Function Questionnaire score	Mohammadzadeh-Moghadam et al., 2015
	Crocetin	*In vitro*	• Superoxide anion and hydrogen peroxide, lipid peroxidation • Acrosome-reacted spermatozoa and blastocyst production	Sapanidou et al., 2016

Contd.

Table 1 *Contd.*

Activity	*Extracts/ bioactive compounds*	*Model used*	*Effect observed*	*Reference*
Cardiovascular activity Hypolipidemic	Crocin	Mice	• Harmful effects of exposure to electromagnetic fields	Vafaei et al., 2020
	Saffron	CT with 56 women	Improved the hot flash-related daily interference and Hamilton depression rating scales	Kashani et al., 2018
	Saffron	CT with 82 women	• Greene Climacteric Scale psychological score, anxiety, and depression scores • Positive and Negative Affect Schedule negative affect score	Lopresti and Smith, 2021
	Saffron	Clinical trial	Improved the female sexual function index domains	Kashani et al., 2022
	Kaempferol and crocin from saffron petals	*In vitro*	• Intracellular ROS formation • Cell viability	Zeka et al., 2020
	Saffron stigma	Isolated rabbit heart	• Myocardial function	Chahine et al., 2013
	Crocin	Rats	Improved the cotinine concentration, inflammation, and antioxidant system	Dianat et al., 2018
	Crocin	Apolipoprotein E knockout mice	• Atherosclerotic area and HIF-1α expression • eNOS expression	Makaritsis et al., 2022
	Saffron	CT with 63 subjects	• McNew, physical, and social domains • The quality of life	Ahmadikhatir et al., 2022
	Crocin from *C. sativus*	Cell culture and quail	• Apoptosis and SMC proliferation • Ox-LDL	He et al., 2005
	Crocin from *C. sativus*	Rats	• Triglyceride (TG), total cholesterol (TC), LDL-C, and VLDL-C	Sheng et al., 2006

Contd.

Table 1 *Contd.*

Activity	*Extracts/ bioactive compounds*	*Model used*	*Effect observed*	*Reference*
	Crocetin from *C. sativus*	Quails	• TC level, aortic plaque formation, MDA, and nitric oxide	He et al., 2007
Liver protection	Hydroalcoholic extract of *C. sativus*	Rats	• Inflammation. Improved ALT, aspartate aminotransferase (AST) and bilirubin. • Total protein and albumin	Omidi et al., 2014
	Saffron and crocin	High-fat-fed obese rats	Improved aspartate transaminase, alanine transaminase (ALT), alkaline phosphatase, and albumin	Mashmoul et al., 2016
	Polyphenols from petals	Wistar rats	Reduced hepatic injury indicators. • Arbitral liver damage	Wali et al., 2020
Pancreas protection	Crocin	Male Swiss albino Mice	• Pancreatic edema, amylase, lipase, and oxidative stress • Nrf-2. • IL-6, TNF-α, nitrotyrosine, and NFκB.	Godugu et al., 2020
Antifibrotic effects	Crocin	Mice	• Serum ALT, AST, and total bilirubin. • Serum TC and TG • Glutathione depletion, SOD and MDA • Nuclear factor-κB and cyclooxygenase-2 expressions	Algandaby, 2018
Kidney protection	Crocin	Rats	• Leukocyte infiltration, intercellular adhesion molecule-1 and TNF-α expression • MDA and antioxidant potential	Yarijani et al., 2017

Contd.

Table 1 *Contd.*

Activity	*Extracts/ bioactive compounds*	*Model used*	*Effect observed*	*Reference*
Anti-diabetic	Crocin from *C. sativus*	Rats	• MDA and xanthine oxidase activities • Glutathione • Blood urea nitrogen and creatinine	Altinoz et al., 2015
	Saffron	CT with 64 women	• Plasma glucose, HbA1c, LDL-C, and LDL/HDL ratio	Moravej Aleali et al., 2019
	Crocin	Rats	• Antioxidant state and hepatic health	Margaritis et al., 2020
	Crocin	CT with 50 patients	• Serum high-sensitive C-reactive protein, TNF-α, NF-κB, and IL-6	Behrouz et al., 2021
	Saffron	CT with 60 patients	• Blood glucose, TNF-α and IL-6 expressions	Mobasseri et al., 2020
	Saffron	CT with 70 patients	• Fasting blood sugar. Improved the lipid profiles	Tajaddini et al., 2023
Anti-obesity	Crocin	Mice	• Weight gain, fat accumulation, and insulin resistance	Liu et al., 2021
	Saffron extract	CT with 60 women	• Snacking and weight gain	Gout et al., 2010
	Saffron	CT with 80 subjects	• Waist circumference and MDA	Ebrahimi et al., 2019a,b
Anti-hepatotoxic	Crocin from *C. sativus*	Rats	• MDA, Bax/Bcl-2 ratio and caspase activation	Lari et al., 2015
	Saffron	CT with 70 overweight/ obese patients with type 2 diabetes	• Hyperglycemia and hyperlipidemia • Fasting plasma glucose, TG, insulin, AST, and ALT • Sleep and life quality	Tajaddini et al., 2021
Diuretic activity	Dried stigma of *C. sativus*	Rats	• Electrolytes input and excretion	Shariatifar et al., 2014

Contd.

Table 1 *Contd.*

Activity	*Extracts/ bioactive compounds*	*Model used*	*Effect observed*	*Reference*
Ergogenic effect	Dried saffron stigma	CT with 28 males	• Muscle mitochondrial biogenesis and motor cortex. • Muscle blood perfusion and aid in oxygen transport	Meamarbashi and Rajabi, 2016
Antimicrobial activity	Safranal		Antifungal activity	Carradori et al., 2016
	Saffron petal extract	*In vitro* and *In vivo*	• Mycelial growth and spore germination • Severity of grey mold disease and green mold	Naim et al., 2022
Atheroprotective effect	Aqueous extract of saffron	ApoE knockout mice	• Aortic stenosis, macrophage content Improved IL-6, TNF-α, MCP-1, MMP-2, -3, -9 and MMP-2/TIMP-2 ratio	Christodoulou et al., 2018
Anti-inflammatory effect	Ethanolic extract of petals	Cell lines	• Nitric oxide production and IL-6 release • Osteoclast-related genes expression	Orabona et al., 2022
	Crocin	Rat coronary atherosclerosis model	• Blood lipid, lipogenesis, inflammation. • Macrophage polarization Inhibits the NF-κB p65 expression.	Li et al., 2018
	Saffron	CT with 76 patients	• High-sensitive C-reactive protein, leptin, and MDA • Total antioxidant capacity	Pour et al., 2020
	Crocetin	Mice	• AST, ALT activities Improved the lipid and antioxidant system • TNF-α, IL-6, IL-1β • HO-1 and Nrf2 expressions	Xu et al., 2021

Contd.

Table 1 *Contd.*

Activity	*Extracts/ bioactive compounds*	*Model used*	*Effect observed*	*Reference*
	Saffron	Rats	Improved weight gain, food intake, TG, FBS, ALT, and AST	Alipour et al., 2023
	Saffron or Crocin	Mice	• Adhesion band frequency and oxidative stress	Arjmand et al., 2021
	Saffron	Mice	Improved the gross and histopathological characteristics of the colonic mucosa	Banskota et al., 2021
	Crocetin	Rats	Improved the players of the inflammatory system	Peng et al., 2023a,b
	Saffron	CT with 80 patients	• Clinical colitis activity index score, antioxidant capacity	Tahvilian et al., 2021
	Saffron	CT in hypertensive men	• Inflammation	Mojtahedi et al., 2022
Arthritis treatment	Safranal	Rats	Improved the arthritis-associated clinical parameters	Cellat et al., 2022
	Saffron	CT with 66 women	• The number of tenders, swollen joints, Physician Global Assessment, and erythrocyte sedimentation rate • High-sensitivity C-reactive protein, TNF-ã, IFN-ã, MDA • Antioxidant capacity	Hamidi et al., 2020
Visual functions	Saffron	Mice	• The fluorescein score of the eyes, TNF-α, IL-1β and IL-6	Yousefi-Manesh et al., 2022
	Saffron	CT with 100 adults	• Mean best-corrected visual acuity • Mean-pooled multifocal electroretinogram	Broadhead et al., 2019
	Saffron	CT with 34 adults	• Intraocular pressure	Bonyadi et al., 2014

Contd.

Table 1 *Contd.*

Activity	*Extracts/ bioactive compounds*	*Model used*	*Effect observed*	*Reference*
	Crocin	CT with 60 adults	• HbA1c and central macular thickness • Best-corrected visual acuity	Sepahi et al., 2018
	Saffron	Mice	• Neuroinflammation	Di Paolo et al., 2023
Breathing functions	Saffron	CT with 80 adults	Improved asthma symptoms and severity • Blood pressure, TG, LDL-C	Zilaee et al., 2019
	Compounds in *C. sativus*	Genetic algorithm-modified neural network	Improved the accuracy for anti-heat shock protein, high sensitivity C-reactive protein, forced expiratory volume and vital capacity	Hosseini et al., 2020
Sleep Quality	Crocetin	CT with 30 adults	Improved sleep quality	Umigai et al., 2018
	Saffron	CT with 63 adults	Improved the sleep-associated indexes	Lopresti et al., 2020
	Saffron	CT with 120 adults	Improved sleep quality and insomnia symptoms	Lopresti et al., 2021
Brain functions	Saffron	Wistar rats	• BDNF, serotonin and neurotrophin-3 Improved memory	Akbari-Fakhrabadi et al., 2021
Protection from toxins	Crocin	Male albino rats	• Toxic effects	Khalaf and El-Mansy, 2019
	Crocin	Male albino rats	• Serum MDA levels • Total antioxidant capacity	Mohamed Mowafy et al., 2021

Saffron enhanced pasta's textural and physicochemical properties, maintaining increased antioxidant activity even after 3 minutes of cooking. Sensory analysis showed higher preference for saffron-enriched pasta (Armellini et al., 2018). Additionally, saffron serves as a functional ingredient in dairy, desserts, cereals, bakery items, and beverages (Ali et al., 2022).

Saffron benefits physically active males by enhancing exercise enjoyment and heart rate variability (Lopresti & Smith, 2022). In Wistar rats, saffron extract increased mtDNA copy number, NRF-1 gene expression, endurance capacity, and antioxidant levels while reducing oxidative stress and inflammation markers (Akbari-Fakhrabadi et al., 2019).

A saffron extract and avocado oil-based anti-wrinkle cream demonstrated effectiveness and safety in facial skin rejuvenation during a 12-week clinical study with 20 volunteers (Naeimifar et al., 2020)[1]. Additionally, saffron supplementation (150 mg/day for six weeks) improved happiness levels in untrained male participants undergoing resistance training, increasing anandamide, 2-arachidonoylglycerol, serotonin, dopamine, and β-endorphin levels (Moghadam et al., 2021). Saffron, as a nutraceutical or supplement, may help alleviate COVID-19 symptoms in pre- and post-infection management. Its antidepressant properties support recovery by addressing anxiety, depression, sleep disturbances, and PTSD. Additionally, it aids immunity and mental well-being in the general population (Husaini et al., 2021; Mentis et al., 20212).

6. Harmful Effects

Crocin and safranal administration during embryogenesis in .BALB/c pregnant mice led to reduced fetal length and weight, with observed skeletal, mandible, and calvaria malformations, as well as growth retardation (Moallem et al., 2016). A survey of 350 pregnant and lactating women found that C. sativus was among the herbs avoided during pregnancy (Eid & Jaradat, 2020).

Conclusion and Future Directions

Saffron shows promising effects on diabetes, obesity, depression, cognitive function, and sexual dysfunction. Reported side effects include nausea, dry mouth, poor appetite, and headache, but no severe adverse reactions. While saffron appears safe, further large-scale clinical trials are needed to confirm its efficacy.

Saffron may alleviate oxidative stress by influencing malondialdehyde and total antioxidant capacity. While preclinical studies highlight its antioxidant, hypolipidemic, antidiabetic, cardioprotective, and anti-inflammatory effects, clinical trials show moderate cardiovascular benefits. Limitations in trial design, including variability in doses, cohorts, and duration, emphasize the need for more rigorous studies. Lifestyle diseases like diabetes contribute to inflammation, mitochondrial dysfunction, ROS, and advanced glycation end-product synthesis, increasing Alzheimer's risk. Saffron may help mitigate both conditions, supporting disease prevention and healthcare. However, limited clinical evidence restricts its use as an herbal therapy, necessitating further studies on pharmacokinetics, bioavailability, and gut microbiota impact.

References

Abbaszade-Cheragheali, A., Beheshti, F., Kakhki, S., Khatibi, S.R., Dehnokhalaji, F. et al. (2022). Crocin, the main active saffron (*Crocus sativus* L.) constituent, as a potential candidate to prevent anxiety and depressive-like behaviors induced by unpredictable chronic mild stress. *Neurosci. Lett*, *791*: 136912.

Abedi, A., Ghobadi, H., Sharghi, A., Iranpour, S., Fazlzadeh, M. et al. (2023). Effect of saffron supplementation on oxidative stress markers (MDA, TAC, TOS, GPx, SOD, and pro-oxidant/antioxidant balance): An updated systematic review and meta-analysis of randomized placebo-controlled trials. *Front. Med (Lausanne)*, *10*: 1071514.

Abu-Izneid, T., Rauf, A., Khalil, A.A., Olatunde, A., Khalid, A. et al. (2022). Nutritional and health beneficial properties of saffron (*Crocus sativus* L): a comprehensive review. *Crit. Rev. Food Sci. Nutr*, *62*(10): 2683–2706.

Agha-Hosseini, M., Kashani, L., Aleyaseen, A., Ghoreishi, A., Rahmanpour, H. et al. (2008). *Crocus sativus* L. (saffron) in the treatment of premenstrual syndrome: a double-blind, randomised and placebo-controlled trial. *BJOG*, *115*(4): 515–519.

Ahmadikhatir, S., Ostadrahimi, A., Safaiyan, A., Ahmadikhatir, S. and Farrin, N. et al. (2022). Saffron (*Crocus sativus* L.) supplements improve quality of life and appetite in atherosclerosis patients: A randomized clinical trial. *J. Res. Med. Sci*, *27*: 30.

Ahmadpanah, M., Ramezanshams, F., Ghaleiha, A., Akhondzadeh, S., Bahmani, D.S. et al. (2019). *Crocus sativus* L. (saffron) versus sertraline on symptoms of depression among older people with major depressive disorders-a double-blind, randomized intervention study. *Psychiatry Res*, *282*: 112613.

Akbari-Fakhrabadi, M., Najafi, M., Mortazavian, S., Memari, A.H., Shidfar, F. et al. (2021). Saffron (*Crocus sativus* L.), Combined with Endurance Exercise, Synergistically Enhances BDNF, Serotonin, and NT-3 in Wistar Rats. *Rep. Biochem. Mol. Biol*, *9*(4): 426–434.

Akbari-Fakhrabadi, M., Najafi, M., Mortazavian, S., Rasouli, M., Memari, A.H. et al. (2019). Effect of saffron (*Crocus sativus* L.) and endurance training on mitochondrial biogenesis, endurance capacity, inflammation, antioxidant, and metabolic biomarkers in Wistar rats. *J. Food Biochem*, *43*(8): e12946.

Akhondzadeh, S., Sabet, M.S., Harirchian, M.H., Togha, M., Cheraghmakani, H. et al. (2010). Saffron in the treatment of patients with mild to moderate Alzheimer's disease: a 16-week, randomized and placebo-controlled trial. *J. Clin. Pharm. Ther*, *35*: 581–588.

Akhondzadeh, B.A., Ghoreishi, S.A., Noorbala, A.A., Akhondzadeh, S.H. and Rezazadeh, S.H. et al. (2008). Petal and stigma of *Crocus sativus* L. in the treatment of depression: A pilot double-blind randomized trial. *J. Med. Plants*, *7*: 29–36.

Alavizadeh, S.H. and Hosseinzadeh, H. (2014). Bioactivity assessment and toxicity of crocin: a comprehensive review. *Food Chem. Toxicol*, *64*: 65–80.

Algandaby, M.M. (2018). Antifibrotic effects of crocin on thioacetamide-induced liver fibrosis in mice. *Saudi J. Biol. Sci*, *25*(4): 747–754.

Ali, A., Yu, L., Kousar, S., Khalid, W., Maqbool, Z. et al. (2022). Crocin: Functional characteristics, extraction, food applications and efficacy against brain related disorders. *Front. Nutr*, *9*: 1009807.

Alipour, R., Aryaeian, N., Hajiluian, G., Soleimani, M. and Barati, M. et al. (2023). The Effect of the Saffron Intervention on NAFLD Status and Related Gene Expression in a Rat Model. *Med. J. Islam Repub. Iran*, *37*: 28.

Altinoz, E., Oner, Z., Elbe, H., Cigremis, Y. and Turkoz, Y. et al. (2015). Protective effects of saffron (its active constituent, crocin) on nephropathy in streptozotocin-induced diabetic rats. *Hum. Exp. Toxicol*, *34*(2): 127–134.

Amin, A., Bajbouj, K., Koch, A., Gandesiri, M. and Schneider-Stock, R. et al. (2015). Defective autophagosome formation in p53-null colorectal cancer reinforces crocin-induced apoptosis. *Int. J. Mol. Sci*, *16*: 1544–1561.

Arjmand, M.H., Hashemzehi, M., Soleimani, A., Asgharzadeh, F., Avan, A. et al. (2021). Therapeutic potential of active components of saffron in post-surgical adhesion band formation. *J. Tradit. Complement. Med*, *11*(4): 328–335.

Armellini, R., Peinado, I., Pittia, P., Scampicchio, M., Heredia, A. et al. (2018). Effect of saffron (*Crocus sativus* L.) enrichment on antioxidant and sensorial properties of wheat flour pasta. *Food Chem*, *254*: 55–63.

Asadi, F., Jamshidi, A.H., Khodagholi, F., Yans, A., Azimi, L. et al. (2015). Reversal effects of crocin on amyloid β-induced memory deficit: Modification of autophagy or apoptosis markers. *Pharmacol. Biochem. Behav*, *139*(Pt A): 47–58.

Asadollahi, M., Nikdokht, P., Hatef, B., Sadr, S.S., Sahraei, H. et al. (2019). Protective properties of the aqueous extract of saffron (*Crocus sativus* L.) in ischemic stroke, randomized clinical trial. *J. Ethnopharmacol*. *238*: 111833.

Asbaghi, O., Sadeghian, M., Sadeghi, O., Rigi, S., Tan, S.C. et al. (2021). Effects of saffron (*Crocus sativus* L.) supplementation on inflammatory biomarkers: A systematic review and meta-analysis. *Phytother. Res*, *35*(1): 20–32.

Ashtiani, A.R., Amirhossein, L.S., Jadidi, A., Ghasami, K., Hezave, A.K. et al. (2020). The effect of novel simple saffron syrup on fatigue reduction in patients with multiple sclerosis. *J. Basic Clin. Physiol. Pharmacol*, *31*(6): doi: 10.1515/jbcpp-2020-0063

Azami, S., Shahriari, Z., Asgharzade, S., Farkhondeh, T., Sadeghi, M. et al. (2021). Therapeutic Potential of Saffron (*Crocus sativus* L.) in Ischemia Stroke. *Evid. Based Complement. Alternat. Med*, *2021*: 6643950.

Bahrami, F., Pour, F.J., Hassanpour, M., Saki, M., Ebrahimzadeh, F. et al. (2023). The effect of saffron and corrective exercises on depression and quality of life in women with multiple sclerosis: A randomized controlled clinical trial. *Mult Scler Relat Disord*, *79*: 105038.

Bakshi, H.A., Zoubi, M.S.A., Hakkim, F.L., Aljabali, A.A.A., Rabi, F.A. et al. (2020). Dietary Crocin is Protective in Pancreatic Cancer while Reducing Radiation-Induced Hepatic Oxidative Damage. *Nutrients*, *12*(6): 1901.

Bangratz, M., Ait Abdellah, S., Berlin, A., Blondeau, C., Guilbot, A. et al. (2018). A preliminary assessment of a combination of rhodiola and saffron in the management of mild-moderate depression. *Neuropsychiatr Dis Treat*, *14*: 1821–1829.

Banskota, S., Brim, H., Kwon, Y.H., Singh, G., Sinha, S.R. et al. (2021). Saffron Pre-Treatment Promotes Reduction in Tissue Inflammatory Profiles and Alters Microbiome Composition in Experimental Colitis Mice. *Molecules*, *26*(11): 3351.

Basker, D. and Negbi, M. (1983). The use of saffron. *Econ Bot*, 37: 228–236.

Behrouz, V., Sohrab, G., Hedayati, M. and Sedaghat, M. (2021). Inflammatory markers response to crocin supplementation in patients with type 2 diabetes mellitus: A randomized controlled trial. *Phytother Res*, *35*(7): 4022–4031.

Blasco-Fontecilla, H., Moyano-Ramírez, E., Méndez-González, O., Rodrigo-Yanguas, M., Martin-Moratinos, M. et al. (2022). Effectivity of Saffron Extract (Saffr'Activ) on Treatment for Children and Adolescents with Attention Deficit/Hyperactivity Disorder (ADHD): A Clinical Effectivity Study. *Nutrients*, *14*(19): 4046.

Bonyadi, M.H.J., Yazdani, S. and Saadat, S. (2014). The ocular hypotensive effect of saffron extractin primary open angle glaucoma: A pilot study. *BMC Complement. Altern. Med*, *14*: 399.

Broadhead, G.K., Grigg, J.R., McCluskey, P., Hong, T., Schlub, T.E. et al. (2019). Saffron therapy for the treatment of mild/moderate age-related macular degeneration: a randomised clinical trial. *Graefes Arch Clin Exp Ophthalmol. 257*(1): 31–40.

Carradori, S., Chimenti, P., Fazzari, M., Granese, A. and Angiolella, L. et al. (2016). Antimicrobial activity, synergism and inhibition of germ tube formation by Crocus sativus-derived compounds against *Candida* spp. *J Enzyme Inhib Med Chem*, 31(sup2): 189–193.

Cellat, M., İşler, C.T., Kutlu, T., Kuzu, M., Etyemez, M. et al. (2022). Investigation of the effects of safranal on the experimentally created rheumatoid arthritis model in rats. *J Biochem Mol Toxicol. 36*(9): e23140.

Cerdá-Bernad, D., Valero-Cases, E., Pastor, J.J. and Frutos, M.J. (2022). Saffron bioactives crocin, crocetin and safranal: effect on oxidative stress and mechanisms of action. *Crit Rev Food Sci Nutr*, *62*(12): 3232–3249.

Chahine, N., Hanna, J., Makhlouf, H., Duca, L., Martiny, L et al. (2013). Protective effect of saffron extract against doxorubicin cardiotoxicity in isolated rabbit heart. *Pharm Biol*, *51*(12): 1564–1571.

Christodoulou, E., Kadoglou, N.P.E., Stasinopoulou, M., Konstandi, O.A., Kenoutis, C. et al. (2018). *Crocus sativus* L. aqueous extract reduces atherogenesis, increases atherosclerotic plaque stability and improves glucose control in diabetic atherosclerotic animals. *Atherosclerosis*, *268*: 207–214.

Correia, A.G.D.S., Alencar, M.B., Dos Santos, A.N., da Paixão, D.C.B., Sandes, F.L.F. et al. (2023). Effect of saffron and fenugreek on lowering blood glucose: A systematic review with meta-analysis. *Phytother Res*, *37*(5): 2092–2101.

Corso, L., Cavallero, A., Baroni, D., Garbati, P., Prestipino, G. et al. (2016). Saffron reduces ATP-induced retinal cytotoxicity by targeting P2X7 receptors. *Purinergic Signal*, *12*(1): 161–174.

D'Alessandro, A.M., Mancini, A., Lizzi, A.R., De Simone, A., Marroccella, C.E. et al. (2013). *Crocus sativus* stigma extract and its major constituent crocin possess significant antiproliferative properties against human prostate cancer. *Nutr Cancer*, *65*(6): 930–942.

Dante, G. and Facchinetti, F. (2011). Herbal treatments for alleviating premenstrual symptoms: a systematic review. *J Psychosom Obstet Gynaecol*, *32*(1): 42–51.

Das, I., Chakrabarty, R.N. and Das, S. (2004). Saffron can prevent chemically induced skin carcinogenesis in Swiss albino mice. *Asian Pac J Cancer Prev*, *5*(1): 70–76.

Delam, H., Keshtkaran, Z., Shokrpour, N., Eidi, A. and Bazrafshan, M.R. et al. (2023). The effect of *Crocus sativus* L. (saffron) herbal tea on happiness in postmenopausal women: a randomized controlled trial. *BMC Complement Med Ther*, *23*(1): 176.

Di Paolo, M., Corsi, F., Cerri, C., Bisti, S., Piano, I. et al. (2023). A Window to the Brain: The Retina to Monitor the Progression and Efficacy of Saffron Repron® Pre-Treatment in an LPS Model of Neuroinflammation and Memory Impairment. *Pharmaceuticals (Basel)*, *16*(9): 1307.

Dianat, M., Radan, M., Badavi, M., Mard, S.A., Bayati, V. et al. (2018). Crocin attenuates cigarette smoke-induced lung injury and cardiac dysfunction by anti-oxidative effects: the role of Nrf2 antioxidant system in preventing oxidative stress. *Respir Res*, *19*(1): 58.

D'Onofrio, G., Nabavi, S.M., Sancarlo, D., Greco, A. and Pieretti, S. et al. (2021). *Crocus sativus* L. (Saffron) in Alzheimer's Disease Treatment: Bioactive Effects on Cognitive Impairment. *Curr Neuropharmacol. 19*(9): 1606–1616.

Dwyer, A.V., Whitten, D.L. and Hawrelak, J.A. (2011). Herbal medicines, other than St. John's Wort, in the treatment of depression: a systematic review. *Altern Med Rev*, *16*(1): 40–49.

Ebrahimi, F., Aryaeian, N., Pahlavani, N., Abbasi, D., Hosseini, A.F. et al. (2019a). The effect of saffron (*Crocus sativus* L.) supplementation on blood pressure, and renal and liver function in patients with type 2 diabetes mellitus: A double-blinded, randomized clinical trial. *Avicenna J Phytomed*, *9*(4): 322–333.

Ebrahimi, F., Sahebkar, A., Aryaeian, N., Pahlavani, N., Fallah, S. et al. (2019b). Effects of Saffron Supplementation on Inflammation and Metabolic Responses in Type 2 Diabetic Patients: A

Randomized, Double-Blind, Placebo-Controlled Trial. *Diabetes Metab Syndr Obes*, *12*: 2107–2115.

Eid, A.M. and Jaradat, N. (2020). Public Knowledge, Attitude, and Practice on Herbal Remedies Used During Pregnancy and Lactation in West Bank Palestine. *Front Pharmacol*, *11*: 46.

Esmaealzadeh, D., Ghalibaf, A.M., Shariati-Rad, M., Rezaee, R., Razavi, B.M. et al. (2023). Pharmacological effects of Safranal: An updated review. *Iran J Basic Med Sci*, *26*(10): 1131–1143.

Falsini, B., Piccardi, M., Minnella, A., Savastano, C., Capoluongo, E. et al. (2010). Influence of saffron supplementation on retinal flicker sensitivity in early age-related macular degeneration. *Investig. Ophthalmol. Vis. Sci*, *51*: 6118–6124.

Festuccia, C., Mancini, A., Gravina, G.L., Scarsella, L., Llorens, S. et al. (2014). Antitumor effects of saffron-derived carotenoids in prostate cancer cell models. *Biomed Res Int*, 2014: 135048.

Fotoohi, A., Moloudi, M.R., Hosseini, S., Hassanzadeh, K., Feligioni, M. et al. (2021). A Novel Pharmacological Protective Role for Safranal in an Animal Model of Huntington's Disease. *Neurochem Res*, *46*(6): 1372–1379.

Ghaffari, S. and Roshanravan, N. (2019). Saffron; An updated review on biological properties with special focus on cardiovascular effects. *Biomed Pharmacother*, *109*: 21–27.

Ghajar, A., Neishabouri, S.M., Velayati, N., Jahangard, L., Matinnia, N. et al. (2017). *Crocus sativus* L. versus Citalopram in the Treatment of Major Depressive Disorder with Anxious Distress: A Double-Blind, Controlled Clinical Trial. *Pharmacopsychiatry*, *50*(4): 152–160.

Giakoumettis, D., Pourzitaki, C., Vavilis, T., Tsingotjidou, A., Kyriakoudi, A. et al. (2019). *Crocus sativus* L. Causes a Non Apoptotic Calpain Dependent Death in C6 Rat Glioma Cells, Exhibiting a Synergistic Effect with Temozolomide. *Nutr Cancer*, *71*(3): 491–507.

Giannoulaki, P., Kotzakioulafi, E., Chourdakis, M., Hatzitolios, A. and Didangelos, T. et al. (2020). Impact of *Crocus sativus* L. on Metabolic Profile in Patients with Diabetes Mellitus or Metabolic Syndrome: A Systematic Review. *Nutrients*, *12*(5): 1424.

Godugu, C., Pasari, L.P., Khurana, A., Anchi, P., Saifi, M.A. et al. (2020). Crocin, an active constituent of *Crocus sativus* ameliorates cerulein induced pancreatic inflammation and oxidative stress. *Phytother Res*, *34*(4): 825–835.

Gout, B., Bourges, C. and Paineau-Dubreuil, S. (2010). Satiereal, a *Crocus sativus* L. extract, reduces snacking and increases satiety in a randomized placebo-controlled study of mildly overweight, healthy women. *Nutr. Res*, *30*: 305–313.

Gudarzi, S., Jafari, M., Jahromi, G.P., Eshrati, R., Asadollahi, M. et al. (2022). Evaluation of modulatory effects of saffron (*Crocus sativus* L.) aqueous extract on oxidative stress in ischemic stroke patients: a randomized clinical trial. *Nutr Neurosci*. *25*(6): 1137–1146.

Hamidi, Z., Aryaeian, N., Abolghasemi, J., Shirani, F., Hadidi, M. et al. (2020). The effect of saffron supplement on clinical outcomes and metabolic profiles in patients with active rheumatoid arthritis: A randomized, double-blind, placebo-controlled clinical trial. *Phytother Res*, *34*(7): 1650–1658.

Hasani-Ranjbar, S., Jouyandeh, Z. and Abdollahi, M. (2013). A systematic review of anti-obesity medicinal plants – an update. *J Diabetes Metab Disord*, *12*(1): 28.

Hausenblas, H.A., Heekin, K., Mutchie, H.L. and Anton, S. (2015). A systematic review of randomized controlled trials examining the effectiveness of saffron (*Crocus sativus* L.) on psychological and behavioral outcomes. *J Integr Med*. *13*(4): 231–240.

He, S.Y., Qian, Z.Y., Tang, F.T., Wen, N., Xu, G.L. et al. (2005). Effect of crocin on experimental atherosclerosis in quails and its mechanisms. *Life Sci*, *77*: 907–921.

He, S.Y., Qian, Z.Y., Wen, N., Tang, F.T., Xu, G.L. et al. (2007). Influence of crocetin on experimental atherosclerosis in hyperlipidamic-diet quails. *Eur. J. Pharmacol*, *554*: 191–195.

Heitmar, R., Brown, J. and Kyrou, I. (2019). Saffron (*Crocus sativus* L.) in Ocular Diseases: A Narrative Review of the Existing Evidence from Clinical Studies. *Nutrients*, *11*(3): 649.

Heydari, M., Zare, M., Badie, M.R., Watson, R.R., Talebnejad, M.R. et al. (2023). Crocin as a vision supplement. *Clin Exp Optom*, *106*(3): 249–256.

Hoshyar, R., Bathaie, S.Z. and Sadeghizadeh, M. (2013). Crocin triggers the apoptosis through increasing the Bax/Bcl-2 ratio and caspase activation in human gastric adenocarcinoma, AGS, cells. *DNA Cell Biol*, *32*: 50–57.

Hosseini, S.A., Jamshidnezhad, A., Zilaee, M., Dehaghi, B.F., Mohammadi, A. et al. (2020). Neural Network-Based Clinical Prediction System for Identifying the Clinical Effects of Saffron (*Crocus sativus* L) Supplement Therapy on Allergic Asthma: Model Evaluation Study. *JMIR Med Inform*, *8*(7): e17580.

Husaini, A.M., Jan, K.N. and Wani, G.A. (2021). Saffron: A potential drug-supplement for severe acute respiratory syndrome coronavirus (COVID) management. *Heliyon*, *7*(5): e07068.

Jackson, P.A., Forster, J., Khan, J., Pouchieu, C., Dubreuil, S. et al. (2021). Effects of Saffron Extract Supplementation on Mood, Well-Being, and Response to a Psychosocial Stressor in Healthy Adults: A Randomized, Double-Blind, Parallel Group, Clinical Trial. *Front Nutr*, *7*: 606124.

Jalali, F. and Hashemi, S.F. (2018). The Effect of Saffron on Depression among Recovered Consumers of Methamphetamine Living with HIV/AIDS. *Subst Use Misuse*, *53*(12): 1951–1957.

Jiang, X., Li, Y., Feng, J.L., Nik-Nabil, W.N., Wu, R. et al. (2020). Safranal Prevents Prostate Cancer Recurrence by Blocking the Re-activation of Quiescent Cancer Cells via Downregulation of S-Phase Kinase-Associated Protein 2. *Front Cell Dev Biol*, *8*: 598620.

Kamalipour, M. and Akhondzadeh, S. (2011). Cardiovascular effects of saffron: an evidence-based review. *J Tehran Heart Cent*, *6*(2): 59–61.

Karamacoska, D., Chan, D.K.Y., Leung, I., Liu, J.X., Brodaty, H. et al. (2023). Study protocol for a phase III randomised controlled trial of Sailuotong (SLT) for vascular dementia and Alzheimer's disease with cerebrovascular disease. *PLoS One*, *18*(3): e0265285.

Karimi-Nazari, E., Nadjarzadeh, A., Masoumi, R., Marzban, A., Mohajeri, S.A. et al. (2019). Effect of saffron (*Crocus sativus* L.) on lipid profile, glycemic indices and antioxidant status among overweight/obese prediabetic individuals: A double-blinded, randomized controlled trial. *Clin Nutr ESPEN*, *34*: 130–136.

Kashani, L., Aslzadeh, S., Shokraee, K., Shamabadi, A., Najafabadi, B.T. et al. (2022). *Crocus sativus* (saffron) in the treatment of female sexual dysfunction: a three-center, double-blind, randomized, and placebo-controlled clinical trial. *Avicenna J Phytomed*. *12*(3): 257–268.

Kashani, L., Esalatmanesh, S., Eftekhari, F., Salimi, S., Foroughifar, T. et al. (2018). Efficacy of *Crocus sativus* (saffron) in treatment of major depressive disorder associated with post-menopausal hot flashes: a double-blind, randomized, placebo-controlled trial. *Arch Gynecol Obstet*, *297*(3): 717–724.

Kashani, L., Eslatmanesh, S., Saedi, N., Niroomand, N., Ebrahimi, M. et al. (2017). Comparison of Saffron versus Fluoxetine in Treatment of Mild to Moderate Postpartum Depression: A Double-Blind, Randomized Clinical Trial. *Pharmacopsychiatry*, *50*(2): 64–68

Kashani, L., Raisi, F., Saroukhani, S., Sohrabi, H., Modabbernia, A. et al. (2013). Saffron for treatment of fluoxetine-induced sexual dysfunction in women: randomized double-blind placebo-controlled study. *Hum Psychopharmacol*, *28*(1): 54–60.

Khaksarian, M., Behzadifar, M., Behzadifar, M., Alipour, M., Jahanpanah, F. et al. (2019). The efficacy of *Crocus sativus* (Saffron) versus placebo and Fluoxetine in treating depression: a systematic review and meta-analysis. *Psychol Res Behav Manag*, *12*: 297–305.

Khalaf, H.A. and El-Mansy, A.A.E. (2019). The possible alleviating effect of saffron on chlorpyrifos experimentally induced cardiotoxicity: Histological, immunohistochemical and biochemical study. *Acta Histochem*, *121*(4): 472–483.

Khazdair, M.R., Boskabady, M.H., Hosseini, M., Rezaee, R. and Tsatsakis, A. et al. (2015). The effects of *Crocus sativus* (saffron) and its constituents on nervous system: A review. *Avicenna J Phytomed*. *5*(5): 376–391.

Lari, P., Abnous, K., Imenshahidi, M., Rashedinia, M., Razavi, M. et al. (2015). Evaluation of diazinon-induced hepatotoxicity and protective effects of crocin. *Toxicol Ind Health*, *31*(4): 367–376.

Lashay, A., Sadough, G., Ashrafi, E., Lashay, M., Movassat, M. et al. (2016). Short-term outcomes of saffron supplementation in patients with age-related macular degeneration: A double-blind, Placebo-controlled, randomized trial. *Med. Hypothesis Discov. Innov. Ophthalmol*, *5*: 32.

Li, J., Lei, H.T., Cao, L., Mi, Y.N., Li, S. et al. (2018). Crocin alleviates coronary atherosclerosis via inhibiting lipid synthesis and inducing M2 macrophage polarization. *Int Immunopharmacol*, *55*: 120–127.

Li Puma, S., Landini, L., Macedo, S.J. Jr., Seravalli, V., Marone, I.M. et al. (2019). TRPA1 mediates the antinociceptive properties of the constituent of *Crocus sativus* L., safranal. *J Cell Mol Med*, *23*(3): 1976–1986.

Lin, S., Li, Q., Jiang, S., Xu, Z., Jiang, Y. et al. (2021). Crocetin ameliorates chronic restraint stress-induced depression-like behaviors in mice by regulating MEK/ERK pathways and gut microbiota. *J Ethnopharmacol*, *268*: 113608.

Liu, X., Wang, Z., Song, X., Chang, X., Zu, E. et al. (2021). Crocetin Alleviates Ovariectomy-Induced Metabolic Dysfunction through Regulating Estrogen Receptor β. *J Agric Food Chem*, *69*(49): 14824–14839.

Llorens, S., Mancini, A., Serrano-Díaz, J., D'Alessandro, A.M. et al. (2015). Effects of Crocetin Esters and Crocetin from *Crocus sativus* L. on Aortic Contractility in Rat Genetic Hypertension. *Molecules*, *20*(9): 17570–17584.

Lopresti, A.L., Smith, S.J. and Drummond, P.D. (2021). An investigation into an evening intake of a saffron extract (affron®) on sleep quality, cortisol, and melatonin concentrations in adults with poor sleep: a randomised, double-blind, placebo-controlled, multi-dose study. *Sleep Med*, *86*: 7–18.

Lopresti, A.L., Smith, S.J., Metse, A.P. and Drummond, P.D. (2020). Effects of saffron on sleep quality in healthy adults with self-reported poor sleep: a randomized, double-blind, placebo-controlled trial. *J Clin Sleep Med*, *16*(6): 937–947.

Lopresti, A.L. and Smith, S.J. (2022). An examination into the mental and physical effects of a saffron extract (affron®) in recreationally-active adults: A randomized, double-blind, placebo-controlled study. *J Int Soc Sports Nutr*, *19*(1): 219–238.

Lopresti, A.L. and Smith, S.J. (2021). The Effects of a Saffron Extract (affron®) on Menopausal Symptoms in Women during Perimenopause: A Randomised, Double-Blind, Placebo-Controlled Study. *J Menopausal Med*, *27*(2): 66–78.

Makaritsis, K.P., Kotidis, C., Papacharalampous, K., Kouvaras, E., Poulakida, E. et al. (2022). Mechanistic insights on the effect of crocin, an active ingredient of saffron, on atherosclerosis in apolipoprotein E knockout mice. *Coron Artery Dis*, *33*(5): 394–402.

Mancini, A., Serrano-Díaz, J., Nava, E., D'Alessandro, A.M., Alonso, G.L. et al. (2014). Crocetin, a carotenoid derived from saffron (*Crocus sativus* L.), improves acetylcholine-induced vascular relaxation in hypertension. *J Vasc Res*, *51*(5): 393–404.

Marangoni, D., Falsini, B., Piccardi, M., Ambrosio, L., Minnella, A.M. et al. (2013). Functional effect of saffron supplementation and risk genotypes in early age-related macular degeneration: A preliminary report. *J. Transl. Med*, *11*: 228.

Margaritis, I., Angelopoulou, K., Lavrentiadou, S., Mavrovouniotis, I.C., Tsantarliotou, M. et al. (2020). Effect of crocin on antioxidant gene expression, fibrinolytic parameters, redox status and blood biochemistry in nicotinamide-streptozotocin-induced diabetic rats. *J Biol Res (Thessalon)*, *27*: 4.

Mashmoul, M., Azlan, A., Mohtarrudin, N., Mohd-Yusof, B.N., Khaza'ai, H. et al. (2016). Protective effects of saffron extract and crocin supplementation on fatty liver tissue of high-fat diet-induced obese rats. *BMC Complement Altern Med*, *16*(1): 401.

Meamarbashi, A. and Rajabi, A. (2016). Potential Ergogenic Effects of Saffron. *J Diet Suppl*. *13*(5): 522–529.

Mentis, A.A., Dalamaga, M., Lu, C. and Polissiou, M.G. (2021). Saffron for "toning down" COVID-19-related cytokine storm: Hype or hope? A mini-review of current evidence. *Metabol Open*, *11*: 100111.

Moallem, S.A., Afshar, M., Etemad, L., Razavi, B.M. and Hosseinzadeh, H. et al. (2016). Evaluation of teratogenic effects of crocin and safranal, active ingredients of saffron, in mice. *Toxicol Ind Health*. *32*(2): 285–291.

Moallem, S.A., Hariri, A.T., Mahmoudi, M. and Hosseinzadeh, H. (2014). Effect of aqueous extract of *Crocus sativus* L. (saffron) stigma against subacute effect of diazinon on specific biomarkers in rats. *Toxicol Ind Health*, *30*(2): 141–146.

Mobasseri, M., Ostadrahimi, A., Tajaddini, A., Asghari, S., Barati, M. et al. (2020). Effects of saffron supplementation on glycemia and inflammation in patients with type 2 diabetes mellitus: A randomized double-blind, placebo-controlled clinical trial study. *Diabetes Metab Syndr*, *14*(4): 527–534.

Modabbernia, A. and Akhondzadeh, S. (2013). Saffron, passionflower, valerian and sage for mental health. *Psychiatr Clin North Am*, *36*(1): 85–91.

Modabbernia, A., Sohrabi, H., Nasehi, A.A., Raisi, F., Saroukhani, S. et al. (2012). Effect of saffron on fluoxetine-induced sexual impairment in men: randomized double-blind placebo-controlled trial. *Psychopharmacology (Berl)*, *223*(4): 381–388.

Modaghegh, M.H., Shahabian, M., Esmaeili, H.A., Rajbai, O. and Hosseinzadeh, H. et al. (2008). Safety evaluation of saffron (*Crocus sativus*) tablets in healthy volunteers. *Phytomedicine*, *15*(12): 1032–1037.

Moghadam, B.H., Bagheri, R., Roozbeh, B., Ashtary-Larky, D., Gaeini, A.A. et al. (2021). Impact of saffron (*Crocus sativus* Linn) supplementation and resistance training on markers implicated in depression and happiness levels in untrained young males. *Physiol Behav*, *233*: 113352.

Mohamadpour, A.H., Ayati, Z., Parizadeh, M.R., Rajbai, O. and Hosseinzadeh, H. et al. (2013). Safety Evaluation of Crocin (a constituent of saffron) Tablets in Healthy Volunteers. *Iran J Basic Med Sci*, *16*(1): 39–46.

Mohamed-Mowafy, S., Hegazy, A.A., Mandour, D.A. and Abd El-Fatah, S.S. (2021). Impact of copper oxide nanoparticles on the cerebral cortex of adult male albino rats and the potential protective role of crocin. *Ultrastruct Pathol*, *45*(4-5): 307–318.

Mohammadzadeh-Moghadam, H., Nazari, S.M., Shamsa, A., Kamalinejad, M., Esmaeeli, H. et al. (2015). Effects of a Topical Saffron (*Crocus sativus* L) Gel on Erectile Dysfunction in Diabetics: A Randomized, Parallel-Group, Double-Blind, Placebo-Controlled Trial. *J Evid Based Complementary Altern Med*, *20*(4): 283–286.

Mojtahedi, S., Hooshmand-Moghadam, B., Rosenkranz, S., Shourideh, Z., Amirshaghaghi, F. et al. (2022). Improvement of inflammatory status following saffron (*Crocus sativus* L.) and resistance training in elderly hypertensive men: A randomized controlled trial. *Exp Gerontol*, *162*: 111756.

Monchaux De Oliveira, C., De-Smedt-Peyrusse, V., Morael, J., Vancassel, S., Capuron, L. et al. (2021). Prevention of Stress-Induced Depressive-like Behavior by Saffron Extract is Associated with Modulation of Kynurenine Pathway and Monoamine Neurotransmission. *Pharmaceutics*, *13*(12): 2155.

Moravej-Aleali, A., Amani, R., Shahbazian, H., Namjooyan, F., Latifi, S.M. et al. (2019). The effect of hydroalcoholic Saffron (*Crocus sativus* L.) extract on fasting plasma glucose, HbA1c, lipid profile, liver, and renal function tests in patients with type 2 diabetes mellitus: A randomized double-blind clinical trial. *Phytother Res*, *33*(6): 1648–1657.

Moshiri, E., Basti, A.A., Noorbala, A.A., Jamshidi, A.H., Abbasi, S.H. et al. (2006). *Crocus sativus* L. (petal) in the treatment of mild-to-moderate depression: a double-blind, randomized and placebo-controlled trial. *Phytomedicine*, *13*(9-10): 607–611.

Mousavi, S.M., Mokhtari, P., Asbaghi, O., Rigi, S., Persad, E et al. (2022). Does saffron supplementation have favorable effects on liver function indicators? A systematic review and meta-analysis of randomized controlled trials. *Crit Rev Food Sci Nutr*, *62*(23): 6315–6327.

Munirah, M.P., Norhayati, M.N. and Noraini, M. (2022). *Crocus sativus* for Insomnia: A Systematic Review and Meta-Analysis. *Int J Environ Res Public Health*, *19*(18): 11658.

Naeimi, M., Shafiee, M., Kermanshahi, F., Khorasanchi, Z., Khazaei, M. et al. (2019). Saffron (*Crocus sativus*) in the treatment of gastrointestinal cancers: Current findings and potential mechanisms of action. *J Cell Biochem*, *120*(10): 16330–16339.

Naeimifar, A., Ahmad-Nasrollahi, S., Samadi, A., Talari, R., Ale-Nabi, S.S. et al. (2020). Preparation and evaluation of anti-wrinkle cream containing saffron extract and avocado oil. *J Cosmet Dermatol*, *19*(9): 2366–2373.

Naim, N., Fauconnier, M.L., Ennahli, N., Tahiri, A., Baala, M. et al. (2022). Chemical Composition Profiling and Antifungal Activity of Saffron Petal Extract. *Molecules*, *27*(24): 8742.

Nanda, S. and Madan, K. (2021). The role of Safranal and saffron stigma extracts in oxidative stress, diseases and photoaging: A systematic review. *Heliyon*, *7*(2): e06117.

Noorbala, A., Akhondzadeh, S., Tahmacebi-Pour, N. and Jamshidi, A. (2005). Hydro-alcoholic extract of *Crocus sativus* L. versus fluoxetine in the treatment of mild to moderate depression: A double-blind, randomized pilot trial. *J. Ethnopharmacol*, *97*: 281–284.

Omidi, A., Riahinia, N., Montazer-Torbati, M.B. and Behdani, M.A. (2014). Hepatoprotective effect of *Crocus sativus* (saffron) petals extract against acetaminophen toxicity in male Wistar rats. *Avicenna J Phytomed*, *4*(5): 330–336.

Orabona, C., Orecchini, E., Volpi, C., Bacaloni, F., Panfili, E. et al. (2022). *Crocus sativus* L. Petal Extract Inhibits Inflammation and Osteoclastogenesis in RAW 264.7 Cell Model. *Pharmaceutics*, *14*(6): 1290.

Pagano, C., Ceccarini, M.R., Faieta, M., di Michele, A., Blasi, F. et al. (2022). Starch-based sustainable hydrogel loaded with *Crocus sativus* petals extract: A new product for wound care. *Int J Pharm*, *625*: 122067.

Peng, J., Deng, H., Du, B., Wu, P., Duan, L. et al. (2023b). Saffron Petal, an Edible Byproduct of Saffron, Alleviates Dextran Sulfate Sodium-Induced Colitis by Inhibiting Macrophage Activation and Regulating Gut Microbiota. *J Agric Food Chem*, *71*(28): 10616–10628.

Peng, S., Wu, Y. and Wang, Y. (2023a). Crocetin Suppresses Uterine Ischemia/Reperfusion-induced Inflammation and Apoptosis through the Nrf-2/HO-1 Pathway. *Curr Mol Med*, *23*(6): 527–535.

Piccardi, M., Marangoni, D., Minnella, A.M., Savastano, M.C., Valentini, P. et al. (2012). A Longitudinal follow-up study of saffron supplementation in early age-related macular degeneration: Sustained benefits to central retinal function. *Evid. Based Complement. Altern. Med*, 2012: 429124.

Pitsikas, N. and Tarantilis, P.A. (2017). Crocins, the active constituents of *Crocus sativus* L., counteracted apomorphine-induced performance deficits in the novel object recognition task, but not novel object location task, in rats. *Neurosci Lett*. *644*: 37–42.

Pitsikas, N. and Tarantilis, P.A. (2021). Crocins, the Bioactive Components of *Crocus sativus* L., Counteract the Disrupting Effects of Anesthetic Ketamine on Memory in Rats. *Molecules*, *26*(3): 528.

Pitsikas, N. and Tarantilis, P.A. (2018). Effects of the active constituents of *Crocus sativus* L. crocins and their combination with memantine on recognition memory in rats. *Behav Pharmacol*, *29*(5): 400–412.

Pitsikas, N. and Tarantilis, P.A. (2020). The GABAA-Benzodiazepine Receptor Antagonist Flumazenil Abolishes the Anxiolytic Effects of the Active Constituents of *Crocus sativus* L. Crocins in Rats. *Molecules*, *25*(23): 5647.

Pitsikas, N. (2016). Constituents of Saffron (*Crocus sativus* L.) as Potential Candidates for the Treatment of Anxiety Disorders and Schizophrenia. *Molecules*, *21*(3): 303.

Pitsikas, N. (2021). *Crocus sativus* L. Extracts and Its Constituents Crocins and Safranal; Potential Candidates for Schizophrenia Treatment? *Molecules*, *26*(5): 1237.

Pitsikas, N. (2015). The Effect of *Crocus sativus* L. and Its Constituents on Memory: Basic Studies and Clinical Applications. *Evid Based Complement Alternat Med*, 2015: 926284.

Pontifex, M.G., Connell, E., Le Gall, G., Pourtau, L., Gaudout, D. et al. (2022). Saffron extract (Safr'Inside™) improves anxiety related behaviour in a mouse model of low-grade inflammation through the modulation of the microbiota and gut derived metabolites. *Food Funct*, *13*(23): 12219–12233.

Pour, F.K., Aryaeian, N., Mokhtare, M., Mirnasrollahi-Parsa, R.S., Jannani, L. et al. (2020). The effect of saffron supplementation on some inflammatory and oxidative markers, leptin, adiponectin, and body composition in patients with nonalcoholic fatty liver disease: A double-blind randomized clinical trial. *Phytother Res*, *34*(12): 3367–3378.

Rahaiee, S., Moini, S., Hashemi, M. and Shojaosadati, S.A. (2015). Evaluation of antioxidant activities of bioactive compounds and various extracts obtained from saffron (*Crocus sativus* L.): a review. *J Food Sci Technol*, *52*(4): 1881–1888.

Rahmani, J., Bazmi, E., Clark, C. and Nazari, S.S.H. (2020). The effect of Saffron supplementation on waist circumference, HA1C, and glucose metabolism: A systematic review and meta-analysis of randomized clinical trials. *Complement Ther Med*, *49*: 102298.

Rajabi, F., Rahimi, M., Sharbafchizadeh, M.R. and Tarrahi, M.J. (2020). Saffron for the Management of Premenstrual Dysphoric Disorder: A Randomized Controlled Trial. *Adv Biomed Res*, *9*: 60.

Riazi, A., Panahi, Y., Alishiri, A.A., Hosseini, M.A., Zarchi, A.A. et al. (2017). The impact of saffron (*Crocus sativus*) supplementation on visual function in patients with dry age-related macular degeneration. *Ital. J. Med*, *11*: 758.

Rossi, G., Placidi, M., Castellini, C., Rea, F.D., Andrea, S. et al. (2021). Crocetin Mitigates Irradiation Injury in an *In Vitro* Model of the Pubertal Testis: Focus on Biological Effects and Molecular Mechanisms. *Molecules*, *26*(6): 1676.

Sadat-Rafiei, S.K., Abolghasemi, S., Frashidi, M., Ebrahimi, S., Gharei, F. et al. (2023). Saffron and Sleep Quality: A Systematic Review of Randomized Controlled Trials. *Nutr Metab Insights*, *16*: 11786388231160317.

Salahshoor, M.R., Azari, P., Roshankhah, S., Zokaei, A. and Foroughinia, A. et al. (2022). The effect of Crocin on sperm parameters disorders induced by peripheral administration of nitrosamines in male rats. *J Med Life*, *15*(3): 362–367.

Samarghandian, S. and Borji, A. (2014). Anticarcinogenic effect of saffron (*Crocus sativus* L.) and its ingredients. *Pharmacognosy Res*, *6*(2): 99–107.

Samarghandian, S. and Shabestari, M.M. (2013). DNA fragmentation and apoptosis induced by safranal in human prostate cancer cell line. *Indian J Urol*, *29*(3): 177–183.

Samarghandian, S., Afshari, J.T. and Davoodi, S. (2011). Suppression of pulmonary tumor promotion and induction of apoptosis by *Crocus sativus* l. *Extraction. Appl. Biochem. Biotechnol*, *164*: 238–247.

Sapanidou, V., Taitzoglou, I., Tsakmakidis, I., Kourtzelis, I., Fletouris, D. et al. (2016). Protective effect of crocetin on bovine spermatozoa against oxidative stress during *in vitro* fertilization. *Andrology*, *4*(6): 1138–1149.

Saxena, K., Kurian, S., Kumar, R., Arnold, L.E. and Simkin, D.R. et al. (2023). Mood Disorders in Youth: Complementary and Integrative Medicine. *Child Adolesc Psychiatr Clin N Am*, *32*(2): 367–394.

Schmidt, M., Betti, G., and Hensel, A. (2007). Saffron in phytotherapy: Pharmacology and clinical uses. *Wien. Med. Wochenschr*, *157*: 315–319.

Sepahi, S., Mohajeri, S.A., Hosseini, S.M., Khodaverdi, E., Shoeibi, N. et al. (2018). Effects of crocin on diabetic maculopathy: A placebo-controlled randomized clinical trial. *Am. J. Ophthalmol*, *190*: 89–98.

Shafiee, M., Arekhi, S., Omranzadeh, A. and Sahebkar, A. (2018). Saffron in the treatment of depression, anxiety and other mental disorders: Current evidence and potential mechanisms of action. *J Affect Disord*, *227*: 330–337.

Shahbaz, K., Chang, D., Zhou, X., Low, M., Seto, S.W. et al. (2022). Crocins for Ischemic Stroke: A Review of Current Evidence. *Front Pharmacol*, *13*: 825842.

Shakeri, M., Tayer, A.H., Shakeri, H., Jahromi, A.S., Moradzadeh, M. et al. (2020). Toxicity of Saffron Extracts on Cancer and Normal Cells: A Review Article. *Asian Pac J Cancer Prev*, *21*(7): 1867–1875.

Shamsa, A., Hosseinzadeh, H., Molaei, M., Shakeri, M.T. and Rajabi, O. et al. (2009). Evaluation of *Crocus sativus* L. (saffron) on male erectile dysfunction: a pilot study. *Phytomedicine*, *16*: 690–693.

Shariatifar, N., Shoeibi, S., Sani, M.J., Jamshidi, A.H., Zarei, A. et al. (2014). Study on diuretic activity of saffron (stigma of *Crocus sativus* L.) Aqueous extract in rat. *J Adv Pharm Technol Res*, *5*(1): 17–20.

Sheng, L., Qian, Z., Zheng, S. and Xi, L. (2006). Mechanism of hypolipidemic effect of crocin in rats: Crocin inhibits pancreatic lipase. *Eur. J. Pharmacol*, *543*: 116–122.

Tahmasbi, F., Araj-Khodaei, M., Mahmoodpoor, A. and Sanaie, S. (2022). Effects of saffron (*Crocus sativus* L.) on anthropometric and cardiometabolic indices in overweight and obese patients: A systematic review and meta-analysis of randomized controlled trials. *Phytother Res*, *36*(9): 3394–3414.

Tahvilian, N., Masoodi, M., Faghihi-Kashani, A., Vafa, M., Aryaeian, N. et al. (2021). Effects of saffron supplementation on oxidative/antioxidant status and severity of disease in ulcerative colitis patients: A randomized, double-blind, placebo-controlled study. *Phytother Res*. *35*(2): 946–953.

Tajaddini, A., Roshanravan, N., Mobasseri, M., Aeinehchi, A., Sefid-MooyeAzar, P. et al. (2021). Saffron improves life and sleep quality, glycaemic status, lipid profile and liver function in diabetic patients: A double-blind, placebo-controlled, randomised clinical trial. *Int J Clin Pract*, *75*(8): e14334.

Tajaddini, A., Roshanravan, N., Mobasseri, M., Al-Qaim, Z.H., Hadi, A. et al. (2023). The effect of saffron (*Crocus sativus* L.) on glycemia, lipid profile, and antioxidant status in patients with type-2 diabetes mellitus: A randomized placebo-controlled trial. *Phytother Res*, *37*(2): 388–398.

Talaei, A., Moghadam, M.H., SajadiTabassi, S.A. and Mohajeri, S.A. (2015). Crocin, the main active saffron constituent, as an adjunctive treatment in major depressive disorder: a randomized, double-blind, placebo-controlled, pilot clinical trial. *J Affect Disord*, *174*: 51–56.

Tiwari, A., Modi, S.J., Girme, A. and Hingorani, L. (2023). Network pharmacology-based strategic prediction and target identification of apocarotenoids and carotenoids from standardized Kashmir saffron (*Crocus sativus* L.) extract against polycystic ovary syndrome. *Medicine (Baltimore)*, *102*(32): e34514.

Tsolaki, M., Karathanasi, E., Lazarou, I., Dovas, K., Verykouki, E. et al. (2016). Efficacy and Safety of *Crocus sativus* L. in Patients with Mild Cognitive Impairment: One Year Single-Blind Randomized, with Parallel Groups, Clinical Trial. *J Alzheimers Dis*, *54*(1): 129–133.

Umigai, N., Takeda, R. and Mori, A. (2018). Effect of crocetin on quality of sleep: A randomized, double-blind, placebo-controlled, crossover study. *Complement Ther Med*, *41*: 47–51.

Vafaei, S., Motejaded, F. and Ebrahimzadeh-Bideskan, A. (2020). Protective effect of crocin on electromagnetic field-induced testicular damage and heat shock protein A2 expression in male BALB/c mice. *Iran J Basic Med Sci*, *23*(1): 102–110.

Verma, S.K. and Bordia, A. (1998). Antioxidant property of Saffron in man. *Indian J Med Sci*, *52*: 205-207.

Wali, A.F., Pillai, J.R., Al Dhaheri, Y., Rehman, M.U., Shoaib, A. et al. (2020). *Crocus sativus* L. Extract Containing Polyphenols Modulates Oxidative Stress and Inflammatory Response against Anti-Tuberculosis Drugs-Induced Liver Injury. *Plants (Basel)*, *9*(2): 167.

Wang, F., Li, W.L., Shen, L.J., Jiang, T.T., Xia, J.J., You, D.L., Hu, S.Y., Wang, L. and Wu, X. (2022). Crocin Alleviates Intracerebral Hemorrhage-Induced Neuronal Ferroptosis by Facilitating Nrf2 Nuclear Translocation. *Neurotox Res*. *40*(2): 596–604.

Wauquier, F., Boutin-Wittrant, L., Pourtau, L., Gaudout, D., Moras, B. et al. (2022). Circulating Human Serum Metabolites Derived from the Intake of a Saffron Extract (Safr'InsideTM) Protect Neurons from Oxidative Stress: Consideration for Depressive Disorders. *Nutrients*, *14*(7): 1511.

Xu, Z., Lin, S., Gong, J., Feng, P., Cao, Y. et al. (2021). Exploring the Protective Effects and Mechanism of Crocetin From Saffron Against NAFLD by Network Pharmacology and Experimental Validation. *Front Med (Lausanne)*, *8*: 681391.

Yarijani, Z.M., Pourmotabbed, A., Pourmotabbed, T. and Najafi, H. (2017). Crocin has anti-inflammatory and protective effects in ischemia-reperfusion induced renal injuries. *Iran J Basic Med Sci*, *20*(7): 753–759.

Yoshino, F., Yoshida, A., Umigai, N., Kubo, K. and Lee, M.C. et al. (2011). Crocetin reduces the oxidative stress induced reactive oxygen species in the stroke-prone spontaneously hypertensive rats (shrsps) brain. *J. Clin. Biochem. Nutr*, *49*: 182–187.

Winterhalter, P. and Straubinger, M. (2000). Saffron—Renewed interest in an ancient spice. *Food Rev. Int*, *16*: 39–59.

Yousefi-Manesh, H., Aghamollaei, H., Dehpour, A.R., Sheibani, M., Tavangar, S.M. et al. (2022). The role of saffron in improvement of ocular surface disease in a mouse model of Lacrimal Gland Excision-induced dry eye disease. *Exp Eye Res*, *221*: 109127.

Zaazaa, L., Mrabti, H.N., Ed-Dra, A., Bendahbia, K., Hami, H. et al. (2021). Determination of Mineral Composition and Phenolic Content and Investigation of Antioxidant, Antidiabetic, and Antibacterial Activities of *Crocus sativus* L. Aqueous Stigmas Extracts. *Adv Pharmacol Pharm Sci*, 2021: 7533938.

Zamani, M., Zarei, M., Nikbaf-Shandiz, M., Gholami, F., Hosseini, A.M. et al. (2022). The effects of saffron supplementation on cardiovascular risk factors in adults: A systematic review and dose-response meta-analysis. *Front Nutr*, *9*: 1055517.

Zeinali, M., Zirak, M.R., Rezaee, S.A., Karimi, G. and Hosseinzadeh, H. et al. (2019). Immunoregulatory and anti-inflammatory properties of *Crocus sativus* (Saffron) and its main active constituents: A review. *Iran J Basic Med Sci*, *22*(4): 334–344.

Zeka, K., Marrazzo, P., Micucci, M., Ruparelia, K.C., Arroo, R.R.J. et al. (2020). Activity of Antioxidants from *Crocus sativus* L. Petals: Potential Preventive Effects towards Cardiovascular System. *Antioxidants (Basel)*. *9*(11): 1102.

Zeka, K., Ruparelia, K.C., Sansone, C., Macchiarelli, G., Continenza. M.A. et al. (2018). New Hydrogels Enriched with Antioxidants from Saffron Crocus Can Find Applications in Wound Treatment and/or Beautification. *Skin Pharmacol Physiol*, *31*(2): 95–98.

Zhao, C., Kam, H.T., Chen, Y., Gong, G., Hoi, M.P. et al. (2021). Crocetin and Its Glycoside Crocin, Two Bioactive Constituents From *Crocus sativus* L. (Saffron), Differentially Inhibit Angiogenesis by Inhibiting Endothelial Cytoskeleton Organization and Cell Migration Through VEGFR2/SRC/FAK and VEGFR2/MEK/ERK Signaling Pathways. *Front Pharmacol*, *12*: 675359.

Zheng, S., Qian, Z., Tang, F. and Sheng, L. (2005). Suppression of vascular cell adhesion molecule-1 expression by crocetin contributes to attenuation of atherosclerosis in hypercholesterolemic rabbits. *Biochem Pharmacol*, *70*: 1192–1199.

Zilaee, M., Hosseini, S.A., Jafarirad, S., Abolnezhadian, F., Cheraghian, B. et al. (2019). An evaluation of the effects of saffron supplementation on the asthma clinical symptoms and asthma severity in patients with mild and moderate persistent allergic asthma: a double-blind, randomized placebo-controlled trial. *Respir Res*, *20*(1): 39

7 Saffron Transcriptomics

Neda Arad[1], Parisa Bolouri[2], Fikrettin Sahin[2], Mitra Aelaei[3], and Seyed Alireza Salami[4*]

Saffron, often revered as the "golden spice", is among the most prized and sought-after spices worldwide. Extracted from the delicate stigmas of *Crocus sativus* L., it is celebrated for its distinct taste, vibrant hue, and extensive health benefits. Despite its acclaim, many aspects of saffron remain enigmatic. As a sterile triploid species within the Crocus genus of the Iridaceae family, its genetic relationship with diploid relatives is still unclear. With a large genome rich in repetitive sequences, transcriptome analysis is crucial for deciphering saffron's genetic complexity. This approach reveals its molecular mechanisms, growth dynamics, and flowering processes while identifying key *genes* responsible for its unique traits. Investigating synthesis pathways within the transcriptome provides insights into enzyme catalysis, potentially enabling biotechnological advancements to enhance saffron yield. A deeper understanding of saffron's transcriptome not only enriches knowledge of its biology but also offers opportunities to refine cultivation techniques and optimize production. As an omics-based approach, transcriptomics facilitates comprehensive genetic analysis, ultimately improving saffron's trade value by enhancing both quantity and quality.

1. Introduction

1.1 Saffron (C. sativus): A Historical and Agronomic Overview

Saffron (*C. sativus*), a perennial stemless herb from the Iridaceae family, reproduces asexually via corm division due to its triploid chromosome structure (Hu et al., 2020, Singh et al., 2023). Its sterile flowers feature six tepals, three stamens, and three stigmas, with the stigma—known as 'Red Gold'—valued for its medicinal, culinary, and dyeing applications (Marchioni et al., 2019, Singh et al., 2023). Thriving in arid climates, saffron is widely cultivated in regions like Iran, providing economic benefits to local communities (Rezvani-Moghaddam, 2020, Darabi et

[1] School of Plant Sciences, The University of Arizona, Tucson, USA.
[2] Department of Genetics and Bioengineering, Yeditepe University, Istanbul/Türkiye.
[3] Department of Horticulture, Faculty of Agriculture, University of Zanjan, Zanjan, Iran.
[4] Department of Horticultural Science, Faculty of Agriculture, University of Tehran, Karaj, Iran.
* Corresponding author: asalami@ut.ac.ir

DOI: 10.1201/9781003403760-7

al., 2023). Its origins remain debated, with Iran and Greece as possible birthplaces (Kazemi-Shahandashti et al., 2022). Historically, saffron has served as food, medicine, perfume, and dye, particularly in the pharma, agro-food, and cosmetic industries (Taheri-Dehkordi et al., 2020, Wenger, 2022).

Saffron's high value stems from its rarity, labor-intensive harvesting, and rich nutritional and therapeutic properties (Hosseini et al., 2018, Kour et al., 2022). Its antioxidant effects derive from phenolic and flavonoid compounds (Karimi et al., 2010, Armellini et al., 2018). Crocin, picrocrocin, and safranal in the stigmas define its color, taste, and aroma, with environmental conditions and cultivation strategies significantly influencing their quality (Yilmaz et al., 2010, Moradi et al., 2022). Saffron thrives in specific climates, with optimal flowering at 23–27°C in summer and 15–20°C in winter (Kumar et al., 2022). Water availability is crucial, particularly in arid regions (Bengouga et al., 2020). Major producers include Iran, India, Spain, Greece, and Afghanistan,with Khorasan Province in Iran and Kashmir in India as key hubs (Mykhailenko et al., 2022, Kour et al., 2022). Yield and quality depend on environmental factors affecting gene regulation, particularly in apocarotenoid biosynthesis. Higher air temperatures and moderate rainfall during blooming enhance stigma production (Cardone et al., 2019). Additionally, Total Daily Light Integral (TDLI) influences flowering, stigma quality, and leaf shape (Gao et al., 2023). Advancements in next-generation sequencing (NGS) and omics-based technologies have revolutionized saffron breeding by enabling the selection of elite genotypes for optimized cultivation. Genomic, transcriptomic, proteomic, and metabolomic studies have decoded key regulators of saffron biology, revealing *genes* and pathways involved in secondary metabolite biosynthesis, including crocins, crocetin, safranal, and picrocrocin. Transcriptome analysis of stigma and petals further clarifies the molecular basis of saffron's aroma, flavor, and color.

1.2 Saffron's Genetic Complexity: Dive into Its Enigmatic DNA Structure

Saffron (*C. sativus*) possesses a triploid genome of 10.5 Gb with a chromosome configuration of $2n=3x=24$ (Chichiricc, 1984, Schmidt et al., 2019). Its karyotype consists of eight triplet chromosomes: subacrocentric (chromosome 1 and 2), metacentric (chromosome 3, 4, and 8), and submetacentric (chromosome 6 and 7) (Schmidt et al., 2019, Husaini et al., 2022).

Saffron reproduces asexually via corms, resulting in nearly identical makeup due to its sterility (Rashed-Mohassel, 2020, Husaini et al., 2022, Ghaffari, 2022). Its large genome and triploidy complicate genetic studies, limiting research on its genome and chromosomes. Despite available techniques for protein structure prediction, comprehensive saffron analysis remains elusive (Jyoti et al., 2022). Functional genomics holds promise for identifying *genes* essential for taste and stigma pigment production in *C. sativus* (Husaini et al., 2022).

2. Decoding the Transcriptome: Insights and Implications in Cellular Biology

The transcriptome represents the complete set of RNA molecules in a cell, crucial for understanding gene expression, regulation, and biological processes (Dastkae et al., 2020, de Sousa et al., 2019, Pandita & Pandita, 2023). It reveals gene activation patterns, novel *genes*, and alternative splicing events (Jain et al., 2016, Pandita & Pandita, 2023). Expressed Sequence Tags (ESTs) databases from saffron stigmas aid genomic studies (D'Agostino et al., 2007). However, transcriptome reconstruction using short reads presents computational challenges (Honass et al., 2016).

2.1 Unlocking Saffron's Secrets: The Pivotal Role of Transcriptome Analysis

Transcriptome analysis is crucial for deciphering saffron's complex genome, revealing molecular mechanisms behind its flavor, color biosynthesis, and genomic structure within the Iridaceae family (Fiore et al., 2010). It identifies *genes* influencing saffron's unique traits, enabling genetic modifications to enhance yield (Husaini et al., 2022). Apocarotenoid biosynthesis, a key function in saffron, may be regulated by microRNAs (Taheri-Dehkordi et al., 2021). Additionally, transcriptome studies highlight flower-regulating *genes* essential for optimizing saffron yield (Qian et al., 2019). Saffron transcriptional analysis begins with RNA extraction from plant tissue, followed by high-throughput sequencing to generate short reads, which are aligned to a reference genome or assembled de novo (Jain et al., 2016). RNA integrity is crucial for accurate transcriptome assembly (Schroeder et al., 2006). Advanced sequencing methods like SMRT and long-read sequencing enhance transcriptome precision, enabling full-length transcript detection and alternative splicing analysis (Eid et al., 2009). The assembled transcriptome is annotated against gene databases to identify functional components and biological roles (Trapnell et al., 2012). This process underpins saffron biology research, including gene expression analysis, functional gene characterization, and genetic enhancement strategies (Kumar & Blaxter, 2010).

2.2 Deciphering the Floral Code: Transcriptional Insights into Saffron Flowering

Transcriptional analysis is crucial for understanding saffron's flowering regulation, especially as the industry faces increased flower withering, rot, and delayed blooming, impacting stigma yield and quality. Investigating molecular mechanisms governing flowering transitions is essential for overcoming these challenges (Hu et al., 2020).

Hu et al. (2020) and Tan et al. (2019) used SMRT sequencing and Illumina RNA-seq to profile flowering and non-flowering saffron transcriptomes. They identified 75,351 full-length unigenes, along with SSRs, lncRNAs, and alternative splicing events. Comparative gene expression analysis revealed 62 flower-associated *genes* linked to vernalization, photoperiod, gibberellins, autonomous, and age-related pathways. Real-Time qPCR validated the high expression of PB.20221.2 and PB.38952.1 in flowering samples, showing transcript accumulation during bloom development. Qian et al. (2019) highlighted the crucial role of flower regulatory *genes* in saffron yield and flowering patterns. Using SMRT sequencing, they mapped a detailed genetic blueprint of flowering dynamics, benefiting both saffron and the broader Iridaceae family.

Identifying flower-related *genes* and their regulatory network clarifies saffron crocus flowering mechanisms. Researchers propose core *genes*—FD, FT, SPL1, SOC1, LFY, AP1, and AP2—as key regulators of flowering and transition.

Research on the Saffron Flowering Regulator (SFR) has improved saffron's resistance to rotting and withering, leading to healthier blooms and higher yields. Manipulating SFR marks a breakthrough in crop refinement and genetic engineering, offering insights into *genes* and pathways that enhance saffron's resilience and productivity (Bhandari, 2015).

2.3 Decoding Saffron's Genetic Symphony: The Role of Transcription Factors

Transcription factors (TFs) regulate gene expression in saffron, with MYB identified as a key family through PlnTFDB analysis (Hu et al., 2020, Ahrazem et al., 2019). Among 5726 distinct proteins, MYB and MYB-related TFs dominate, comprising 11.86% of all TFs. Other major TF classes include bHLH, NAC, C2H2, FAR1, C3H, ERF, WRKY, bZIP, and B3.

Jain et al. (2016) highlighted MYB, MYB-related bHLH, and WRKY transcription factors as key regulators of apocarotenoid biosynthesis in saffron. Previous transcriptomic studies identified 3819 and 2601 TFs, maintaining a consistent hierarchy of dominant families (Baba et al., 2015). Additionally, C2H2, C3H, and AN20/AN1 TFs may influence apocarotenoid biosynthesis, with CsSAP09 highly expressed in stigma at anthesis and induced under abiotic stress (Malik & Asharaf, 2017). Saffron's genetic profile includes MADS-box and TFL transcription factors, with MYB playing a key role in pigment synthesis. The TFL gene, cloned and characterized in saffron, is regulated by hormonal and environmental factors, influencing flowering and reproduction (Tsaftaris et al., 2010, Busconi et al., 2021).

2.4 Unveiling the Mystery: The Role of miRNAs in Saffron's Genetic Landscape

miRNAs, compact 20–24 nt non-coding RNAs, regulate gene expression at transcriptional and post-transcriptional levels, though their role in saffron remains largely unexplored (Taheri-Dehkordi et al., 2021). Using SUmiRFind, researchers compared 474,280 non-coding transcripts with 5645 mature plant miRNAs, identifying 8030 matching sequences with minimal mismatches. SUmirFold analysis revealed 175 sequences (2.17%) capable of forming hairpin structures, with 20 meeting miRNA identification criteria. These findings enhance understanding of saffron's genetic regulation and miRNAs involvement in apocarotenoid biosynthesis.

2.5 Decoding the Golden Spice: Transcriptional Insights into Saffron's Medicinal Metabolites

Transcriptional analysis has clarified gene expression patterns in saffron, revealing its role in carotenoid, terpenoid, and polyphenol biosynthesis. Studies have identified therapeutic targets, including antidepressant, anticancer, and neuroprotective agents. Understanding transcriptional pathways enables genetic engineering and environmental optimization to enhance secondary metabolite yields. Key compounds—crocin, picrocrocin, and safranal—are synthesized through enzymatic processes regulated by specific *genes*. Environmental factors like light, temperature, and nutrients influence saffron's transcriptional pathways, aiding cultivation optimization (Husaini et al., 2022). Despite the medicinal value of crocin, safranal, and picrocrocin, their biosynthetic *genes* remain unclear. RNA-seq studies identified regulatory mechanisms in apocarotenoid biosynthesis, showing crocin accumulation increases as the stigma matures (Tan et al., 2019) Key enzymes in crocin synthesis include LYCb, which converts lycopene to β-carotene, and BCH, which hydroxylates β-carotene into zeaxanthin (Ronen et al., 2000, Alquézar et al., 2009, Sathasivam et al., 2021). CCD2 cleaves zeaxanthin into hydroxy β-cyclocitral and crocetin dialdehyde, precursors for picrocrocin and crocin (Frusciante et al., 2014). UDP-glucosyltransferase (UGT) convert hydroxy β-cyclocitral into picrocrocin and safranal, while ALDH transforms crocetin dialdehyde into crocetin (Moraga et al., 2004, Nagatoshi et al., 2021). Crocins are finalized through UGT-mediated glycosylation, with UGT74AD1 and UGT91P3 playing key roles (Liu et al., 2020). UGT91P3, co-expressed with UGT74AD1, was identified as essential for crocin biosynthesis (Lopez-Jimenez et al., 2021). Transcriptome analysis has also pinpointed transcription factors regulating apocarotenoid accumulation in saffron stigma (Husaini et al., 2022).

Despite limited knowledge of the *genes* driving saffron apocarotenoid production, OMICS techniques have significantly advanced our understanding of its genetic networks (Yue et al., 2020). High-throughput approaches have mapped *genes*, biological pathways, and regulatory mechanisms, aiding gene editing and engineering. Genome sequencing has provided insights into saffron's origin and ancestry, while integrating transcriptomic, proteomic, and metabolomic data clarifies its biological processes. These advancements enhance knowledge of apocarotenoid synthesis, stigma biogenesis, corm activity, and flowering development, paving the way for phenome-engineering to improve saffron yield and quality (Tan et al., 2019).

Table 1 Secondary Metabolites of Saffron (*C. sativus*).

Secondary Metabolites	*Features*	*The bioactive role in Saffron*	*Bioactive chemical structure*	*Reference*
Safranal	Volatile compound	Responsible for the distinct aroma of saffron	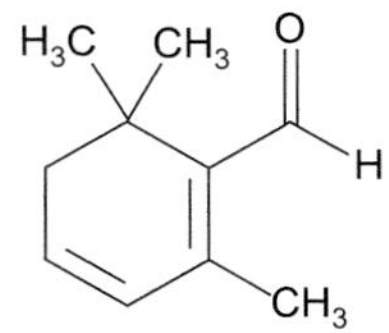	Bolhassani (2018)
Crocin	Water-soluble carotenoid	Gives saffron its vibrant color	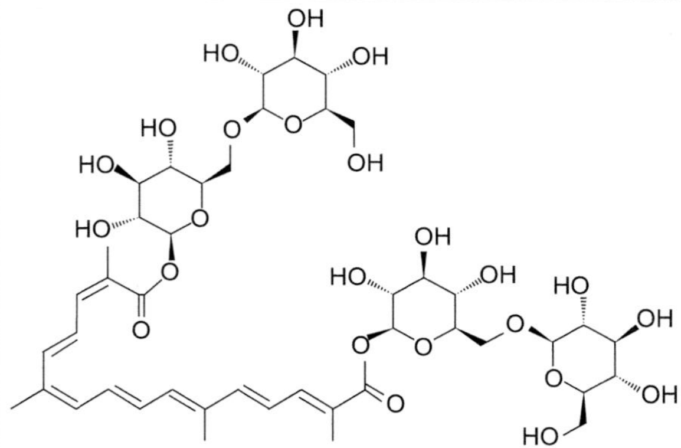	Dasgupta and Klein (2014)
Crocetin	Carotenoid derivative	Contributes to the color and medicinal properties of saffron	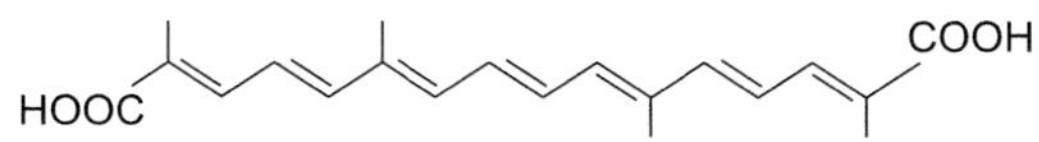	Dasgupta and Klein (2014)

Contd.

Table 1 *Contd.*

Secondary Metabolites	*Features*	*The bioactive role in Saffron*	*Bioactive chemical structure*	*Reference*
Picrocrocin	Bitter compound	Responsible for the bitter taste of saffron	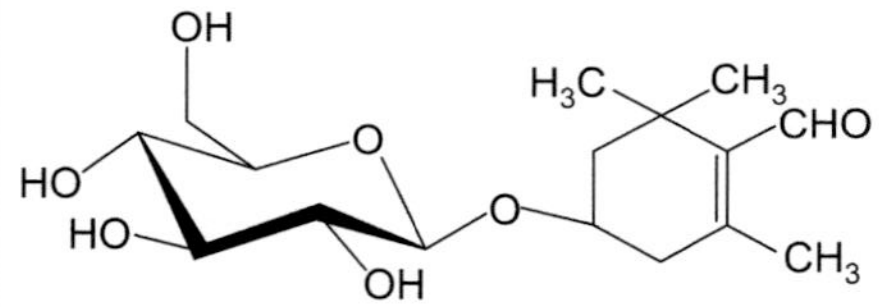	Silva dos Santos et al. (2021)
Kaempferol	Flavonoid compound	Potential health benefits	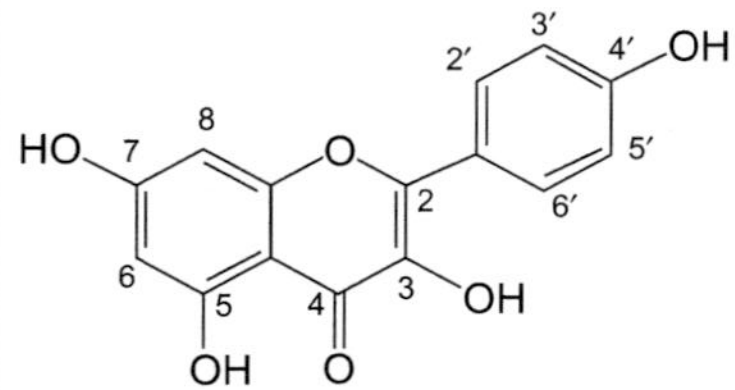	Pandey et al. (2022)
Flavonoids (Quercetin)	The most important flavonoid in saffron	Antioxidant and anti-inflammatory	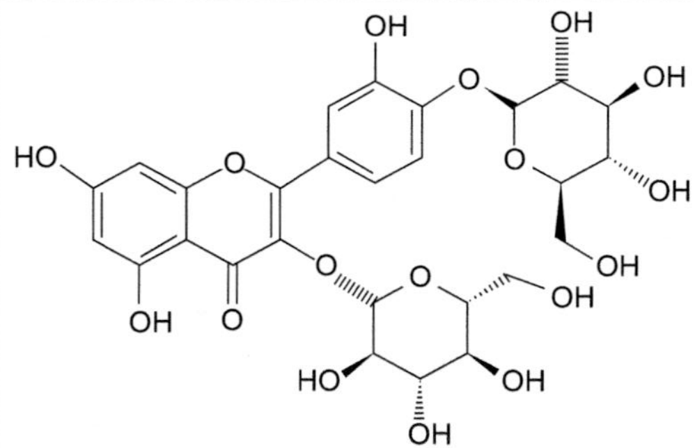	Bakshi et al. (2022)

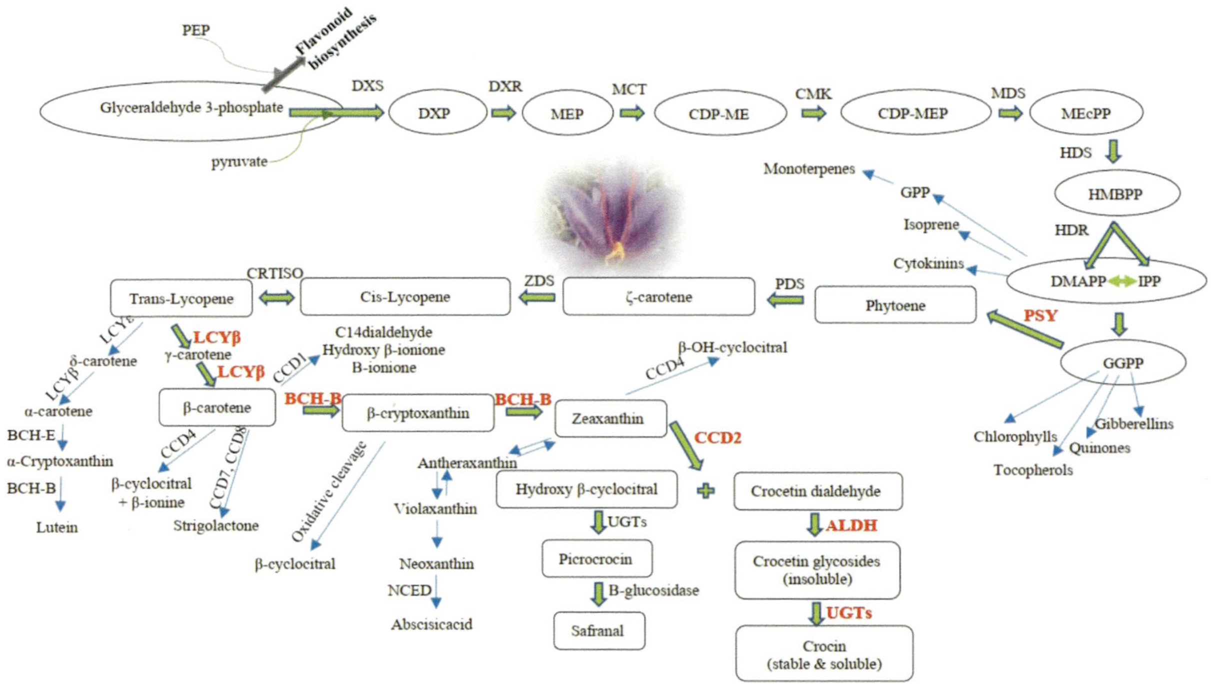

Fig. 1 A schematic diagram of the carotenoid/apocarotenoid pathway, depicting the different *genes* involved in the biosynthetic pathway. The list of the *genes*/enzymes highlighted in red in the figure are: PSY, Phytoene synthase, LCYB, lycopene β-cyclase, BCH, β-carotene hydroxylase, CCD2, crotenoid cleavage dioxygenase, ALDH, aldehyde dehydrogenase and UGT, uridine diphosphate glycosyltransferases (Fallah et al., 2024).

Metabolomic and transcriptomic analyses identified 61,202 unigenes, 28 regulators, and 32 carotenogen *genes* involved in *saffron* apocarotenoid biosynthesis. CsALDH3 was confirmed to oxidize crostin, and UGT709G1 was found essential for picrocrocin biosynthesis, the precursor of safranal (Diretto et al., 2019). Transcriptome sequencing revealed 15 candidate *genes* linked to safranal and crocin production, along with a new branching pathway for geranyl-geranyl pyrophosphate conversion (Jain et al., 2016). Differential transcriptome expression studies showed stigma-specific gene activity in apocarotenoid biosynthesis, enhancing understanding of saffron's molecular mechanisms (Jain et al., 2016).

As saffron transcriptome research advances, its apocarotenoid biosynthesis is becoming clearer. The key challenge now is leveraging these insights to enhance saffron's medicinal potential and yield.

2.6 Saffron's Resilience: Transcriptomic Insights into Stress Response and Light-Induced Metabolism

Saffron faces various biotic and abiotic challenges affecting its yield and quality. Transcriptomic studies have identified molecular responses to drought, salinity, pathogens, and pests. Avila-Sosa et al. (2022) highlighted drought-responsive *genes* involved in osmotic regulation, antioxidant defenses, and stress signaling. Wani et al. (2021) found salinity stress-activated *genes* regulating ion balance, osmotic adjustment, and ROS detoxification. Hu et al. (2020) mapped pathogen-response *genes*, revealing mechanisms for pathogen detection and plant immunity. These findings enhance understanding of saffron's stress-adaptive pathways, aiding resilience strategies. Mapping stress-responsive genetic pathways enables scientists to develop stress-resistant saffron strains through breeding and genetic interventions, refining cultivation strategies to mitigate environmental stressors (Asensi-Fabado et al., 2017, Cerda-Bernad et al., 2022). Light significantly influences saffron metabolism, with monochromatic Blue light enhancing crocin content, while Red and Blue light boost picrocrocin levels. Safranal accumulates most under high Red light conditions (Moradi et al., 2022). At the transcript level, Blue light upregulates crocin biosynthesis *genes* (CsCCD2, CsALDH3I1, CsUGT2), whereas Red and White light increase CsUGT709G1 expression, crucial for picrocrocin and safranal pathways. These findings reinforce the role of specific enzymes and *genes* in saffron's apocarotenoid synthesis, as previously noted by Diretto et al. (2019). Deciphering saffron's transcriptomic responses enhances our understanding of its adaptability, paving the way for optimized cultivation and maximized medicinal and culinary benefits.

Conclusions

Recent research on *Crocus* species has revealed its genetic complexity, genomic structure, and transcriptional activity, particularly in saffron-producing tissues. While key *genes* for flavor, fragrance, carotenoid, and flavonoid synthesis have been identified, the metabolism of primary apocarotenoids in stigmas remains unclear. The role of volatile compounds in saffron's aroma is still enigmatic. Advances in next-generation sequencing and high-resolution mass spectrometry offer a hybrid approach to link genetic determinants with biochemical traits, deepening our understanding of this prized spice.

Reference

Ahrazem, O., Argandoña, J., Fiore, A., Rujas, A., Rubio-Moraga, Á. et al. Multi-species transcriptome analyses for the regulation of crocins biosynthesis in Crocus. *BMC Genomics*. 2019. *27*: 20(1).

Alquézar, B., Zacarías, L. and Rodrigo, M.J. Molecular and functional characterization of a novel chromoplast-specific lycopene β-cyclase from Citrus and its relation to lycopene accumulation. *Journal of Experimental Botany* 2009, 60, 1783–1797.

Armellini, R., Peinado, I., Pittia, P., Scampicchio, M,, Heredia, A., et al. Effect of saffron (*Crocus sativus* L.) enrichment on antioxidant and sensorial properties of wheat flour pasta. *Food Chem.* 2018. 15; *254*: 55–63.

Asensi-Fabado, M.A., Amtmann, A. and Perrella, G. Plant responses to abiotic stress: The chromatin context of transcriptional regulation. Vol. 1860, Biochimica et Biophysica Acta - Gene Regulatory Mechanisms. Elsevier B.V., 2017. pp. 106–122.

Avila-Sosa, R., Nevárez-Moorillón, G.V., Ochoa-Velasco, C.E., Navarro-Cruz, A.R., Hernández-Carranza, P. et al. Detection of Saffron's Main Bioactive Compounds and their Relationship with Commercial Quality. Vol. 11, *Foods. MDPI*, 2022.

Baba, S.A., Mohiuddin, T., Basu, S., Swarnkar, M.K., Malik, A.H. et al. Comprehensive transcriptome analysis of *Crocus sativus* for discovery and expression of *genes* involved in apocarotenoid biosynthesis. *BMC Genomics*. 2015. *15*: 16(1).

Bakshi, R.A., Sodhi, N.S., Wani, I.A., Khan, Z.S., Dhillon, B. et al. Bioactive constituents of saffron plant: Extraction, encapsulation and their food and pharmaceutical applications. Vol. 2, *Applied Food Research*. Elsevier B.V., 2022.

Bengouga, K., Lahmadi, S., Zeguerrou, R., Maaoui, M., Halis, Y. et al. The Saffron (*Crocus sativus* L.) Cultivation Introduction in Mountainous Oases of Algeria. *Acta Horticulturae et Regiotecturae*. 2020. 1; *23*(1): 8–11.

Bhandari, P.R. *Crocus sativus* L. (saffron) for cancer chemoprevention: A mini review. Vol. 5, *Journal of Traditional and Complementary Medicine*. National Taiwan University, 2015. pp. 81–7.

Bolhassani, A. Bioactive Components of Saffron and Their Pharmacological Properties. In: Studies in Natural Products Chemistry. Elsevier B.V., 2018. pp. 289–311.

Busconi, M., Wischnitzki, E., Del Corvo, M., Colli, L., Soffritti, G. et al. Epigenetic Variability Among Saffron Crocus (*Crocus sativus* L.) Accessions Characterized by Different Phenotypes. *Front Plant Sci*. 2021. *4*: 12.

Cardone, L., Castronuovo, D., Perniola, M., Cicco, N., Candido, V. et al. (2019). Evaluation of Corm Origin and Climatic Conditions on Saffron (*Crocus Sativus* L.) Yield And Quality. *J. Sci. Food Agric.*, 13(99), 5858–5869.

Cerdá-Bernad, D., Costa, L., Serra, A.T., Bronze, M.R., Valero-Cases, E. et al. Saffron against Neuro-Cognitive Disorders: An Overview of Its Main Bioactive Compounds, Their Metabolic Fate and Potential Mechanisms of Neurological Protection. Vol. 14, *Nutrients. MDPI*, 2022.

Chichiriccò, G. Karyotype and meiotic behaviour of the triploid *Crocus sativus* L. Caryologia. 1984. 1; *37*(3): 233–239.

D'Agostino, N., Pizzichini, D., Chiusano, M.L., Giuliano, G. (2007). An EST database from saffron stigmas. *BMC Plant Biology* *7*: (53), 1–8.

Darabi, H., Moarrab, Y., Balist, J. and Naroei, B. Resilient plant species selection for urban green infrastructure development in arid regions: a case of Qom, Iran. *Urban Ecosyst.* 2023.

Dasgupta, A. and Klein, K. (2014). Antioxidants in food, vitamins and supplements: prevention and treatment of disease. Academic Press.

Dastkar, E., Soleimani, A., Jafary, H., de Dios Alche, J., Bahari, A. et al. Differential expression of *genes* in olive leaves and buds of ON-versus OFF-crop trees. *Sci Rep.* 2020. *1*: 10(1).

de Sousa, M.C., Gjorgjieva, M., Dolicka, D., Sobolewski, C., Foti, M. et al. Deciphering miRNAs' action through miRNA editing. Vol. 20, *International Journal of Molecular Sciences*. MDPI AG, 2019.

Diretto, G., Ahrazem, O., Rubio-Moraga, Á., Fiore, A., Sevi, F. et al. UGT709G1: a novel uridine diphosphate glycosyltransferase involved in the biosynthesis of picrocrocin, the precursor of safranal in saffron (*Crocus sativus*). *New Phytologist*. 2019. 1; *224*(2): 725–740.

Eid, J., Fehr, A., Gray, J., Luong, K., Lyle, J. et al. Real-Time DNA Sequencing from Single Polymerase Molecules. *New Series*. 2009, 323(5910):133–8.

Fallah, M., Naderi, R. and Salami, S.A. Phytochemical Profile and Gene Expression of Apocarotenoid Biosynthetic Pathway in different Crocus Species and Molecular Phylogeny Studies of Related Genes. 2024. PhD Thesis.

Fiore, A., Pizzichini, D., Diretto, G., Federico Scossa, Spanò, L. et al. Genomics and Transcriptomics of Saffron: New Tools to Unravel the Secrets of an Attractive Spice. 2010. *The Editor*, 25, pp.1–14.

Frusciante, S., Diretto, G., Bruno, M., Ferrante, P., Pietrella, M. et al. Novel carotenoid cleavage dioxygenase catalyzes the first dedicated step in saffron crocin biosynthesis. *Proceedings of the National Academy of Sciences* 2014, 111, 12246–12251.

Gao, D., Ji, X., Yuan, Q., Pei, W., Zhang, X. et al. Effects of total daily light integral from blue and broad-band red LEDs on flowering of saffron (*Crocus sativus* L.). *Sci Rep.* 2023 Dec 1; 13(1).

Ghaffari SM. Cytogenetics of saffron (*Crocus sativus*), A natural autotriploid species, An investigation of the reasons for its sterility. *Iranian Journal of Botany*. 2022 1, *28*(1): 57–63.

Honaas, L.A., Wafula, E.K., Wickett, N.J., Der, J.P., Zhang, Y. et al. Selecting superior de novo transcriptome assemblies: Lessons learned by leveraging the best plant genome. *PLoS One*. 2016. 5;11(1).

Hosseini, A., Razavi, B.M. and Hosseinzadeh, H. Pharmacokinetic Properties of Saffron and its Active Components. Vol. 43, *European Journal of Drug Metabolism and Pharmacokinetics*. Springer-Verlag France, 2018. pp. 383–390.

Hu, J., Liu, Y., Tang, X., Rao, H. and Ren, C. et al. Transcriptome profiling of the flowering transition in saffron (*Crocus sativus* L.). *Sci Rep.* 2020. 15; *10*(1): 9680.

Husaini, A.M., Haq, S.A. ul, Jiménez, A.J.L. Understanding saffron biology using omics- and bioinformatics tools: stepping towards a better Crocus phenome. *Mol Biol Rep.* 2022. 1; *49*(6): 5325–5340.

Jain, M, Srivastava, P.L., Verma, M., Ghangal, R., Garg, R. De novo transcriptome assembly and comprehensive expression profiling in *Crocus sativus* to gain insights into apocarotenoid biosynthesis. *Sci Rep.* 2016. 3; 6.

Jyoti, V., Sheetal, A., Salami, S.A. and Chittaranjan K. *The Saffron Genome.* 2022.

Karimi, E., Oskoueian, E., Hendra, R. and Jaafar, H.Z.E. Evaluation of *Crocus sativus* L. stigma phenolic and flavonoid compounds and its antioxidant activity. *Molecules.* 2010. *15*(9): 6244–56.

Kazemi-Shahandashti, S.S., Mann, L., El-nagish, A., Harpke, D., Nemati, Z. et al. Ancient Artworks and Crocus Genetics Both Support Saffron's Origin in Early Greece. Vol. 13, *Frontiers in Plant Science. Frontiers Media* S.A., 2022.

Kour, K., Gupta, D., Gupta, K., Juneja, S., Kaur, M. et al. Controlling Agronomic Variables of Saffron Crop Using IoT for Sustainable Agriculture. Sustainability (Switzerland). 2022 May 1;14(9).

Kumar, A., Devi, M., Kumar, R. and Kumar, S. Introduction of high-value *Crocus sativus* (saffron) cultivation in non-traditional regions of India through ecological modelling. *Sci Rep.* 2022. 1; 12(1).

Kumar, S. and Blaxter, M.L. Comparing de novo assemblers for 454 transcriptome data. *BMC Genomics*. 2010. 16; 11(1).

Liu, T., Yu, S., Xu, Z., Tan, J., Wang, B. et al. Prospects and progress on crocin biosynthetic pathway and metabolic engineering. *Computational and Structural Biotechnology Journal* 2020, 18, 3278–3286.

López-Jimenez, A.J., Frusciante, S., Niza, E., Ahrazem, O., Rubio-Moraga, Á. et al. A New Glycosyltransferase Enzyme from Family 91, UGT91P3, Is Responsible for the Final Glucosylation Step of Crocins in Saffron (*Crocus sativus* L.). *Int J Mol Sci.* 2021 16; *22*(16): 8815.

Malik, A.H. and Ashraf, N. (2017). Transcriptome wide identification, phylogenetic analysis, and expression profiling of zinc-finger transcription factors from *Crocus sativus* L. *Mol Genet Genom 292*(3): 619–633.

Marchioni, I., Pistelli, L., Ferri, B., Cioni, P.L., Copetta, A., Pistelli, L., et al. Preliminary studies on edible saffron bio-residues during different post-harvest storages [Internet]. Vol. 51, *Bulgarian Chemical Communications*. 2019. Available from: https://www.zafferanoitaliano.it/lo-zafferano-in

Moradi, S., Kafi, M., Aliniaeifard, S., Moosavi-Nezhad, M., Pedersen, C. et al. Monochromatic blue light enhances crocin and picrocrocin content by upregulating the expression of underlying biosynthetic pathway *genes* in saffron (*Crocus sativus* L.). 2022. *Front. Hortic. 1*: 960423.

Moradi, S., Kafi, M., Aliniaeifard, S., Salami, S.A., Shokrpour, M. et al. (2021). Blue light improves photosynthetic performance and biomass partitioning toward harvestable organs in saffron (*Crocus sativus* L.). *Cells 10*: 1994.

Moraga, A.R., Nohales, P.F., Pérez, J.A.F. and Gómez-Gómez, L. Glucosylation of the saffron apocarotenoid crocetin by a glucosyltransferase isolated from *Crocus sativus* stigmas. *Planta* 2004, 219, 955–966.

Mykhailenko, O., Saidov, N.B., Ivanauskas, L. and Georgiyants, V. Model implementation of the legal regulation on medicinal plant cultivation for pharmaceutical purposes. Case study of *Crocus sativus* cultivation in Ukraine. *Botanica.* 2022. *28*(1): 27–38.

Mzabri, I., Rimani, M., Charif, K., Kouddane, N., Berrichi, A. et al. Effect of Shade on Agro-Morphological Parameters and Weed Flora of Saffron (*Crocus sativus* L.) Cultivation in the Semiarid Zone of Eastern Morocco. *Scientific World Journal.* 2022; 2022.

Nagatoshi, M., Terasaka, K., Owaki, M., Sota, M., Inukai, T. et al. UGT75L6 and UGT94E5 mediate sequential glucosylation of crocetin to crocin in Gardenia jasminoides. *FEBS letters* 2012, 586, 1055–1061.

Pandey, S.S., Jain, R., Bhardwaj, P., Thakur, A., Kumari, M. et al. Plant probiotics – Endophytes pivotal to plant health. Vol. 263, *Microbiological Research. Elsevier GmbH*, 2022.

Pandita, D. and Pandita, A. Plant MicroRNAs and Stress Response. *Plant MicroRNAs and Stress Response.* CRC Press, 2023.

Qian, X., Sun, Y., Zhou, G., Yuan, Y., Li, J. et al. Single-molecule real-time transcript sequencing identified flowering regulatory *genes* in *Crocus sativus*. *BMC Genomics*. 2019. 14; 20(1).

Rashed-Mohassel, M.H. Evolution and botany of saffron (*Crocus sativus* L.) and allied species. In: *Saffron: Science, Technology and Health*. Elsevier, 2020. pp. 37–57.

Rezvani-Moghaddam, P. Ecophysiology of saffron. In: *Saffron: Science, Technology and Health.* Elsevier, 2020. pp. 119–37.

Ronen, G., Carmel-Goren, L., Zamir, D. and Hirschberg, J. An alternative pathway to β-carotene formation in plant chromoplasts discovered by map-based cloning of Beta and old-gold color mutations in tomato. *Proceedings of the National Academy of Sciences* 2000, 97, 11102–11107.

Sathasivam, R., Radhakrishnan, R., Kim, J.K. and Park, S.U. An update on biosynthesis and regulation of carotenoids in plants. *South African Journal of Botany* 2021, 140, 290–302.

Schmidt, T., Heitkam, T., Liedtke, S., Schubert, V., Menzel, G. et al. Adding color to a century-old enigma: multi-color chromosome identification unravels the autotriploid nature of saffron (*Crocus sativus*) as a hybrid of wild *Crocus cartwrightianus* cytotypes. *New Phytologist.* 2019. 1; *222*(4): 1965–1980.

Schroeder, A., Mueller, O., Stocker, S., Salowsky, R., Leiber, M. et al. The RIN: An RNA integrity number for assigning integrity values to RNA measurements. *BMC Mol Biol.* 2006. 31; 7.

Silva dos Santos, J., Gonçalves Cirino, J.P., de Oliveira Carvalho, P. and Ortega, M.M. The Pharmacological Action of Kaempferol in Central Nervous System Diseases: A Review. Vol. 11, *Frontiers in Pharmacology*. Frontiers Media S.A., 2021.

Singh, D., Sharma, S., Jose-Santhi, J., Kalia, D., Singh, R.K. et al. Hormones regulate the flowering process in saffron differently depending on the developmental stage. *Front Plant Sci.* 2023; 14.

Taheri-Dehkordi, A., Naderi, R., Martinelli, F. and Salami, S.A. (2020). A robust workflow for indirect somatic embryogenesis and cormlet production in saffron (*Crocus sativus* L.) and its wild allies;*Crocus caspius* and *Cheilocostus speciosus*. *Heliyon* 6:e05841.

Taheri-Dehkordi, A., Naderi, R., Martinelli, F. and Salami, S.A. Computational screening of miRNAs and their targets in saffron (*Crocus sativus* L.) by transcriptome mining. *Planta.* 2021. 1; 254(6).

Tan, H., Chen, X., Liang, N., Chen, R., Chen, J. et al. Transcriptome analysis reveals novel enzymes for apo-carotenoid biosynthesis in saffron and allows construction of a pathway for crocetin synthesis in yeast. *J Exp Bot.* 2019. 24; *70*(18): 4819–4833.

Trapnell, C., Roberts, A., Goff, L., Pertea, G., Kim, D. et al. Differential gene and transcript expression analysis of RNA-seq experiments with TopHat and Cufflinks. *Nat Protoc.* 2012.; *7*(3): 562–78.

Tsaftaris, A., Pasentsis, K., Kalivas, A., Michailidou, S., Madesis, P. et al. Isolation of a CENTRORADIALIS/TERMINAL FLOWER1 homolog in saffron (*Crocus sativus* L.): Characterization and expression analysis. *Mol Biol Rep.* 2012; *39*(8): 7899–7910.

Vahedi, M., Kabiri, M. and Salami, S.A. (2018). Quantitative HPLC-based metabolomics of some Iranian saffron (*Crocus sativus* L.) accessions. *Ind Crop Prod* 118: 26–29.

Vahedi, M., Kalanrtari, S. and Salami, S.A. (2014). Factors affecting callus induction and organogenesis in saffron (*Crocus sativus* L.). *Plant Tissue Cult Biotech 24*(1): 1–9.

Wani, K.I., Naeem, M., Castroverde, C.D.M., Kalaji, H.M., Albaqami, M. et al. Molecular mechanisms of nitric oxide (No) signaling and reactive oxygen species (ros) homeostasis during abiotic stresses in plants. Vol. 22, *International Journal of Molecular Sciences.* MDPI, 2021.

Wenger, T. History of saffron. Longhua Chinese Medicine. 2022; 5:15.

Yilmaz, A., Nyberg, N.T., Mølgaard, P., Asili, J. and Jaroszewski, J.W. 1H NMR metabolic fingerprinting of saffron extracts. *Metabolomics.* 2010. *6*(4): 511–517.

Yue, J., Wang, R., Ma, X., Liu, J., Lu, X. et al. Full-length transcriptome sequencing provides insights into the evolution of apocarotenoid biosynthesis in *Crocus sativus*. *Comput Struct Biotechnol J.* 2020. 1; *18*: 774–783.

Zhang, Y., Jin, J., Zhu, S., Sun, Q., Zhang, Y. et al. Citrus β-carotene hydroxylase 2 (BCH2) participates in xanthophyll synthesis by catalyzing the hydroxylation of β-carotene and compensates for BCH1 in citrus carotenoid metabolism. *Horticulture Research* 2023, 10, uhac290.

8 Saffron Post-Transcriptome

Neda Arad[1], Mitra Aelaei[2], Fahime Salehi[2] and Seyed Alireza Salami[3*]

Saffron (*Crocus sativus* L.) holds deep historical and cultural significance, valued for its color, aroma, and health benefits. Its bioactive compounds—crocin, picrocrocin, and safranal—define its sensory and therapeutic properties. Understanding their biosynthesis and regulation is key to improving saffron's quality and yield. Advances in transcriptomics and post-transcriptomics provide insights into gene-protein-metabolite interactions, despite challenges posed by the lack of a fully annotated genome. MicroRNAs (miRNAs), crucial post-transcriptional regulators, remain underexplored in saffron but influence gene expression under biological and environmental signals. Integrating traditional knowledge with modern genetic engineering and synthetic biology can unlock new cultivation and application possibilities, enhancing saffron's role in pharmaceutical, culinary, and cosmetic industries. Continued research promises innovative strategies to optimize saffron's production and medicinal potential. Keywords: miRNAs, Crocus, Regulatory networks, Apocarotenoids, Post-transcriptome.

1. Introduction

1.1 Saffron (C. sativus) Overview

Saffron (*Crocus sativus* L.) within the Iridaceae family, valued for its culinary and medicinal applications (Vahedi et al., 2014, 2018; Taheri-Dehkordi et al., 2020, 2021; Moradi et al., 2021, 2022). Used for over four millennia, it serves as a spice, dye, and medicinal herb (Molina et al., 2005; Taheri-Dehkordi et al., 2020). The Crocus genus includes ~100 species, many cultivated for ornamental purposes across Europe, North Africa, and China (Harpke et al., 2013; Taheri-Dehkordi et al., 2020; Ruksans, 2023). *C. sativus* is specifically grown for saffron extraction, blooming in autumn with slender dark green leaves and purple flowers, though color variations exist among varieties. Saffron's domestication remains debated, with origins proposed in the Middle East, Asia Minor, or Greek islands. It likely spread

[1] School of Plant Sciences, The University of Arizona, Tucson USA.
[2] Department of Horticulture, Faculty of Agriculture, University of Zanjan, Zanjan, Iran.
[3] Department of Horticultural Science, Faculty of Agriculture, University of Tehran, Karaj, Iran.
* Corresponding author: asalami@ut.ac.ir

DOI: 10.1201/9781003403760-8

to India, China, and the Middle East, with Arabs facilitating its Mediterranean dispersion. Iran, first cultivating saffron during the Kingdom of Media (708–550 B.C.), now dominates global production, contributing over 90% of yield (Vahedi et al., 2018). Cultivation extends across Central Europe, North Africa, western China, and West Asia. Saffron's red three-branched stigma and terminal cream-colored segment hold the highest medicinal value, rich in antioxidant, anticancer, and neuroprotective compounds (Festuccia et al., 2014). In Ayurveda, saffron acts as an adaptogen, aiding resilience against stress, anxiety, and fatigue (Kianbakht & Ghazavi, 2005; Xi et al., 2007). Stigma and petal extracts exhibit anti-inflammatory effects (Amin & Hosseinzadeh, 2012). Crocin and safranal help treat high blood pressure, seizures, genotoxicity, and inflammatory disorders, while also improving retinal blood flow and memory (Nazarian et al., 2021).

Research aims to enhance saffron yield and quality through biostimulants, fertilizers, and molecular breeding. Understanding its biology is crucial for sustainable cultivation. While saffron's bioactive compound biosynthesis remains underexplored, transcriptome and post-transcriptome techniques help identify key molecular regulators. These methods could optimize crocin, picrocrocin, and safranal production. Limited studies on miRNAs suggest their role in apocarotenoid biosynthesis, offering potential for further saffron quality improvements (Taheri-Dehkordi et al., 2021).

2. Saffron's Secondary Metabolic Pathway

Secondary metabolic pathways in plants produce essential compounds for defense, food, medicine, color, taste, and aroma. Extensive research has identified 150+ volatile and non-volatile compounds in saffron, with crocin, picrocrocin, and safranal as its most valuable metabolites (Haghighi et al., 2007; Amin & Hosseinzadeh, 2012). Saffron's secondary metabolites include carotenoids, monoterpenes, phenolics, anthocyanins, and flavonoids (Nair et al., 1995; Cerdá-Bernad et al., 2020). Crocin and picrocrocin, abundant in saffron stigma, originate from C40 carotenoid oxidation, imparting its color and flavor (Kanakis et al., 2004; Melnyk et al., 2010; Gomez-Gomez et al., 2010). Volatile compounds in stigma tissues suggest a complex metabolic interplay, with apocarotenoid degradation contributing to saffron's aroma, including safranal (Dauria et al., 2006).

Crocin, a water-soluble carotenoid glycoside, imparts saffron's golden-yellow to red color, undergoing glucosyltransferase-mediated modifications (Ahrazem et al., 2015; Rahaiee et al., 2015). Unlike other carotenoids, crocins dissolve in water but not in alcohol or ether (Carmona et al., 2006). Crocin esters (16–28%) exist in trans and cis forms, with trans being more valuable (Moratalla-Lopez et al., 2019). Crocin serves as a natural food colorant and exhibits antioxidant and anti-inflammatory properties, though its oral absorption is poor (Lage & Cantrell, 2009; Demurtas et al., 2018; Martin et al., 2002; Soeda et al., 2007).Picrocrocin, exclusive to the Crocus genus, gives saffron its bitter taste, acting as a molecular marker (Sánchez et al., 2011). Safranal, derived from picrocrocin degradation,

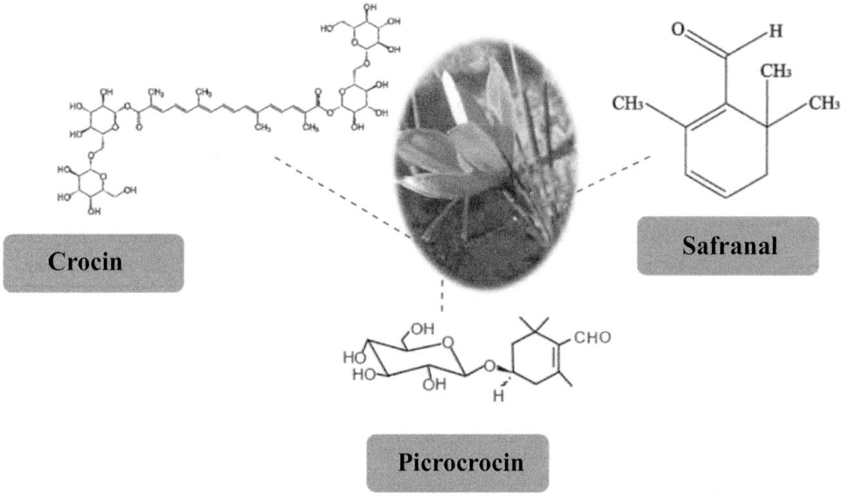

Fig. 1 The structure of three primary metabolites in saffron stigma.

constitutes ~70% of saffron's volatile compounds, contributing to its distinct fragrance and offering antioxidant, antimicrobial, and anti-inflammatory benefits (Kanakis et al., 2007).

Saffron metabolite production depends on genetic background, climate, and cultivation practices. OMICS studies reveal gene networks and metabolic pathways governing apocarotenoid biosynthesis and plant-environment interactions (Jyoti et al., 2022). Transcriptomics and post-transcriptomics highlight saffron's primary metabolites—carbohydrates, amino acids, proteins, fats, and vitamins—essential for growth and survival. Crocin, picrocrocin, and safranal define saffron's unique characteristics and value. Understanding their transcriptional and post-transcriptional regulation is crucial for optimizing production. Metabolomic studies offer insights for metabolic engineering and biotechnological approaches, enhancing their application in food, pharmaceuticals, and cosmetics. The interplay of crocin, picrocrocin, and safranal shapes saffron's distinct sensory profile, reinforcing its culinary and medicinal value. Understanding their metabolic pathways, genes, and miRNA regulation is crucial for expanding saffron's applications in food, health, and cosmetics, highlighting the complexity of plant secondary metabolism.

3. The Role of MicroRNAs In Saffron Plant

Omics technologies—genomics, transcriptomics, proteomics, and metabolomics—have revolutionized plant genome studies, offering insights into gene function, regulatory networks, and metabolic pathways (Fleury et al., 2010). RNA sequencing (RNA-seq) enables precise identification and quantification of gene transcripts, advancing understanding of regulatory and metabolic networks. MicroRNAs (miRNAs), small non-coding RNAs (21-24 nt), act as post-transcriptional gene regulators in eukaryotes, influencing cellular homeostasis and stress responses (Bartel, 2018; Jones-Rhoades et al., 2006). In plants, RNA polymerase II transcribes pri-miRNA, which forms a hairpin structure processed by Dicer-like (DCL) enzymes (Liu et al., 2009). DCL1 cleaves pri-miRNA into pre-miRNA, releasing a miRNA/miRNA duplex*, stabilized by HEN1 methyltransferase (Axtell et al., 2011; Yu, 2005). The guide miRNA binds Argonaute (AGO) proteins, integrating into the RNA-induced silencing complex (RISC) to regulate gene expression via target cleavage or translational inhibition (Budak & Akpinar, 2015; Wu et al., 2009). MiRNAs regulate plant development, from seed germination to maturation (Allen et al., 2005). They influence terpenoid and alkaloid biosynthesis, leaf growth, flowering, and stress responses (Chen et al., 2020; Verma et al., 2020; D'Ario et al., 2017; Sunkar et al., 2012; Pandita & Pandita, 2023). Though few mRNA targets are modulated, their impact is significant. miRNA156 SPL9 controls anthocyanin and flavonoid biosynthesis in Arabidopsis thaliana (Gou et al., 2011). In saffron, miRNAs remain underexplored but are known to modulate gene expression in response to environmental signals (Kurihara & Watanabe, 2010).

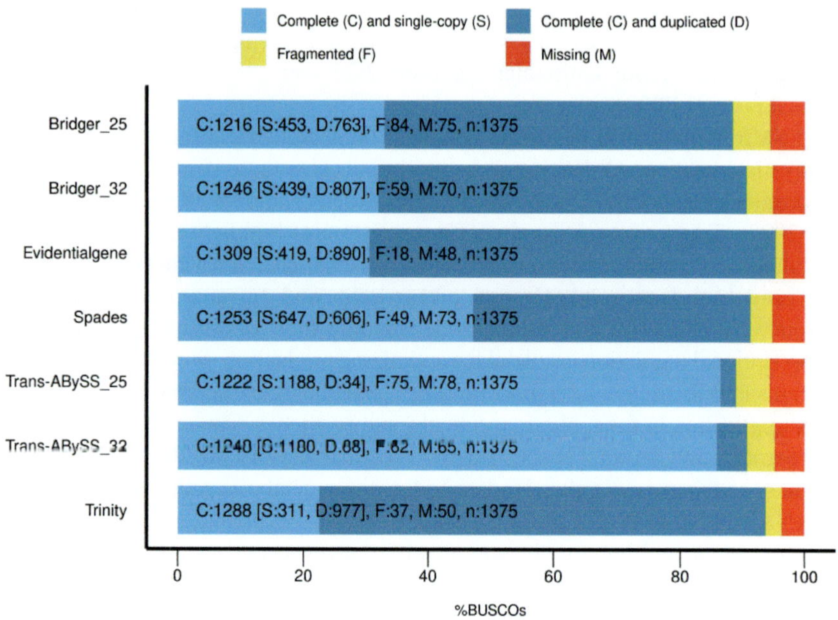

Fig. 2 BUSCO results for different assembly tools. The numbers represent the absolute values of the complete (C) and single-copy (S), complete and duplicated (D), fragmented (F), and missing (M) BUSCOs (Taheri-Dehkordi et al., 2021).

Cloning remains a key method for miRNA discovery in plants, complemented by sequencing technologies and bioinformatics in species like *Taxus mairei*,

Xanthium, and opium poppy (Hao et al., 2012; Fan et al., 2015; Boke et al., 2015). *Despite saffron's* potential, miRNA studies remain limited. Taheri-Dehkordi et al. (2021) identified 66 miRNAs from 19 families, targeting 2880 genes, including CsCCD and CsUGT, crucial for apocarotenoid biosynthesis. Among assemblers, EvidentialGene had the highest BUSCO completeness (1319/1375), outperforming Trinity (1288) and Trans-ABySS (92.08% RMBT).

A total of 66 miRNAs from 19 families were identified in saffron, with miR156 being the most frequent, followed by miR164, miR319, and mir396 (Taheri-Dehkordi et al., 2021). These findings enhance understanding of post-transcriptional regulation in saffron, particularly its apocarotenoid biosynthetic pathway.

4. Implementing Post-transcriptomics in Saffron Research

Post-transcriptomics research enhances understanding of saffron's bioactive compounds, particularly crocin, picrocrocin, and safranal, known for their antioxidant, anti-inflammatory, analgesic, and neuroprotective properties. A high-quality draft genome assembled via NGS technology provides insights into saffron's genetic traits and biosynthetic pathways, aiding metabolite regulation (Ambardar et al., 2021, 2022). Transcriptomic analysis has been crucial in identifying genes involved in secondary metabolite biosynthesis in saffron. Next-generation sequencing of stigmas and petals has mapped key genes regulating metabolite production (Baba et al., 2015). Studies on gene expression dynamics across developmental stages provide insights into temporal regulation (Castillo et al., 2005). At the post-transcriptome level, miRNAs and their targets further clarify apocarotenoid biosynthetic pathway control in saffron.

Proteomics reveals protein products involved in saffron's metabolic pathways, offering insights into enzymatic processes governing bioactive compound synthesis. High-throughput mass spectrometry enables comprehensive proteome profiling, identifying key enzymes and post-translational modifications crucial for secondary metabolite biosynthesis and regulation (Baba et al., 2015). Metabolomics offers crucial insights into saffron's metabolic processes, enabling a deeper understanding of secondary metabolite biosynthesis and regulation. Comparative analysis across developmental stages and growth conditions helps elucidate these mechanisms (Baba et al., 2015). Integrating interactomics enhances understanding of saffron's secondary metabolite biosynthesis and regulation. Protein-protein and metabolite-protein interaction analyses reveal the complex metabolic orchestration within saffron (Ahrazem et al., 2016). Multi-omics has transformed plant research, revealing intricate metabolic pathways in saffron. As the field advances, these insights may optimize metabolite production, enhancing applications in food, pharmaceuticals, and cosmetics.

A deep understanding of saffron's secondary metabolic pathways is crucial for both academic research and practical applications. It underscores the complexity and richness of plant metabolism, offering valuable compounds with diverse

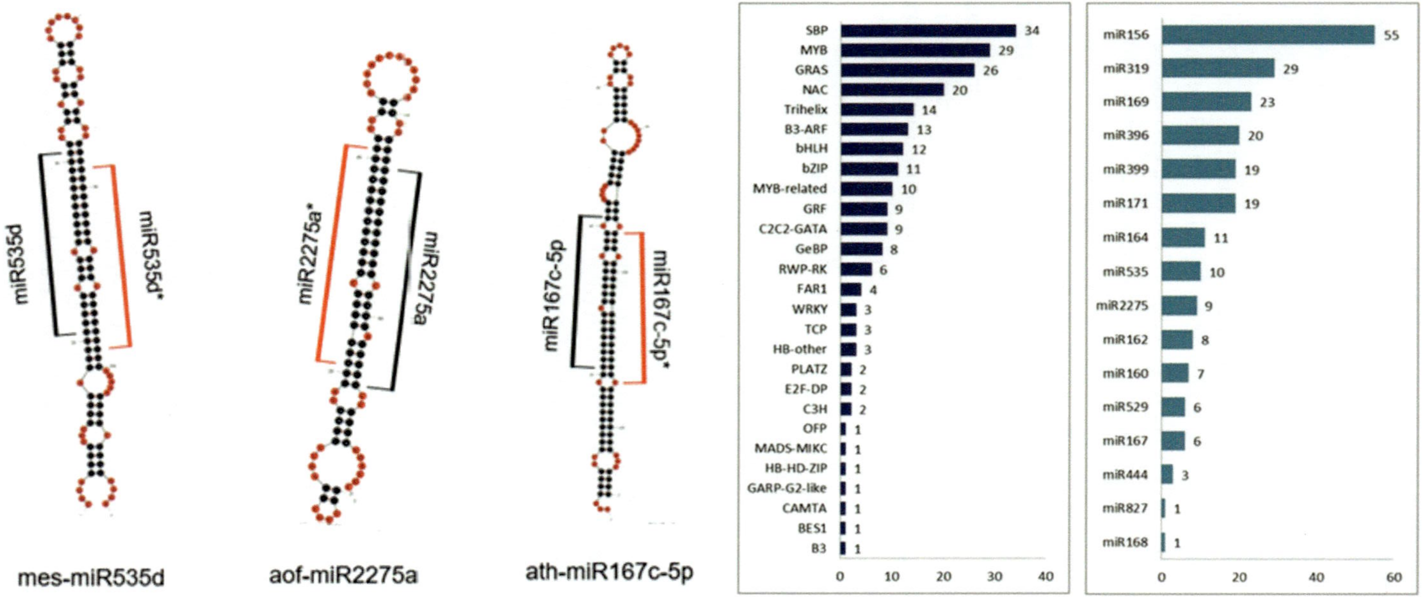

Fig. 3 Stem-loop structures of several predicted mi-RNAs in saffron and frequency of different transcription factor (T.F.) families identified as miRNA targets and identified miRNAs targeting T.F.s in the saffron transcriptome (Taheri-Dehkordi et al., 2021).

benefits. Continued research will unlock new opportunities to harness these metabolites across various industries.

5. Challenges in Saffron Post-transcriptomics

Post-transcriptomics holds great promise for saffron research but faces challenges due to limited data and the complexity of its genome, transcriptome, proteome, and metabolome. The stigma, saffron's source, produces diverse genes, proteins, and metabolites, complicating efforts to identify, quantify, and regulate key components in crocin, picrocrocin, and safranal biosynthesis. Saffron's metabolite biosynthesis is influenced by developmental stage, environment, and genetic variations, complicating analysis. The lack of a fully annotated genome hinders gene identification, making research difficult. Additionally, saffron's perennial nature and specific climatic needs pose challenges for controlled experimentation.

Advancements in omics technologies and bioinformatics are key to overcoming saffron research challenges. Mass spectrometry improvements enhance proteome and metabolome analysis, while sequencing advancements aid genome completion. Enhanced bioinformatics tools refine large-scale data interpretation. Developing experimental systems, including *in vitro* culture and vertical farming, enables controlled saffron studies (Taheri-Dehkordi et al., 2020; Moradi et al., 2021, 2022). Despite challenges in deciphering saffron's biosynthesis and regulation, technological and experimental advancements offer promising solutions, unlocking its culinary and therapeutic potential.

6. Advances in Omics Technologies for Saffron Research

Advancements in omics technologies have enhanced research on saffron's transcriptome, post-transcriptome, proteome, metabolome, and interactome. High-resolution mass spectrometry, next-generation sequencing, and bioinformatics enable deeper, more precise analysis. NGS has transformed genomics and transcriptomic, enabling rapid, cost-effective sequencing of DNA, RNA, and miRNAs at an unprecedented scale. High-resolution sequencing reveals mRNA and miRNA modifications and interactions, crucial for understanding their roles in crocin, picrocrocin, and safranal biosynthesis. Bioinformatics advancements enable efficient data management, analysis, and interpretation in high-throughput omics studies. Sophisticated algorithms reveal gene-protein-metabolite interactions, clarifying saffron's biosynthesis and regulation. Computational modeling predicts network modifications, guiding experimental design. Advancing post-transcriptomics in saffron requires a multidisciplinary approach, integrating biology, chemistry, and bioinformatics. Collaborative efforts will unlock saffron's secrets, maximizing its culinary and medicinal potential.

7. Future Perspectives for Saffron Research

Post-transcriptomics enhances understanding of saffron stigma metabolism and its regulatory mechanisms. Integrating transcriptomics, proteomics, metabolomics, and interactomics offers a comprehensive view of crocin, picrocrocin, and safranal biosynthesis and regulation.

Understanding saffron's biology and biosynthetic pathways can drive metabolite production strategies. Identifying key genes, enzymes, and proteins enables genetic manipulation and breeding to enhance yield. Optimizing growth conditions further maximizes metabolite production. The future of saffron research relies on post-transcriptomic advancements, driving breakthroughs in saffron biology. These insights will enhance biotechnology, improving farming efficiency and enabling therapeutic applications. Post-transcriptomics holds promise for unlocking saffron's culinary and therapeutic potential. Advancing research will deepen appreciation of this remarkable plant, driving innovation in its applications.

8. Industrial Applications of Post-transcriptomics in Saffron

Post-transcriptomic research in saffron not only advances academic knowledge but also transforms industrial applications. It aids in optimizing cultivation, improving quality, and fostering synthetic biology innovations for enhanced metabolite production. Saffron cultivation is labor-intensive, requiring hand-picking and drying of stigmas. Post-transcriptomic insights can optimize planting densities, indoor lighting, and watering to enhance metabolite production. Understanding biosynthesis timing aids in ideal harvest periods for maximum potency and quality.

Saffron quality depends on key metabolite concentrations. Post-transcriptomic insights into biosynthesis pathways and regulation can guide strategies to enhance metabolite levels, improving overall quality. Post-transcriptomic insights enable synthetic biology approaches for saffron metabolite production. By identifying key genes and pathways, scientists can engineer yeast or bacteria to synthesize these compounds, offering a sustainable, scalable alternative free from geographical and seasonal constraints. Despite challenges,post-transcriptomic research in saffron offers significant academic and industrial benefits. Advancing molecular insights can drive innovations in production, quality, and therapeutic applications, unlocking saffron's full potential.

Conclusion

Post-transcriptomics in saffron research deciphers biosynthetic pathways and regulatory networks despite challenges like proteome complexity, incomplete genome, and cultivation constraints. Proteomics reveals key enzymes, metabolomics profiles metabolites, and interactomics clarifies protein-metabolite interactions, enhancing regulatory insights.

miRNAs play a crucial role in gene regulation, yet their impact on apocarotenoid biosynthesis remains underexplored. Expanding research could uncover control mechanisms for valuable metabolite production, optimizing cultivation and harvesting. Insights may also enable synthetic biology approaches, using microbial hosts for sustainable metabolite synthesis. Advancing post-transcriptomics will accelerate gene, enzyme, and pathway discovery, driving biotechnological innovations in agriculture, nutrition, and therapeutics. Saffron research holds vast potential for medicinal and industrial applications.

Reference

Ahrazem, O., Rubio-Moraga, A., Argandola-Picazo, J., Castillo, R. and Gomez-Gomez, L. et al. 2016. Intron retention and rhythmic diel pattern regulation of carotenoid cleavage dioxygenase 2 during crocetin biosynthesis in saffron. *Plant Mol. Biol*, *91*(3): 355–374.

Ahrazem, O., Rubio-Moraga, A., Berman, J., Capell, T., Christou, P. et al. 2015. The carotenoid cleavage dioxygenase CCD2 catalyzes the synthesis of crocetin in spring crocuses and saffron is a plastidial enzyme. *New Phytol*, *209*(2): 650–663.

Allen, E., Xie, Z., Gustafson, A.M. and Carrington, J. 2005. MicroRNA-directed phasing during trans-acting siRNA biogenesis in plants. *Cell*, *121*: 207–221.

Amin, B. and Hosseinzadeh, H. 2012. Evaluation of aqueous and ethanolic extracts of saffron, *Crocus sativus* L. and its constituents, safranal and crocin in allodynia and hyperalgesia induced by chronic constriction injury model of neuropathic pain *in rats, Fitoterapia*, *83*(5): 888–895.

Axtell, M.J., Westholm, J.O., Lai, E.C. (2011). Vive la différence: biogenesis and evolution of microRNAs in plants and animals. *Genome sBiol, 12*: 221.

Baba, S.A., Mohiuddin, T., Basu, S., Swarnkar, M.K., Malik, A.H. et al. Comprehensive transcriptome analysis of *Crocus sativus* for discovery and expression of genes involved in apocarotenoid biosynthesis. *BMC Genomics*. 2015. 15; 16(1).

Bartel, D.P. 2018. Metazoan microRNAs. *Cell. 173*: 20–51.

Boke, H., Ozhuner, E., Turktas, M., Parmaksiz, I., Ozcan, S. et al., 2015. Regulation of the alkaloid biosynthesis by miRNA in opium poppy. *Plant Biotechnol. J. 13*: 409–420.

Budak, H, Akpinar, B.A. 2015. Plant miRNAs: biogenesis, organization and origins. *Funct Integr Genomics 15*: 523–531.

Carmona, M., Zalacain, A., Sanchez, A.M., Novella, J.L. and Alonso, G.L. et al. 2006. Crocetin esters, picrocrocin and its related compounds present in *Crocus sativus* stigmas and Gardenia jasminoides fruits. Tentative identification of seven new compounds by LC-ESIMS. *Journal of Agriculture and Food Chemistry*. *54*: 973–979.

Castillo, R., Fernandez, J.A., Gomez-Gomez, L. 2005. Implications of Carotenoid Biosynthetic Genes in Apocarotenoid Formation during the Stigma Development of *Crocus sativus* and Its Closer Relatives. *Plant Physiol. 139*: 674–689.

Cerdá-Bernad, D., Valero-Cases, E., Pastor, J. and José Frutos, M. 2020. Saffron bioactives crocin, crocetin and safranal: effect on oxidative stress and mechanisms of action. *Critical Reviews in Food Science and Nutrition*, *62*: 12–20.

Chen, C., Zhong, Y., Yu, F. and Xu, M. 2020. Deep sequencing identifies miRNAs and their target genes involved in the biosynthesis of terpenoids in *Cinnamomum camphora. Ind Crops Prod*, *145*: 111853.

D'Ario, M., Griffiths-Jones, S. and Kim, M. 2017. Small RNAs: big impact on plant development. *Trends Plant Sci*, *22*: 1056–1068.

Dauria, M., Mauriello, G., Racioppi, R. and Rana, G.L. 2006. Use of SPME-GC–M.S. in the study of time evolution of the constituents of saffron aroma, modifications of the composition during storage. *Journal of Chromatographic Science*, *44*: 18–21.

Demurtas, O.S., Frusciante, S., Ferrante, P., Diretto, G., Hosseinpour Azad, N. et al. 2018. Candidate enzymes for saffron crocin biosynthesis are localized in multiple cellular compartments. *Plant Physiol*, *177*(3): 990–1006.

Fan, R., Li, Y., Li, C. and Zhang, Y. 2015. Differential microRNA analysis of glandular trichomes and young leaves in *Xanthium strumarium* L. Reveals their putative roles in regulating terpenoid biosynthesis. *PLoS One*, *10*: e0139002.

Festuccia, C., Mancini, A., Gravina, G.L., Scarsella, L., Llorens, S. et al., 2014. Anti-tumor effects of saffron-derived carotenoids in prostate cancer cell models. *BioMed Research International*, pp 1–12.

Fleury, D., Jefferies, S., Kuchel, H. and Langridge P. 2010. Genetic and genomic tools to improve drought tolerance in wheat. *J. Exp. Bot*, *4*: 152.

Gao, D., Ji, X., Yuan, Q., Pei, W., Zhang, X. et al. Effects of total daily light integral from blue and broad-band red LEDs on flowering of saffron (*Crocus sativus* L.). *Sci Rep.* 2023 Dec 1; 13(1).

Gomez-Gomez, L., Rubio-Moraga, A. and Ahrazem, O. 2010. Understanding carotenoid metabolism in saffron stigmas: Unravelling aroma and colour formation. Func. *Plant Sci. & Biotechnol*, *4* (2): 56–63.

Gou, J.Y., Felippes, F.F., Chang-Jun, L., Detlef, W. and Jia-Wei, W. et al. 2011. Negative regulation of anthocyanin biosynthesis in Arabidopsis by a miR156-targeted SPL transcription factor. *Plant Cell*, *23*:1512–1522.

Haghighi, B., Feizy, J. and Hemati kakhki, A. 2007. L.C. Determination of Adulterated Saffron Prepared by Adding Styles Colored with Some Natural Colorants, *Chromatographia*, *66*: 325–332.

Hao, D.C., Yang, L., Xiao, P.G. and Liu, M. 2012. Identification of Taxus microRNAs and their targets with high-throughput sequencing and degradome analysis. *Physiol. Plant*, *146*: 388–403.

Harpke, D., Meng, S. and Rutten, T. 2013. Phylogeny of Crocus (Iridaceae) based on one chloroplast and two nuclear loci: ancient hybridization and chromosome number evolution. *Mol. Phylogenet. Evol*, *66*: 617–627.

Jeong, D.H., German, M.A. and Rymarquis, L.A. 2010. Abiotic stress-associated miRNAs: detection and functional analysis. *Plant microRNAs.* Springer, pp 203–230.

Jones-Rhoades, M., Bartel, D.P. and Bartel, B. 2006. MicroRNAs and Their Regulatory Roles in Plants. *Plant Biol*, *19*: 53.

Jyoti, V., Sheetal, A., Salami, S.A., Chittaranjan, K. The Saffron Genome. 2022.

Kanakis, C.D., Daferera, D.J., Tarantilis, P.A. and Polissiou, M.G. 2004. Qualitative determination of volatile compounds and quantitative evaluation of safranal and 4-hydroxy-2, 6, 6-trimethyl-1-cyclohexene-1-carboxaldehyde (HTCC) in Greek saffron, *Journal of Agricultural Food Chemistry*, *52*: 4515–4521.

Kanakis, C.D., Tarantilis, P.A., Tajmir-Riahi, H.A. and Polissiou, M.G. 2007. Crocetin, dimethylcrocetin and safranal bind human serum albumin: stability and antioxidative properties. *J. Agric. Food Chem.* *55*(3): 970–977.

Kianbakht, S. and Ghazavi, A. 2005. Evaluation of immunological and hematological effects of saffron in men. Ethnopharmacologia, *36*:78–83.

Kurihara, Y. and Watanabe Y. 2010. Processing of miRNA precursors. *Plant microRNAs*. Springer, pp 231–241.

Lage, M. and Cantrell. C. 2009. Quantification of saffron (*Crocus sativus* L.) metabolites crocins, picrocrocin and safranal for quality determination of the spice grown under different environmental Moroccan conditions. *Scientia Horticulturae*, *121*(3): 366–373.

Liu, Q., Feng, Y. and Zhu, Z. 2009. Dicer-like (DCL) proteins in plants. *Funct Integr Genomics*, *9*: 277–286.

Martin, G., Goh, E. and Neff, A.W. 2002. Evaluation of the developmental toxicity of crocetin on Xenopus. *Food and Chemical Toxicology*, *40*: 959–964.

Melnyk, J.P., Wang, S. and Marcone. M.F. 2010. Chemical and biological properties of the world's most expensive spice: Saffron. *Food Research International*, *43*(8): 1981–1989.

Molina, R.V., Valero, M., Navarro, Y., Guardiola, J.L. and Garcia-Luice, A. et al. 2005. Temperature effects on flower formation in saffron (*Crocus sativus* L.). *Scientia Horticulture*. *103*: 361–379.

Moradi, S., Kafi, M., Aliniaeifard, S., Moosavi-Nezhad, M., Pedersen, C. et al. Monochromatic blue light enhances crocin and picrocrocin content by upregulating the expression of underlying biosynthetic pathway genes in saffron (*Crocus sativus* L.). 2022. Front. *Hortic. 1*: 960423.

Moradi, S., Kafi, M., Aliniaeifard, S., Salami, S.A., Shokrpour, M. et al. (2021). Blue light improves photosynthetic performance and biomass partitioning toward harvestable organs in saffron (*Crocus sativus* L.). *Cells 10*: 1994.

Moratalla-López, N., José Bagur, M., Lorenzo, C., Martínez-Navarro, M.E., Rosario Salinas, M. et al., 2019. Bioactivity and Bioavailability of the Major Metabolites of *Crocus sativus* L. *Flower. Molecules*, *24*(15): 28–27.

Nair, S.C., Kurumboor, S.K. and Hasegawa, J.H. 1995. Saffron chemoprevention in biology and medicine: a review. *Cancer Biother*, *10*(4): 25764.

Nazarian, R., Samim, Noorahmad, H. and Qasem Jami, M. 2021. Determination of saffron (*Crocus sativus* L.) quality characteristics in Herat-Afghanistan, *Journal of Horticulture and Postharvest Research*, *4*:11–18.

Pandita, D. and Pandita, A. Plant MicroRNAs and Stress Response. *Plant MicroRNAs and Stress Response*. CRC Press; 2023.

Rahaiee, S., Moini, S., Hashemi, M. and Shojaosadati. S.A. 2015. Evaluation of antioxidant activities of bioactive compounds and various extracts obtained from saffron (*Crocus sativus* L.): A review. *Journal of Food Science and Technology*, *52*(4): 1881–1888.

Ruksans, J. The World of Crocuses. 2017. Latvian Acady of Sciences.

Sánchez, A.M., Maggi, L., Carmona, M. and Alonso, G.L. 2011. Authentication of saffron spice (*Crocus sativus* L.). In Progress in Authentication of Food and Wine; Ebeler, S.E., Takeoka, G.R., Winterhalter, P., Eds.; *American Chemical Society: Washington, DC,* USA, *1081*: 309–331.

Soeda, S., Ochiai, T., Shimeno, H., Saito, H., Abe, K. et al. 2007. Pharmacological activities of crocin in saffron. *Journal of Natural Medicines*. *61*: 102–111.

Sunkar, R., Li, Y.F. and Jagadeeswaran, G. 2012. Functions of microRNAs in plant stress responses. *Trends Plant Sci*, *17*: 196–203.

Taheri-Dehkordi, A., Naderi, R., Martinelli, F., Salami, S.A. (2020) A robust workflow for indirect somatic embryogenesis and cormlet production in saffron (*Crocus sativus* L.) and its wild allies;*Crocus caspius* and *Cheilocostus speciosus*. Heliyon *6*: e05841.

Taheri-Dehkordi, A., Naderi, R., Martinelli, F., Salami, S.A. Computational screening of miRNAs and their targets in saffron (*Crocus sativus* L.) by transcriptome mining. *Planta*. 2021. 1; 254(6).

Vahedi, M., Kabiri, M. and Salami, S.A. (2018) Quantitative HPLC-based metabolomics of some Iranian saffron (*Crocus sativus* L.) accessions. *Ind Crop Prod 118*: 26–29.

Vahedi, M., Kalanrtari, S. and Salami, S.A. (2014). Factors affecting callus induction and organogenesis in saffron (*Crocus sativus* L.). *Plant Tissue Cult Biotech 24*(1):1–9.

Verma, P., Singh, N. and Khan, S.A. 2020. TIAs pathway genes and asso- ciated miRNA identification in Vinca minor: supporting aspi-dosperma and eburnamine alkaloids linkage via transcriptomic analysis. *Physiol Mol Biol Plants*, *26*:1695–1711.

Wu, L., Zhang, Q. and Zhou, H. (2009). Rice microRNA effector complexes and targets. *Plant Cell 21*: 3421–3435.

Xi, L., Aian, Z., Xu, G., Zheng, S.H., Sun, S. et al., 2007. Beneficial impact of concern, a Carotenoid from saffron, on insulin sensitivity in fructose-fed rats. *Journal of Nutritional Biochemistry*, *18*: 64–72.

Yu, B. 2005. Methylation as a crucial step in plant microRNA biogenesis. *Science 307*: 932–935.

9 Metabolomic Analysis of Saffron: An Overview of the Current Status

Sairam Reddy Palicherla[1], Mithun Chakraborty[1*] and Rambabu Pogiri[1]

Saffron (*Crocus sativus* L.) is the world's most expensive spice, valued for its pharmacological effects and rich phytochemical profile. It contains over 150 secondary metabolites, including carotenoids, monoterpenes, and isopherones. Key carotenoid derivatives—crocin, crocetin, picrocrocin, and safranal—define its color, taste, and aroma. Kaempferol glycosides, found in flowers and leaves, offer antioxidant benefits, making saffron useful in food, cosmetics, and pharmaceuticals. Its quality depends on genetics, agronomic conditions, geography, and processing.

Precision agriculture and tissue culture enhance secondary metabolite production in saffron. Vertical farming boosts corm yield up to threefold and improves stigma quality. Light spectra and nutrient recipes regulate flower production and metabolite composition. Saffron quality is traditionally assessed using gas chromatography, spectroscopy, and liquid chromatography. Advanced methods include HPLC–PDA–ESI–MS, NMR, MIR, FT–IR, FT–NIR, and DRIFTS for precise metabolite detection. AI-driven authentication employs CVS, e-tongue, and e-nose for sensory analysis. Metabolomics with PCA, HCA, SVM, and SIMCA predicts quality and metabolite concentration with over 98% accuracy.

1. Introduction

Saffron (*Crocus sativus* L.) is the world's most expensive spice, valued for its phytochemical richness and pharmacological benefits. Its stigma contains crocin and crocetin esters (color), picrocrocin (taste), and safranal (aroma) (Azarabadi & Özdemir, 2018). Additionally, kaempferols and glycosides offer antioxidant properties, supporting applications in food, cosmetics, and pharmaceuticals (Moshiri et al., 2015).

Carotenoid and apocarotenoid biosynthesis occurs in Crocus species, including C. ancyrensis, which accumulates picrocrocin, crocetin, and crocins (Castillo et al., 2005). In *C. sativus* and *C. ancyrensis*, zeaxanthin biosynthesis, a precursor of apocarotenoids, is developmentally regulated (Castillo et al., 2005; Rubio-Moraga et al., 2009). *C. sativus* expresses four phytoene synthase isoforms, with CsPSY2 predominantly active in stigmas, driving apocarotenoid biosynthesis (Ahrazem

[1] Urban Kisaan Farms Pvt. Ltd., Hyderabad, India.
* Corresponding author: mithun@urbankisaan.com; https://orcid.org/0000-0002-4926-2648

DOI: 10.1201/9781003403760-9

et al., 2019). Chemical analysis of *C. sativus* stigmas identifies ~150 compounds, including carotenoids (crocetin, crocin, β-carotene, lycopene, zeaxanthin), monoterpene aldehydes (picrocrocin, safranal), terpene alcohols, monoterpenoids, and isopherones (Bathaie & Mousavi, 2010; Mir et al., 2018). Beyond stigmas, stamens, petals, and styles contain alkaloids, flavonoids, polyphenols, tannins, and essential nutrients (Guijarro-Díez et al., 2017). Crocetin/crocins ($C_{44}H_{64}O_{24}$) are unique carotenoids with lipophilic and water-soluble properties, attributed to their carboxyl group. This makes saffron a popular food colorant. Crocin, a crocetin ester, exists in trans and cis isomers, with the trans-isomer having higher pharmacological value (Moratalla-López et al., 2016; García-Rodríguez et al., 2021). It constitutes ~0.3% of saffron stigma weight (Melnyk et al., 2010; Anjum et al., 2015).

2. Genetics and Metabolites

C. sativus is a sterile triploid (2n = 3x = 24), causing chromosome mispairing, abnormal pollen, and self-sterility (Chichiriccò, 1984; Rashed-Mohassel, 2020). Initially considered autotriploid, recent findings suggest alloploidy, with C. cartwrightianus likely contributing two genomes, though the second parent remains unknown (Fernández et al., 2004).

Saffron's clonal nature suggests a single polyploidization and domestication event, with all plants being genetically identical clones. The sole cultivar, "Saffron," varies only in commercial ecotypes. Vegetative propagation via daughter corms maintains uniformity but lacks genetic variation, making classical breeding and biodiversity conservation crucial (Fernández, 2004; Gresta et al., 2008; Nehvi et al., 2010). Genetic erosion threatens saffron, but clonal selection and wild relatives offer potential for trait enhancement via cross-pollination (Fernández, 2006).

3. Epigenetic and the Effect on Metabolites

Chemical mutagens have been used to induce morphological and phytochemical variations in *C. sativus* due to its lack of genetic diversity (Zaffar et al., 2003; Khan et al., 2011; Kashtwari et al., 2018). Samadi et al. (2022) found that colchicine and EMS treatments led to stigmas with 2–6 threads, higher ALDH, BGL, and CCD2 gene expression, increased picrocrocin, and decreased crocetin content.

A study of 112 accessions using MS-AFLP fingerprinting revealed low genetic but high epigenetic variation, linked to geographic origin (Busconi et al., 2015). Later, MRE-seq analysis identified differentially methylated regions, confirming epigenetic variability among accessions (Busconi et al., 2021). DNA methylation, influenced by environmental conditions, likely drives phenotypic differences. Co-cultivation may have led to converging methylation profiles over time. Additionally, high SNP variability was detected across five accessions with distinct phenotypic traits.

3.1 Microbiome and the Effect on Metabolites

Wani et al. (2016) identified 294 fungal endophytes from saffron corms in Jammu and Kashmir, classifying them into 36 genotypes across 19 genera. The microbiome was dominated by *Phialophora mustea* and *Cadophora malorum*, which produce auxin (IAA) to stimulate growth. Endophyte diversity was higher in the dormant stage than in the vegetative stage. Caser et al. (2019) found that arbuscular mycorrhizal symbionts (*Rhizophagus intraradices*, *Funneliformis mosseae*) act as bio-stimulants, enhancing antioxidant activity and increasing picrocrocin, crocin II, and quercitrin levels in saffron. Ramandi et al. (2023) examined the saffron microbiome's role in yield and quantity, identifying dominant culturable bacteria in high- and low-yielding fields. Key rhizospheric and endophytic bacteria—*Cupriavidus metallidurans*, Bacillus sp., Solibacillus sp., Planococcus sp., *Serratia oryzae*, and *S. odorifera*—were found to enhance saffron growth and flowering. Saffron's quality and metabolite composition vary by geographical origin, edapho-climatic conditions, and cultivation practices. Epigenetic variations, especially DNA methylation, influence phenotypic differences across climates. The plant microbiome, including endophytic and rhizospheric microbes, enhances growth, stress resilience, and bioactive compound production. These factors are crucial for crop improvement and conservation.

4. Environmental Influences on Metabolite Production

Saffron (*Crocus sativus* L.) is native to the Mediterranean and parts of Asia, cultivated across Iran, India, China, Pakistan, Japan, Nepal, Turkey, Azerbaijan, Greece, Italy, Spain, France, Switzerland, Morocco, the U.S., Peru, and Colombia. It thrives in temperate, semi-arid to arid climates with cold winters, hot summers, and low precipitation (Gresta et al., 2008; Menia et al., 2018). Saffron tolerates –18°C to 40°C (Kouzegaran et al., 2020). Its secondary metabolite composition, shaped by environmental factors, determines quality and yield. Saffron's secondary metabolite accumulation is influenced by growing conditions and environmental factors like altitude, soil type, nutrients, water availability, photoperiod, and temperature (Siracusa et al., 2013; Rahimi et al., 2017; Cardone et al., 2020). Biotic and abiotic stress triggers metabolic shifts, enhancing secondary metabolite production while reducing plant dry matter. Soil composition significantly influences saffron metabolite production. Sandy soil favors flower and stigma yield, while clay soil enhances stigma weight. Loam-clay-sandy soils with high organic matter, potassium, and phosphorus boost crocins and kaempferol 3-sophoroside 7-glucoside but reduce safranal levels (Chaouqi et al., 2023). Low iron content improves saffron's coloring strength. Optimized fertilizer management enhances saffron yield and quality. Leaf nitrogen directly correlates with yield, while phosphorus influences safranal and crocin levels (Naghdi Badi et al., 2011; Kheiry et al., 2018). Plant growth regulators like gibberellic acid, potassium nitrate, and salicylic acid influence saffron's reproductive and vegetative traits, enhancing picrocrocin and safranal levels.

Saffron's secondary metabolite production is shaped by altitude, soil properties, nutrient management, and plant growth regulators. Optimizing these factors enhances yield and quality.

5. Metabolite Production Through Controlled Environment Agriculture (CEA)

Saffron cultivation is declining in traditional regions due to climate change impacts on yield. Controlled Environment Agriculture (CEA) offers a solution, optimizing indoor conditions to enhance yield and secondary metabolite production.

Soilless systems improve saffron flowering and yield. Perlite-based cultivation outperforms peat-perlite mixtures (Maggio et al., 2006). Greenhouse and growth chambers double saffron yield compared to field cultivation (Molina et al., 2004). Controlled microclimate and nutrition enhance quality and productivity (Mollafilabi, 2014). Light plays a key role in phytochemical synthesis in plants (Jung et al., 2021). LEDs with specific wavelengths help researchers analyze light spectra effects on growth and physiology. Kajikawa et al. (2018) found that a low R/FR ratio during propagule development enhances photosynthate translocation, producing carbohydrate-rich corms that boost crocin content in saffron.

Moradi et al. (2021) found that blue light directs biomass toward corms and floral organs, while red light enhances leaf growth. Later, Moradi et al. (2022) reported that monochromatic blue light maximizes crocetin ester (trans-isomer), red-blue light boosts picrocrocin, and high red light proportion increases safranal. Blue light also upregulates crocin biosynthesis genes. Applying GA3 and GABA alongside specific light spectra improves saffron flower yield and enhances phytochemical composition in saffron stigmas (Eftekhari et al., 2023).

Pharma-grade saffron is crucial for treating macular diseases like RPE degeneration and AMD. Nardi et al. (2022) optimized hydroponic cultivation, enhancing yield and bioactive metabolite accumulation (crocins, picrocrocin). Their study confirms hydroponically grown saffron matches pharma-grade quality and efficacy. Controlled Environment Agriculture (CEA) optimizes saffron cultivation by precisely managing environmental and nutritional factors, enhancing crocin, picrocrocin, and safranal content. IoT-based frameworks enable real-time monitoring, advancing artificial cultivation for year-round, location-independent saffron production (Kour et al., 2022).

***In-vitro* culture for direct organogenesis and secondary metabolites synthesis**

Plant tissue culture is a key tool for producing commercially valuable secondary metabolites (SMs). In saffron, it enables micropropagation, pathogen-free plant production, superior clone selection, somaclonal variation-based crop improvement, and gene transformation. Advances in tissue culture now facilitate artificial production of crocin, safranal, picrocrocin, and crocetin. Studies have optimized *in-vitro* conditions for generating stigma-like structures (SLS) from *C. sativus* stigma explants, which resemble natural saffron stigmas and contain crocin,

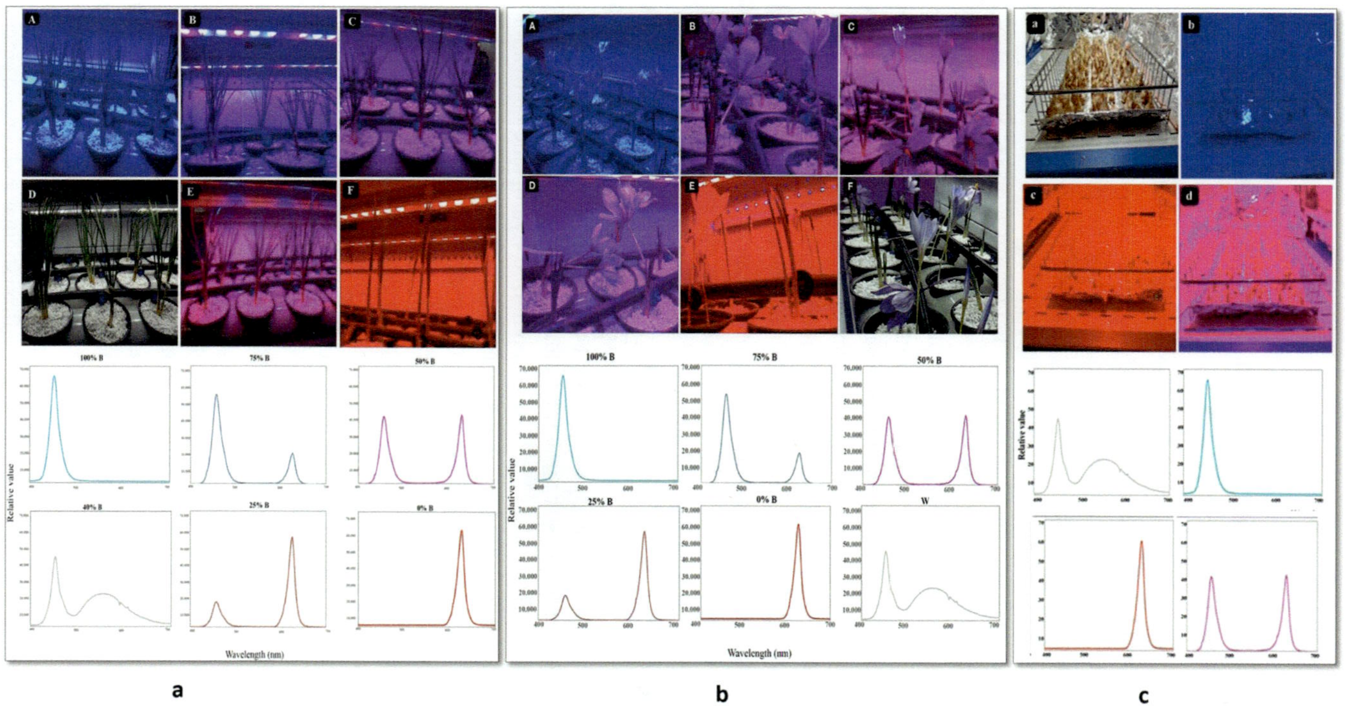

Fig. 1 **(a)** Saffron plants grown under varying Blue (B) to Red (R) light ratios. Blue light promotes the production of high-quality daughter corms (Moradi et al., 2021). **(b)** Monochromatic blue light yields the highest crocin, while red-blue light combination maximizes picrocrocin, and high red light proportion enhances safranal production (Moradi et al., 2022). **(c)** A red-blue light combination improves stigma quality and increases flower yield (Eftekhari et al., 2023).

crocetin, picrocrocin, and safranal. HPLC analysis confirmed SLS metabolite levels comparable to natural stigmas (Sano & Himeno, 1987; Sarma et al., 1991; Loskutov et al., 1999). SLS have also been successfully generated from flower, petal explants, and half ovaries, while roots, corms, petals, and leaves showed reduced SLS development (Himeno et al., 1988; Kohda et al., 1993). Hori et al. (1988) induced calli from saffron pistils, producing crocins and crocetin-di-ester, suggesting high auxin levels aid callus induction and organ differentiation enhances pigment production. Research has successfully produced stigma-like structures (SLS) from floral explants, identifying key pigments (Han & Zhang, 1993; Zeng et al., 2003). Callus cultures from floral buds generated crocin, crocetin, picrocrocin, and safranal (Visvanath et al., 1990; Chen et al., 2003). However, SLS from ovary explants contained significantly lower crocin (6×) and picrocrocin (11×) levels than natural stigmas, with weaker sensory properties (Sarma et al., 1991).

Visvanath et al. (1990) found higher picrocrocin levels in callus and filamentous structures than in natural stigmas, while safranal content remained comparable. However, crocin levels were lower in tissue-cultured saffron. Ovary development stage is critical for stigma proliferation, as noted in multiple studies (Bhagyalakshmi et al., 1997; Castellar & Ibora, 1997. Bo et al. (2005) used cell culture techniques to produce crocin, generating yellow-orange callus cultures from leaf sheaths, increasing biomass fivefold in 25 days. Chen et al. (2003) found that NAA and 6-BA boosted callus biomass, while IAA and 6-BA in a dual-stage culture significantly enhanced crocin production compared to a single-stage system. Chib et al. (2020) developed a callus production and somatic embryogenesis system, achieving 4% transformation efficiency via Agrobacterium-mediated genetic transformation in saffron. The study also established CRISPR/Cas9 integration, advancing gene-editing strategies for saffron improvement. Plant tissue culture effectively enhances secondary metabolite production in saffron. Research focuses on optimizing *in vitro* conditions for stigma-like structure (SLS) generation, considering ovary development stage and explant type. Two-stage culture systems improve crocin yield over single-stage methods.

Metabolic engineering through transgenic approach in heterologous systems

Crocins, abundant in *C. sativus* stigmas, also occur in Crocus tepals, Gardenia jasminoides fruits, and Buddleja davidii flowers (Rubio-Moraga et al., 2013; Ahrazem et al., 2018). In saffron and Buddleja, crocins derive from zeaxanthin cleavage, while Gardenia synthesizes them from lycopene and β-carotene. Due to high demand, research explores crocin biosynthesis in alternative plant systems. Frusciante et al. (2014) identified CCD2 enzyme cleaving zeaxanthin (C7′–C8′/C7–C8), forming crocetin and HTCC, a safranal precursor in C. sativus and Buddleja. Ahrazem et al. (2015) demonstrated that CaCCD2 cleaves zeaxanthin to produce crocetin in E. coli and transgenic rice cell lines. Rice embryos transformed with CaCCD2 cDNA exhibited carotenoid accumulation in white, yellow, and orange lines. Later, Ahrazem et al. (2016) identified CCD2 (AIG94929) as the key enzyme in saffron crocetin biosynthesis, cloning CaCCD1, CaCCD2, CaCCD4a/b, and CaCCD4c for characterization. Genetic engineering of biosynthetic pathways enables secondary metabolite production in heterologous systems.

Fig. 2 Production of Crocins and Picrocrocin in Heterologous Plant Systems. 2: Crocetin biosynthesis in rice callus stably expressing CaCCD2, coloured phenotypes of nine independent rice callus lines (Ahrazem et al., 2015). 3a: Cleavage of maize kernel carotenoids by transiently expressed CCD2. Pigmentation of maize kernels after 48 h of agroinfiltration with pBI-GUS, pBI-CCD2, and pBI-ZCD; 3b: *In-vitro* cleavage assay. *In vitro* cleavage of zeaxanthin by E. coli extracts (Frusciante et al., 2014). 4a: Crocetin/crocin production in transplastomic tobacco plants expressing BoCCD4-3. 4b: Crocetin/crocin production in transgenic Microtom fruits expressing TP:BoCCD4-3 and CsUGT74AD1 (Frusciante, et al., 2022). 5: Internal phenotype of wild type (WT) (cv. Desirée) potato tubers and O6 transgenic (cv. Desirée) potato tuber pool (Gómez, et al., 2022). 6: Tobacco (N. tabacum) plants sixth leaves from wild-type (Wt) control, T2 homozygous lines expressing CsCCD2L and coexpressing CsCCD2L + BrCrtZ + AtOrMut genes. (Ahrazem, et al., 2022). 7a: Agro-infiltrated N. benthamiana leaf transiently expressing GjCCD4a or sCCD2L; 7b: Agrobacterium mediated multi-gene transformation generates various engineered citrus callus lines with distinct color phenotypes and carotenoid profiles (Zheng et al., 2022).

Research has demonstrated crocins accumulation in *Nicotiana benthamiana* leaves with CsCCD2L gene introduction. *N. tabacum* showed a 3.5-fold increase in crocin levels when co-expressed with β-carotene hydroxylase (BCH) (Diretto et al., 2019; Lopez-Jimenez et al., 2021). Martí et al. (2020) successfully expressed CsCCD2L and BdCCD4.1 in *N. benthamiana*. Ahrazem et al. (2022a) confirmed CsCCD2L expression in *N. benthamiana* and *N. tabacum*, leading to crocins accumulation. Frusciante et al. (2022) identified BoCCD4–3, capable of cleaving lycopene, β-carotene, and zeaxanthin for crocetin dialdehyde/crocin production. Transient expression of BoCCD4–3 in *N. benthamiana* resulted in 0.026 mg/g DW crocins, while constitutive expression with CsUGT74AD1 in tomato transgenic plants yielded 0.110 mg/g DW crocins. Ahrazem et al. (2022b) successfully introduced CsCCD2L, CsUGT2 (CsUGT74AD1), and CsUGT709G1 into tomato plants using fruit-specific (pE8, p2A11) and constitutive (CaMV p35S) promoters. The transgenic tomatoes accumulated 14.80 mg/g crocins and 2.92 mg/g picrocrocin DW, maintaining normal plant growth.

Gómez-Gómez et al. (2022) developed transgenic potato lines expressing CsCCD2L, CsUGT74AD1, and CsUGT709G1 under the patatin promoter, yielding crocins (0.019–0.36 mg/g) and picrocrocin (0.105–0.80 mg/g). These tubers exhibited a 4-fold increase in antioxidant capacity over wild-type potatoes. Boiled potatoes showed the highest bio-accessibility (97.3%), followed by raw (80.9%), baked (78.9%), and fried (76.1%).

Zheng et al. (2022) explored carotenoid and apocarotenoid biosynthesis using CsCCD2L (*C. sativus*) and GjCCD4a (*G. jasminoides*). Co-expression of GjCCD4a, tpCrtB, and rice β-carotene hydroxylase enabled crocins production in N. benthamiana leaves and citrus callus. Agro-infiltration of CsCCD2L yielded 0.67 mg/g DW crocins, while GjCCD4a with intermediate genes achieved 1.6 mg/g DW crocins.

6. Conventional Metabolites Detection Techniques

Saffron quality depends on its chemical profile, shaped by geography, climate, agronomic practices, soil, and processing (Caballero-Ortega, 2007; Liu, 2018). Due to its high value, adulteration is common. ISO 3632 sets global saffron quality standards, revised in 2003, 2010, and 2011, with the latest update as ISO 3632/TS.

Researchers have analyzed saffron's main components, their deterioration during storage, and adulteration detection (Ordoudi & Tsimidou, 2004; Petrakis et al., 2015, 2017a; Shawky et al., 2020; Amirvaresi et al., 2021a). Saffron quality is assessed by crocin/crocetin (color), picrocrocin (taste), and safranal (aroma). Techniques include UV-Vis, FS, MS, FT-IR, FT-NIR, NMR, and Raman spectroscopy (Verma et al., 2010; Moratalla-López et al., 2019; Bakshi et al., 2022).

Avila-Sosa et al. (2022) identified saffron's bioactive compounds by geographic origin, detailing analytical techniques for quality assessment and adulteration detection. UV-Vis spectroscopy, per ISO 3632, is commonly used for quality validation. Zalacain et al. (2005) pioneered FT-NIR spectrometry for saffron identification. Petrakis et al. (2017b) used FTIR spectroscopy to detect 6 plant-based adulterants, identifying contamination as low as 5-20% (w/w). FTIR and pattern-recognition techniques have distinguished authentic saffron from colorant-added and plant-based adulterants (Karimi et al., 2016; Varliklioz et al., 2017; Amirvaresi et al., 2021b). Anastasaki et al. (2010) applied MIR spectroscopy with multivariate analysis, achieving optimal saffron differentiation at the 2000–700 cm^{-1} fingerprint region. HPLC is a fast, robust method for saffron quality assessment and adulterant detection, commonly used to quantify crocin, picrocrocin, and safranal (Maggi et al., 2009; Nescatelli et al., 2017; Ordoudi et al., 2018). HPLC-DAD enhances sensitivity and specificity, improving genuine vs. adulterated saffron discrimination (Carmona et al., 2006). Mass spectrometry has been widely used to analyze saffron components in stigmas, crude extracts, and Gardenia jasminoides fruits via HPLC-UV-VIS-PDA-MS. Coupling MS detectors with HPLC/DAD enhances saffron compound quantification (Tarantilis et al., 1995; Carmona et al., 2006; Moras et al., 2018). NMR spectroscopy effectively analyzes saffron components, distinguishing crocetin ester isomers and detecting adulterants at low concentrations (Ordoudi et al., 2018). Combined with chemometric techniques, it authenticates saffron samples by geographic origin (Rubert et al., 2016). Raman spectroscopy is a rapid, non-destructive method for saffron analysis, identifying crocin and picrocrocin and distinguishing authentic vs. adulterated samples (Anastasaki et al., 2010; Dai et al., 2020). Mass spectrometry, UV-Vis, FT-NIR, FT-IR, fluorescence, NMR, and Raman spectroscopy are widely used for saffron quality assessment and adulteration detection. These methods, applied individually or in combination, enhance accuracy and reliability. Ongoing research aims to develop faster, cost-effective, and highly sensitive techniques for quality control, ensuring genuine saffron products.

7. Recent Non-conventional Detection Techniques

Conventional chromatography and spectroscopy methods effectively identify saffron but require expertise, lengthy sample preparation, high costs, and complex technology management. With increasing saffron adulteration due to rising costs and demand, food and pharmaceutical industries seek low-cost, rapid, non-destructive, and automated detection techniques. Modern approaches like computer vision and spectral imaging offer reliable solutions. Machine vision and image processing require extensive saffron adulterant data, prompting researchers to employ machine learning for constructing saffron sample databases (pure and adulterated). Kiani et al. (2016) developed a portable e-nose system for identifying different types of saffron stigma based on to identify saffron stigma types based on VOCs, using metal oxide semiconductor gas sensors. They applied unsupervised learning (HCA, PCA) and supervised learning (MLP neural networks, PLS), finding MLP achieved superior prediction accuracy (100% efficacy, $R^2 = 0.989$). Later, Kiani et al. (2017)

integrated a computer vision system (CVS) with multivariate algorithms (PCA, HCA, SVM) for saffron adulterant detection based on color and aroma. The ANN-MLP model achieved 89% color prediction accuracy and 100% for aroma.

Kiani et al. (2023) developed a portable hyperspectral imaging (HSI) system with chemometric techniques for rapid, non-destructive saffron origin recognition and crocins content determination. Using the Savitzky–Golay (SG) algorithm, 160 saffron samples from two locations were preprocessed and classified via PCA (unsupervised) and PLS-DA, RBF, MLP (supervised). PLS-DA achieved 100% clustering accuracy, while RBF demonstrated the highest crocins content predictive capability ($R^2 > 0.94$). Minaei et al. (2017) developed a portable computer vision system for saffron color analysis using PCA on ISO 3632 color parameters (R, Y, I, Cr). They applied PLS, MLR, and MLP neural networks, achieving a high correlation ($R^2=0.89$) with 96.67% efficacy. Akbari-Adergani et al. (2022) integrated image processing with the support vector machine (SVM) method for rapid, non-destructive saffron adulteration detection. Using various SVM kernels, images were classified into authentic and adulterated saffron, evaluated via statistical indicators (accuracy, precision, sensitivity, specificity, AUC). The SVM cubic kernel model achieved 97–99% accuracy for authentic saffron and 83–97.5% for adulterated samples. Alighaleh et al. (2022) employed deep neural networks (DNN) with RGB images under varied conditions to detect saffron adulterants. Five adulterant classes and two genuine saffron classes optimized CNN models, evaluating VGG11, ResNet18, ResNet50, Inception V3, and DarkNet53 using batch normalization. Their S5 structure achieved 99.8% accuracy with a processing time of 36.26 ms. Fattahi et al. (2023) utilized ion mobility spectrometry (IMS) with multivariate analysis for rapid detection of four synthetic saffron-like colorants (E102, E110, E124, E127). The PLSR model, built on discriminative features, and the PCA model achieved 92.28% accuracy in distinguishing original saffron color from synthetic adulterants.

Guan et al. (2023) utilized machine vision and visible-near infrared spectroscopy to determine saffron origin and crocin content. The KNN model achieved 90.32% recognition accuracy for geographical origin, while spectroscopy effectively predicted crocin levels, with correlation coefficients of 0.9883 (crocin I) and 0.9982 (crocin II). Amin et al. (2023) developed a saffron adulteration prediction systems (SAPS) using visual computing and data-driven algorithms (SaffNet). They created a Kashmiri Saffron image dataset (Saff-Kash) and designed SaffNet using a gradient boosting ensemble-based model for authentic saffron detection. Saff-Kash outperformed SVM and KNN models, achieving 98% accuracy. Momeny et al. (2023) developed a deep learning-based machine vision system for saffron adulteration detection and quality assessment. Using a dataset of 1,869 images, they designed the LAII-v4 CNN for fraud detection and grading via smartphone images. The model, incorporating batch normalization, dropout, and leaky ReLU, achieved 99.5% classification accuracy.

Conclusion

Saffron is a highly valuable spice with potent medicinal properties. Its metabolite content varies due to genetic diversity, influenced by geographic origin, environment, and human intervention. Epigenetic factors like DNA methylation and histone modifications also affect its composition. *In-vitro* culture, genetic engineering, and controlled environment agriculture offer strategies to enhance saffron quality and yield. Geographical and environmental factors—temperature, light, soil composition, water, and nutrients—significantly influence saffron production and metabolite content. Advances in controlled environment agriculture (CEA) enable precise growth condition optimization, while *in-vitro* culture facilitates direct organogenesis and secondary metabolite synthesis. Researchers have used *in-vitro* techniques to enhance crocins and picrocrocin production, with genetic engineering further improving yield and quality in heterologous systems. Conventional chromatography and spectroscopy methods effectively analyze saffron metabolites and detect adulterants but require expertise, complex sample preparation, and high costs. Emerging non-conventional techniques like computer vision and spectral imaging offer rapid, low-cost, non-destructive, automated solutions. Advances in machine learning models, incorporating pure and adulterated saffron databases, will enhance accuracy and future detection systems. Continued research will deepen insights into saffron's metabolite production and expand applications in medicine, food, and industry.

References

Ahrazem, O., Rubio-Moraga, A., López, R.C. and Gómez-Gómez, L. (2010). The expression of a chromoplast-specific lycopene beta cyclase gene is involved in the high production of saffron's apocarotenoid precursors. *Journal of Experimental Botany, 61*(1): pp. 105–119

Ahrazem, O., Rubio-Moraga, A., Jimeno, M.L. and Gómez-Gómez, L. (2015). Structural characterization of highly glucosylated crocins and regulation of their biosynthesis during flower development in Crocus. *Frontiers in Plant Science, 6*: p. 971

Ahrazem, O., Rubio-Moraga, A., Berman, J., Capell, T., Christou, P. et al. (2016). The carotenoid cleavage dioxygenase CCD 2 catalysing the synthesis of crocetin in spring crocuses and saffron is a plastidial enzyme. *New Phytologist. 209*(2): pp 650–663.

Ahrazem, O., Argandoña, J., Fiore, A., Aguado, C., Luján, R. et al. (2018). Transcriptome analysis in tissue sectors with contrasting crocins accumulation provides novel insights into apocarotenoid biosynthesis and regulation during chromoplast biogenesis. *Scientific reports. 8*(1): p. 2843.

Ahrazem, O., Diretto, G., ArgandonaPicazo, J., Fiore, A., Rubio-Moraga, Á. et al. (2019). The specialized roles in carotenogenesis and apocarotenogenesis of the phytoene synthase gene family in saffron. *Frontiers in Plant Science*. 4; *10*: p. 249.

Ahrazem, O., Zhu, C., Huang, X., Rubio-Moraga, A., Capell, T. et al. (2022a). Metabolic engineering of crocin biosynthesis in nicotiana species. *Front. Plant Sci.,* 861140

Ahrazem, O., Diretto, G., Rambla, J.L., Rubio-Moraga, Á., Lobato-Gómez, M. et al. (2022b). Engineering high levels of saffron apocarotenoids in tomato. *Horticulture Research*. 5; 9.

AkbariAdergani, B., Moghadam, M.M., Noghabi, M.K., Mohammadpour, M., Khalilian-Movahhed, M. et al. (2022). Application of Computer Vision on Non-Destructive Detection of the Authentic and Adulterated Saffron. Saffron Agronomy and Technology Vol. 9, No. 4, Winter, pp. 409–429.

Alighaleh, P., Khosravi, H., Rohani, A., HosseinSaeidirad, M., Einafshar, S. et al. (2022). The detection of saffron adulterants using a deep neural network approach based on RGB images taken under uncontrolled conditions. *Expert Systems with Applications*, Vol. *198*(15): 116890

Amin, J., Selwal, A., Sabha, (2023). A SaffNet: an ensemble-based approach for saffron adulteration prediction using statistical image features. *Multimed. Tools Appl.* https://doi.org/10.1007/s11042-023-14934-9

Amirvaresi, A. and Parastar, H. (2021a). External parameter orthogonalization-support vector machine for processing of attenuated total reflectance-mid-infrared spectra: A solution for saffron authenticity problem. *Anal. Chim. Acta 1154*: p. 338308.

Amirvaresi, A., Nikounezhad, N., Amirahmadi, M., Daraei, B., Parastar, H. et al. (2021b). Comparison of Near-Infrared (NIR) and Mid-Infrared (MIR) Spectroscopy Based on Chemometrics for Saffron Authentication and Adulteration Detection. *Food Chem.*, 344, 128647.

Anastasaki, E.G., Kanakis, C.D., Pappas, C., Maggi, L., Zalacain, A. et al. (2010). Quantification of crocetin esters in saffron (*Crocus sativus* L.) using Raman spectroscopy and chemometrics. *Journal of Agricultural and Food Chemistry*. *58*(10): pp. 6011–6017.

Anjum, N., Pal, A. and Tripathi, Y.C. (2015). Phytochemistry and pharmacology of saffron, the most precious natural source of colour, flavour and medicine. *SMU Med J,* 2: pp. 335–347.

Avila-Sosa, R., Nevárez-Moorillón, G.V., Ochoa-Velasco, C.E., Navarro-Cruz, A.R., Hernández-Carranza, P. et al. (2022). Detection of Saffron's Main Bioactive Compounds and Their Relationship with Commercial Quality. *Foods*. *11*(20): 3245.

Azarabadi, N. and Özdemir, F. (2018). Determination of crocin content and volatile components in different qualities of Iranian saffron. *Gıda*. *43*(3): pp. 476–489.

Bakshi, R.A., Sodhi, N.S., Wani, I.A., Khan, Z.S., Dhillon, B. et al. (2022). Bioactive constituents of saffron plant: extraction, encapsulation and their food and pharmaceutical applications. *Applied Food Research.* 18:100076.

Bathaie, S.Z. and Mousavi, S.Z. (2010). New applications and mechanisms of action of saffron and its important ingredients. *Critical Reviews in Food Science and Nutrition*. *50*(8): pp.761–786.

Bhagyalakshmi, N., Ravishankar, G.A. and Venkataraman, L.V. (1997). Progress in saffron biotechnology. *Biotechnological Application of Plant Tissue Culture*, Oxford IBH Publishing Co., New Delhi:101–120.

Bo, Y., Zhigang, G. and Ruizhi, L. (2005). Crocin synthesis mechanism in *Crocus sativus.Tsinghua Science and technology.* *10*(5): pp. 567–572.

Busconi, M., Colli, L., Sánchez, R.A., Santaella, M., De-Los-MozosPascual, M. et al. (2015). AFLP and MS-AFLP analysis of the variation within saffron crocus (*Crocus sativus* L.) germplasm. *PLoS One*. *10*(4): e0123434.

Busconi, M., Wischnitzki, E., Del Corvo, M., Colli, L., Soffritti, G. et al. (2021). Epigenetic variability among saffron crocus (*Crocus sativus* L.) accessions characterized by different phenotypes. *Frontiers in Plant Science. 12*: 642631.

Caballero-Ortega, H., Pereda-Miranda, R. and Abdullaev, F.I. (2007). HPLC quantification of major active components from 11 different saffron (*Crocus sativus* L.) Sources. *Food Chem.*, 100, 1126–1131.

Cardone, L., Castronuovo, D., Perniola, M., Scrano, L., Cicco, N., Candido, V. et al. (2020). The influence of soil physical and chemical properties on saffron (*Crocus sativus* L.) growth, yield and quality. *Agronomy*. *10*(8): 1154.

Carmona, M., Zalacain, A., Sánchez, A.M., Novella, J.L., Alonso, G.L. et al. (2006). Crocetin esters, picrocrocin and its related compounds present in *Crocus sativus* stigmas and Gardenia jasminoides fruits. Tentative identification of seven new compounds by LC-ESI-MS. J. Agric. *Food Chem., 54*: pp. 973–979.

Caser, M., Demasi, S., Victorino, Í.M., Donno, D., Faccio, A. et al. (2019). Arbuscular mycorrhizal fungi modulate the crop performance and metabolic profile of saffron in soilless cultivation. *Agronomy. 9*(5): p. 232.

Castellar, M.R. and Iborra, J.L. (1997). Callus induction from explants of Crocus sativus. *Journal of Plant Biochemistry and Biotechnology*. *6*: pp. 97–100.

Castillo, R., Fernández J.A. and Gómez-Gómez, L. (2005). Implications of carotenoid biosynthetic genes in apocarotenoid formation during the stigma development of *Crocus sativus* and its closer relatives. *Plant Physiology*. *139*(2): pp. 674–689.

Chaouqi, S., Moratalla-López, N., Alonso, G.L., Lorenzo, C., Zouahri, A. et al. (2023). Effect of Soil Composition on Secondary Metabolites of Moroccan Saffron (*Crocus sativus* L.). *Plants*. *12*(4): p. 711.

Chen, S.A., Wang, X., Zhao, B., Yuan, X., Wang, Y. et al. (2003). Production of crocin using *Crocus sativus* callus by two-stage culture system. *Biotechnology Letters*. *25*:1235–1238.

Chib, S., Thangaraj, A., Kaul, S., Dhar, M.K., Kaul, T. et al. (2020). Development of a system for efficient callus production, somatic embryogenesis and gene editing using CRISPR/Cas9 in Saffron (*Crocus sativus* L.). *Plant methods*. *16*(1):16, 47.

Chichiriccò, G. (1984). Karyotype and meiotic behaviour of the triploid *Crocus sativus* L. *Caryologia*. *37*(3): 233–9.

Dai, H., Gao, Q. and He, L. (2020). Rapid determination of saffron grade and adulteration by thin-layer chromatography coupled with Raman spectroscopy. *Food Analytical Methods*. *13*: pp. 2128–2137.

Diretto, G., Ahrazem, O., Rubio-Moraga, Á., Fiore, A., Sevi, F. et al. (2019). UGT709G1: a novel uridine diphosphate glycosyltransferase involved in the biosynthesis of picrocrocin, the precursor of safranal in saffron (*Crocus sativus*). *New Phytologist*. *224*(2): pp. 725–740.

Eftekhari, M., Javid, M.G., Aliniaeifard, S. and Nicola, S. (2023). Alteration of flower yield and phytochemical compounds of saffron (*Crocus sativus* L.) by application of different light qualities and growth regulators. *Horticulturae*. *9*(2): 169.

Fattahi, R., Mani-Varnosfaderani, A., Barzegar, M. and Sahari, M.A. (2023). An ion mobility spectrometry-chemometrics combination approach for assessing adulteration in saffron (*Crocus sativus* L.) with synthetic colorants. *Industrial Crops and Products*. *193*: 116161.

Fernández, J.A. (2004). Biology, biotechnology and biomedicine of saffron. *Recent Research Developments in Plant Science 2*: 127–159.

Fernández, J.A. (2006). Genetic resources of saffron and allies (Crocus spp.). *InII International Symposium on Saffron Biology and Technology 739*: pp. 167–185.

Frusciante, S., Diretto, G., Bruno, M., Ferrante, P., Pietrella, M. et al. (2014). Novel carotenoid cleavage dioxygenase catalyzes the first dedicated step in saffron crocin biosynthesis. Proc. Natl. Acad. Sci. USA 111, 12246–12251.

García-Rodríguez, M.V., Moratalla-López, N., López-Córcoles, H. and Alonso, G.L. (2021). Saffron quality obtained under different forcing conditions, considering various vegetative stages of corms. *ScientiaHorticulturae*. *277*: 109811.

Gómez-Gómez, L., Rubio-Moraga, Á. and Ahrazem, O. (2010). Understanding carotenoid metabolism in saffron stigmas: unravelling aroma and colour formation. *Func Plant Sci Biotech*. *4*: pp. 56–63.

Gómez-Gómez, L., Parra-Vega, V., Rivas-Sendra, A., Seguí-Simarro, J.M., Molina, R.V. et al. (2017). Unraveling massive crocins transport and accumulation through proteome and microscopy tools during the development of saffron stigma. *International Journal of Molecular Sciences*. *18*(1): p.76. https://doi.org/10.3390/ijms18010076

Gómez-Gómez, L., Morote, L., Frusciante, S., Rambla, J.L., Diretto, G. et al. (2022). Fortification and bioaccessibility of saffron apocarotenoids in potato tubers. *Frontiers in Nutrition*, 2961.

Gresta, F., Lombardo, G.M., Siracusa, L. and Ruberto, G. (2008). Effect of mother corm dimension and sowing time on stigma yield, daughter corms and qualitative aspects of saffron (*Crocus sativus* L.) in a Mediterranean environment. *Journal of the Science of Food and Agriculture*. *88*(7): pp. 1144–1150.

Guan, B., Zhou, M., Xiong, J. and Li, K. (2023). Identification of saffron origin and determination of crocin content in saffron based on machine vision and visible-near infrared spectroscopy. In Biophysical Society of GuangDong Province Academic Forum: *Precise Photons and Life Health (PPLH 2022)* Apr 4 (Vol. 12603, pp. 26–31). SPIE.

Guijarro-Díez, M., Castro-Puyana, M., Crego, A.L. and Marina, M.L. (2017). A novel method for the quality control of saffron through the simultaneous analysis of authenticity and adulteration markers by liquid chromatography-(quadrupole-time of flight)-mass spectrometry. *Food Chem., 228*: pp. 403–410.

Han, L.L. and Zhang, X.Y. (1993). Morphogenesis of style stigma like structures from floral explants of *Crocus sativus* L. and identification of the pigments. *Acta Bot. Sin. 35*: 157–160.

Himeno, H., Matsushima, H. and Sano, K. (1988). Scanning electron microscopic study on the in vitro organogenesis of saffron stigma-and style-like structures. *Plant Science*. 1; *58*(1): pp. 93–101.

Hori, H., Enomoto, K. and Nakaya, M. (1988). Induction of callus from pistils of *Crocus sativus* L. and production of colour compounds in the callus. *Plant Tiss. Cult. Lett. 5*: pp. 72–77

ISO 3632/TS-1, 2 Saffron (*Crocus sativus* L.) Part 1: Specifications, Part 2: Test Methods; ISO: Geneva, Switzerland, 2003.

ISO 3632-1 Saffron (*Crocus sativus* L.), Part 1 (Specification); ISO: Geneva, Switzerland, 2011.

ISO 3632-2 Saffron (*Crocus sativus* L.), Part 2 (Test Methods); ISO: Geneva, Switzerland, 2010.

Jung, W.S., Chung, I.M., Hwang, M.H., Kim, S.H., Yu, C.Y. and Ghimire, B.K. (2021). Application of light-emitting diodes for improving the nutritional quality and bioactive compound levels of some crops and medicinal plants. *Molecules. 26*(5): p.1477.

Kajikawa, N., Uno, Y., Kuroki, S., Miyagawa, S., Yamashita, Y., Hamaguchi, Y. et al. (2018). Effect of far-red light on saffron (*Crocus sativus* L.) growth and crocin yields. *Environmental Control in Biology*, *56*(2): pp. 51–57.

Karimi, S., Feizy, J., Mehrjo, F. and Farrokhnia, M. (2016). Detection and quantification of food colorant adulteration in saffron sample using chemometric analysis of FT-IR spectra. *RSC Adv. 6*: pp. 23085–23093.

Kashtwari, M., Wani, A.A., Dhar, M.K., Jan, S., Kamili, A.N. et al. (2018). Development of an efficient *in vitro* mutagenesis protocol for genetic improvement of saffron (*Crocus sativus* L.). *Physiology and Molecular Biology of Plants. 24*: pp. 951–962.

Khan, M.A., Nagoo, S., Naseer, S., Nehvi, F., Zargar, S.M. et al. (2011). Induced mutation as a tool for improving corm multiplication in saffron (*Crocus sativus* L.). *Journal of Phytology*, *3*(7).

Kheiry, A., Parsa, H., Sani Khani, M. and Razavi, F. (2018). Effect of bio-fertilizers and nitrogen on quantitative and qualitative characteristics of tepals in saffron (*Crocus sativus* L.). *Saffron agronomy and technology*. *6*(3): pp. 309–322.

Kiani, S. and Minaei, S. (2016). Potential application of machine vision technology to saffron (*Crocus sativus* L.) quality characterization. *Food Chemistry*. *212*: pp. 392–394.

Kiani, S., Minaei, S. and Ghasemi-Varnamkhasti, M. (2017). Integration of computer vision and electronic nose as non-destructive systems for saffron adulteration detection. *Computers and Electronics in Agriculture. 141*: pp. 46–53.

Kiani, S., Yazdanpanah, H. and Feizy, J. (2023). Geographical origin differentiation and quality determination of saffron using a portable Hyperspectral imaging system. *Infrared Physics and Technology*. 1; *131*: 104634.

Kohda, H., Yamasaki, K., Koyama, A., Miyagawa, H., Fujioka, N. et al. (1993). Process for culturing saffron stigma tissues. United States Patent N 5, 217, 897;

Kour, K., Gupta, D., Gupta, K., Dhiman, G. and Juneja, S. et al. (2022). Smart-hydroponic-based framework for saffron cultivation: a precision smart agriculture perspective. *Sustainability. 14*(3): 1120.

Kouzegaran, S., Baygi, M.M., Babaeian, I. and Khashei-Siuki, A.J.T., (2020). Climatology A Modeling of the saffron yield in Central Khorasan region based on meteorological extreme events. *139*(3): 1207–1217. https://doi.org/10.1007/s00704-019-03028-y

Liu, J., Chen, N., Yang, J., Yang, B., Ouyang, Z. et al. (2018). An integrated approach combining HPLC, GC/MS, NIRS, and chemometrics for the geographical discrimination and commercial categorization of saffron. *Food Chem. 253*: pp. 284–292.

López-Jimenez, A.J., Frusciante, S., Niza, E., Ahrazem, O., Rubio-Moraga, Á. et al. (2021). A new glycosyltransferase enzyme from family 91, UGT91P3, is responsible for the final glucosylation step of crocins in saffron (*Crocus sativus* L.). *International Journal of Molecular Sciences. 22*(16): p. 8815.

Loskutov, A.V., Benginger, C.W. and Ball, T.M. (1999). Optimization of *in vitro* conditions for stigma-like-structure production from half-ovary explants of *Crocus sativus* L. *In Vitro Cell Dev. Biol. (Plant) 35*: pp. 200–205.

Maggi, L., Carmona, M., del Campo, C.P., Kanakis, C.D., Anastasaki, E. and Tarantilis, P.A. (2009). Worldwide market screening of saffron volatile composition. *J. Sci. Food Agric. 89*(11): 1950–1954.

Maggio, A., Raimondi, G., Martino, A. and De Pascale, S. (2006). Soilless cultivation of saffron in Mediterranean environment. *In III International Symposium on Models for Plant Growth, Environmental Control and Farm Management in Protected Cultivation* 718, *29*: pp. 515–522.

Martí, M., Diretto, G., Aragonés, V., Frusciante, S., Ahrazem, O. et al. (2020). Efficient production of saffron crocins and picrocrocin in Nicotianabenthamiana using a virus-driven system. *Metabolic Engineering. 61*: pp. 238–250.

Melnyk, J.P., Wang, S. and Marcone, M.F. (2010). Chemical and Biological Properties of the World's Most Expensive Spice: Saffron. *Food Res. Int. 43*: pp. 1981–1989.

Menia, M., Iqbal, S., Zahida, R., Tahir, S., Kanth, R.H. et al. (2018). Production technology of saffron for enhancing productivity. *Journal of Pharmacognosy and Phytochemistry.7*(1): pp.1033–1039.

Minaei, S., Kiani, S., Ayyari, M. and Ghasemi-Varnamkhasti, M. (2017). A portable computer-vision-based expert system for saffron color quality characterization. *Journal of Applied Research on Medicinal and Aromatic Plants. 7*: pp. 124–130.

Mir, J.I., Ahmed, N., Wafai, A.H., Shafi, W. and Qadri, R.A., (2018). Relative quantification of apocarotenoids in saffron (*Crocus sativus* L.). *ActaHorticulturae 1200*: pp. 41–46. https://doi.org/10.17660/ActaHortic.2018.1200.6

Molina, R.V., Valero, M., Navarro, Y., Garcıa-Luis, A. and Guardiola, J.L. (2004). The effect of time of corm lifting and duration of incubation at inductive temperature on flowering in the saffron plant (*Crocus sativus* L.). *Scientiahorticulturae. 103*(1): pp. 79–91.

Mollafilabi, A., Koocheki, A., RezvaniMoghaddam, P. and NassiriMahallati, M. (2014). Effect of plant density and corm weight on yield and yield components of saffron (*Crocus sativus* L.) under soil, hydroponic and plastic tunnel cultivation. *Saffron Agronomy and Rechnology. 1*(2): pp. 14–28.

Momeny, M., Neshat, A.A., Jahanbakhshi, A., Mahmoudi, M., Ampatzidis, Y. et al. (2023). Grading and fraud detection of saffron via learning-to-augment incorporated Inception-v4 CNN. *Food Control. 147*: 109554.

Moradi, S., Kafi, M., Aliniaeifard, S., Salami, S.A., Shokrpour, M. et al. (2021). Blue light improves photosynthetic performance and biomass partitioning toward harvestable organs in saffron (*Crocus sativus* L.). *Cells. 10*(8): pp. 1994.

Moradi, S., Kafi, M., Aliniaeifard, S., Moosavi-Nezhad, M., Pedersen, C. et al. (2022). Monochromatic blue light enhances crocin and picrocrocin content by upregulating the expression of underlying biosynthetic pathway genes in saffron (Crocus sativus L.). *Front. Hortic. 1*: 960423. doi: 10.3389/fhort.2022.960423

Moras, B., Loffredo, L. and Rey, S. (2018). Quality assessment of saffron (*Crocus sativus* L.) extracts via UHPLC-DAD-MS analysis and detection of adulteration using Gardenia fruit extract (*Gardenia jasminoides* Ellis) *Food Chem., 257*: pp. 325–332. doi: 10.1016/j.foodchem.2018.03.025.

Moratalla-López, N., Lorenzo, C., Alonso, G.L. and Sánchez, A.M. (2016). Kaempferolglycosides in crocus: sources, biosynthesis, and uses. In: Garde-Cerdán T., Gonzalo-Diago A., editors.

Kaempferol: Biosynthesis, Food Sources and Therapeutic Uses. Nova Science Publisher's, Inc.; Hauppauge, NY, USA: pp. 151–195. Biochemistry research trends.

Moratalla-López, N., José Bagur, M., Lorenzo, C., Martínez-Navarro, M.E., Mora-talla-López, N. et al. (2019) Bioactivity and bioavailability of the major metabolites of*Crocus sativus* L. Flower. *Molecules 24*: 2827.

Moshiri, M., Vahabzadeh, M. and Hosseinzadeh, H. (2015), Clinical applications of saffron (Crocus sativus) and its constituents: a review. *Drug research. 65*(06): pp. 287–295.

NaghdiBadi, H,A., Omidi, H., Golzad, A., Torabi, H. and Fotookian, M.H. et al. (2011). Change in crocin, safranal and picrocrocin content and agronomical characters of saffron (*Crocus sativus* L.) under biological and chemical of phosphorous fertilizers. *J. Med. Plants. 4*(40): 58–68.

Nardi, L., Metelli, G., Garegnani, M., Villani, M.E., Massa, S. et al. (2022). Farming for Pharming: Novel Hydroponic Process in Contained Environment for Efficient Pharma-Grade Production of Saffron. *Molecules. 27*(24): p. 8972.

Nehvi, F.A. and Salwee, Y. (2010). Saffron farming in India the Kashmir connection. *Financ Agric. 42*(5): pp. 9–15.

Nescatelli, R., Carradori, S., Marini, F., Caponigro, V., Bucci, R. et al. (2017). Geographical characterization by MAE-HPLC and NIR methodologies and carbonic anhydrase inhibition of Saffron components. *Food Chem. 221*: 855–863.

Ordoudi, S.A. and Tsimidou, M. (2004). Saffron Quality: Effect of Agricultural Practices, Processing and Storage. In Production Practices and Quality Assessment of Food Crops; Dris, R., Jain, S.M., Eds.; Kluwer Academic: Amsterdam, *The Netherland*, Vol. 1, pp. 209–260. ISBN 1402025335.

Ordoudi, S.A., Staikidou, C., Kyriakoudi, A. and Tsimidou, M.Z. (2018). A stepwise approach for the detection of carminic acid in saffron with regard to religious food certification. *Food Chem. 267*: pp. 410–419.

Petrakis, E.A., Cagliani, L.R., Polissiou, M.G. and Consonni, R. (2015). Evaluation of Saffron (*Crocus sativus* L.) Adulteration with Plant Adulterants by 1 H NMR Metabolite Fingerprinting. *Food Chem. 173*: pp. 890–896. DOI: 10.1016/j.foodchem.2014.10. 107.

Petrakis, E.A., Cagliani, L.R., Tarantilis, P.A., Polissiou, M.G., Consonni, R. et al. (2017a). Sudan dyes in adulterated saffron (*Crocus sativus* L.): Identification and quantification by 1H NMR. *Food Chem. 217*: pp. 418–424.

Petrakis, E.A. and Polissiou, M.G., (2017b). Assessing saffron (*Crocus sativus* L.) adulteration with plant-derived adulterants by diffuse reflectance infrared fourier transform spectroscopy coupled with chemometrics. *Talanta 162*: pp. 558–566.

Rahimi, H., Shokrpour, M., TabriziRaeini, L. and Esfandiari, E. (2017). A study on the effects of environmental factors on vegetative characteristics and corm yield of saffron (Crocus sativus). Iran *J. Hortic. Sci. 48*: pp. 45–52.

Ramandi, A., Nourashrafeddin, S.M., Marashi, S.H. and Seifi, A. (2023). Microbiome contributes to phenotypic plasticity in saffron crocus. *World Journal of Microbiology and Biotechnology. 39*(1): p. 9.

Rashed-Mohassel, M.H. (2020). Evolution and botany of saffron (*Crocus sativus* L.) and allied species. In Saffron pp. 37–57. Woodhead Publishing.

Rubert, J., Lacina, O., Zachariasova, M. and Hajslova, J. (2016). Saffron authentication based on liquid chromatography high resolution tandem mass spectrometry and multivariate data analysis. *Food Chem., 204*: pp. 201–209.

Rubio-Moraga, A., Castillo-López, R., Gómez-Gómez, L. and Ahrazem, O. (2009). Saffron is a monomorphic species as revealed by RAPD, ISSR and microsatellite analyses. *BMC Research Notes. 2*(1): p.1–5.

Rubio-Moraga, Á., Gómez-Gómez, L., Trapero, A., Castro-Díaz, N. and Ahrazem, O. (2013). Saffron corm as a natural source of fungicides: The role of saponins in the underground. *Industrial Crops and Products. 49*: pp. 915–921.

Samadi, N., Naghavi, M.R., Moratalla-López, N., Alonso, G.L., Shokrpour, M. et al. (2022). Morphological, molecular and phytochemical variations induced by colchicine and EMS chemical mutagens in *Crocus sativus* L. *Food Chemistry: Molecular Sciences. 4*: 100086.

Sano, K. and Himeno, H. (1987). *In vitro* proliferation of saffron (*Crocus sativus* L.) stigma. *Plant Cell Tissue Organ Cult. 11*: pp.159–166.

Sarma, K.S., Sharada, K., Maesato, K., Hara, T., Sonoda, Y. et al. (1991). Chemical and sensory analysis of saffron produced through tissue cultures of *Crocus sativus. Plant cell, tissue and organ culture. 26*: pp.11–16.

Shawky, E., Abu El-Khair, R.A. and Selim, D.A. (2020). NIR Spectroscopy-Multivariate Analysis for Rapid Authentication, Detection and Quantification of Common Plant Adulterants in Saffron (*Crocus sativus* L.) Stigmas. *LWT–Food Sci. Technol. 122*: 109032. DOI: 10.1016/j.lwt.2020.109032.

Siracusa, L., Gresta, F., Avola, G., Albertini, E., Raggi, L. et al. (2013). Agronomic, chemical and genetic variability of saffron (*Crocus sativus* L.) of different origin by LC-UV–vis-DAD and AFLP analyses. *Genetic Resources and Crop Evolution. 60*: pp.711–721.

Tarantilis, P.A., Tsoupras, G. and Polissiou, M. (1995). Determination of saffron (*Crocus sativus* L.) components in crude plant extract using high-performance liquid chromatography-UV-visible photodiode-array detection-mass spectrometry. *J. Chromatogr*. A *699*: pp.107–118.

VarlikliozEr, S., Eksi-Kocak, H., Yetim, H. and Boyaci, I.H. (2017). Novel spectroscopic method for determination and quantification of saffron adulteration. *Food Anal. Methods* 10: pp. 1547–1555.

Verma, R.S. and Middha, D. (2010). Analysis of saffron (*Crocus sativus* L.) stigma components by LC-MS-MS. *Chromatographia, 71*: pp.117–123.

Visvanath, S., Ravishankar, G.A. and Venkataraman, L.V. (1990). Induction of crocin, crocetin, picrocrocin, and safranal synthesis in callus cultures of saffron: *Crocus sativus* L. *Biotechnology and Applied Biochemistry* (USA).

Wani, Z.A., Mirza, D.N., Arora, P. and Riyaz-Ul-Hassan, S. (2016). Molecular phylogeny, diversity, community structure, and plant growth promoting properties of fungal endophytes associated with the corms of saffron plant: An insight into the microbiome of *Crocus sativus* Linn. *Fungal Biology. 120*(12): pp. 1509–1524.

Zaffar, G., Wani, S.A., Anjum, T. and Zeerak, N.A. (2003). Colchicine induced variability in saffron. I *International Symposium on Saffron Biology and Biotechnology 650*: pp. 277–280.

Zalacain, A., Ordoudi, S.A., Díaz-Plaza, E.M., Carmona, M., Blázquez, I. et al. (2005). Near-Infrared spectroscopy in saffron quality control: Determination of chemical composition and geographical origin. *J. Agric. Food Chem. 53*: pp. 9337–9341.

Zeng, Y., Yan, F., Tang, L. and Chen, F. (2003). Increased crocin production and induction frequency of stigma-like-structure from floral organs of *Crocus sativus* L. by precursor feeding. *Plant Cell, Tissue and Organ Culture, 72*: pp.185–191.

Zheng, X., Mi, J., Balakrishna, A., Liew, K.X., Ablazov, A. et al. (2022). Gardenia carotenoid cleavage dioxygenase 4a is an efficient tool for biotechnological production of crocins in green and non-green plant tissues. *Plant Biotechnol. J. 20*: pp. 2202–2216.

10 Computational Biology of Saffron: An Expensive Spice

Mahantesha B.N. Naika[1*], Mamatha Y.S.[2] and
Dwijesh Chandra Mishra[2]

Advanced omics technologies and next-generation sequencing have transformed biological research, extending beyond model organisms to specialized plants like saffron. Computational biology tools—innovative algorithms, data structures, and visualization software—help decode complex genetic and molecular networks. Systematic perturbations and monitoring refine our understanding of saffron's genes, proteins, and pathways. An integrative systems biology approach, leveraging predictive models, reveals the molecular mechanisms shaping its agricultural and horticultural traits. This chapter explores the fusion of ancient wisdom and modern science to unravel the mysteries of the world's most treasured spice.

1. Introduction

Saffron (*Crocus sativus* L.), known as 'red gold,' is valued for its culinary and medicinal properties. It has a chromosome number of 2n=3x=24, with other Crocus species also documented (Table 1). Derived from flower stigmas, saffron holds significance in both traditional and modern therapeutics (Naghshineh et al., 2015; Singla et al., 2011). Saffron's deep hues, aroma, and health benefits justify its status as the world's most expensive spice (Yadollahi et al., 2006). Beneath its golden threads lies a complex genetic architecture, gradually unraveled through bioinformatics and modern computational tools. Genomics reveals unique genetic expressions shaping saffron's lifecycle and adaptability, particularly its sterile reproductive system relying on vegetative propagation via corms (Husaini et al., 2009). Advancements in transcriptomics reveal how environmental factors like light and temperature influence saffron's gene expression. Proteomics uncovers its dynamic protein interactions, while metabolomics maps the metabolic pathways driving its therapeutic properties (Schmidt et al., 2007; Frusciante et al., 2014; Kyriakoudi et al., 2015). Omics technologies and systems biology approaches further enhance our understanding of plant genome functions and physiological

[1] Centre for Biotechnology Research, Department of Biotechnology and Crop Improvement, College of Horticulture, Bagalkot, University of Horticultural Sciences, Bagalkot, India.
[2] ICAR-Indian Agricultural Statistics Research Institute, PUSA, New Delhi, India.
* Corresponding author: mahantesha.bnn@uhsbagalkot.edu.in

DOI: 10.1201/9781003403760-10

responses (Shameer et al., 2019). Saffron research blends ancient wisdom with modern computational methods, expanding genomic insights and digital archives. Computational biology, alongside botanical studies, illuminates its complex biology. From Ottoman-era cultivation to its status as the world's most expensive spice, saffron's historical and scientific significance endures (Fernández et al., 2009). Its therapeutic potential—addressing neurodegenerative disorders and severe respiratory syndromes like COVID-19—reinforces its value in modern medicine (Hausenblas et al., 2015; Broadhead et al., 2015). Advancements in computational biology and systems science are rapidly expanding saffron research, unlocking new applications and promising further discoveries and innovations.

Fig. 1 (a) Flowering in saffron (b) dried saffron stigmas. (Figure reproduced from Kumar et al. 2022. *Sci Rep* 12, 11925 (2022).

Table 1 Different species of Crocus with chromosome numbers.

Genus	*Chromosome number*	*C value (pg)*	*Reference*
Crocus sativus	2n = 3x = 24	11.37	Brighton, 1977; Mathew, 1977
C. cartwrightianus, *C. hadriaticus,* *C. haussknechtii,* *C. oreocreticus,* *C. thomasii*	2n = 16	15.7 19.4 15.9 19.4 19.0	Brighton et al., 1973; Karamplianis et al., 2013
C. dispathaceus, *C. moabiticus,* *C. naqabensis,*	2n = 14	–	Brighton et al., 1973; Karamplianis et al., 2013
C. asumaniae	2n = 26	–	Brighton et al., 1973
C. mathewii	2n = 70	–	Brighton et al., 1973
Crocus turcicus *C. haussknechtii*	2n = 12	– 15.9	Brighton, 1977
C. pallasii	2n = 12, 14, 16	–	Jacobsen and Orgaard 2003

2. Bioinformatics and Saffron

Bioinformatics plays a crucial role in deciphering saffron's genetic and molecular complexity, mapping gene-protein-metabolite interactions through advanced computational methods. Omics technologies enhance this understanding via databases, machine learning insights, and integrative studies. Table 2 provides key tools and databases for saffron research, aiding deeper exploration.

Table 2 Databases/tools specific to Saffron

S. No.	*Database/Tool*	*Description*	*Reference/DOI*
1	Saffron genes	EST collection from saffron stigmas	Husaini et al. (2009).
2	SaffNet	An ensemble-based approach for saffron adulteration prediction using statistical image features	10.3390/rs13132532
3	Saffron Adulteration Prediction Systems (SAPS)	Automatic Saffron Adulteration Prediction Systems, research issues, and prospective solutions	10.1109/ CCICT53244.2021.00024

2.1 Machine Learning and its Impact on Saffron

Machine learning (ML) and artificial intelligence (AI) are revolutionizing saffron research, enhancing genetic differentiation, economic evaluation, and therapeutic insights. Traditionally difficult to classify due to its sterile triploid nature, saffron identification has improved with data-driven algorithms and Sequence-Related Amplified Polymorphism markers, achieving over 90% accuracy (Beiki, 2014). ML models effectively predict crocin content, a key compound in saffron, with correlation coefficients up to 0.9982 (Guan et al., 2023). Adulteration detection has also advanced through ML-based systems like SAPS (Amin et al., 2021) and SaffNet, the latter achieving 98% accuracy in distinguishing authentic and adulterated saffron (Amin et al., 2023). The "SAFFRON" framework provides adaptive false discovery rate control for genetic studies, enhancing research potential (Ramdas et al., 2018). As ML evolves, its integration into saffron research promises deeper insights into this valuable spice.

2.2 Omics Technologies Applied to Saffron

Saffron, a biological marvel, exhibits complex metabolite accumulation regulated by intricate networks. Omics technologies and bioinformatics provide deep insights, transforming vast datasets into meaningful knowledge (Figure 2). Systems biology integrates transcriptomics, proteomics, and metabolomics for a unified understanding (Kumar et al., 2015). Bioinformatics aids in data generation, analysis, and integration, refining high-throughput insights (Maturana et al., 2019; Ogbe et al., 2016). Essential tools like GenoType, GenoDive, PIECE, MISA,

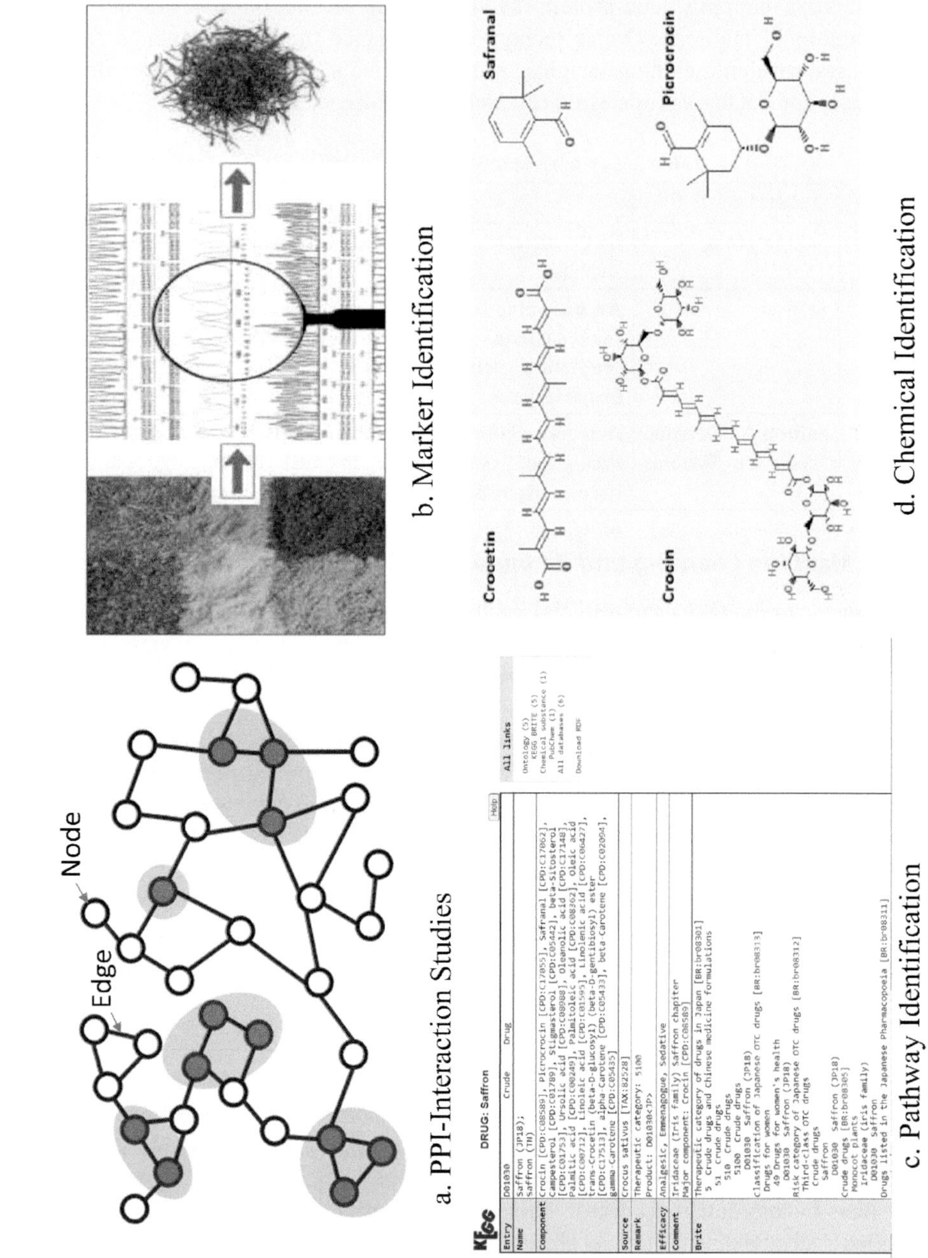

Fig. 2 Integrative approaches in saffron biology

Trinity, and SOAP denovo-Trans enhance genetic diversity and transcriptome analysis (Meirmans & Tienderen, 2004; Ahrazem et al., 2020; Yue et al., 2020; Tan et al., 2019). Algorithms such as RSEM, eXpress, Sailfish, kallisto, BLAST, and Blast2GO facilitate transcript expression and annotation (Conesa et al., 2005; Rhoads et al., 2015; Langmead, 2010). Studies highlight the role of bioinformatics in deciphering glycosidase activity for stigma development and bioactive compound synthesis (Husaini et al., 2018; Husaini et al., 2021), advancing saffron research and applications.

Genomics in Saffron

Saffron's genomic complexity continues to intrigue scientists. As a sterile perennial, it reproduces via corms, yet no Crocus species has undergone full genome sequencing, leaving gaps in genetic understanding. While some studies suggest saffron is monomorphic, requiring sequencing to distinguish isolates (Ambardar et al., 2022), others demonstrate polymorphism through molecular markers. Epigenetic factors, like methylation-sensitive restriction enzyme sequencing (MRE-seq), reveal that genetically similar accessions can show distinct methylation patterns, affecting phenotypes (Husaini et al., 2022). Functional genomics aims to identify genes critical for saffron's pigmentation and flavor compound synthesis (Dhar et al., 2017), while bioinformatics-driven DNA analysis could enhance saffron quality via transgenic methods (Bergomi et al., 2022). Conservation efforts like CROCUSBANK (De-Los-Mozos-Pascual et al., 2009) further advance genetic resource characterization. Table 3 outlines key projects decoding saffron genomics.

Transcriptomics

Transcriptomics has been instrumental in uncovering saffron's genetic dynamics, particularly across cultivation regions like Khorasan (Iran) and Kashmir (India). DNA microarrays and bioinformatics have advanced this research (Husaini et al., 2009). Notably, 6603 high-quality ESTs from saffron stigma cDNA libraries helped establish a comprehensive gene database (D'Agostino et al., 2007). Baba et al.'s transcriptome analysis provided insights into apocarotenoid biosynthesis, generating a TF database for Crocus species (Ahrazem et al., 2019). This study identified numerous saffron-specific transcripts (Jain, 2016). UDP-glucosyltransferase plays a key role in metabolite differentiation among Crocus species (Trapero et al., 2012). The European initiative 'Saffronomics' advances saffron research through genomics and metabolomics (Panchangam et al., 2016). Limited saffron ESTs indicate early-stage transcriptomic and genomic exploration (Hu et al., 2020), but progress in carotenoid/apocarotenoid biosynthesis and gene identification signals promising developments (Sathasivam et al., 2022). Continued research will enhance saffron cultivation and biotechnological applications.

Table 3 Sequence resources available at web resources on saffron.

Project data type	*Scope*	*Accession*	*Cultivar/Variety/Description*	*Submitter*
RefSeq Genome	Monoisolate	PRJNA531999	*Crocus sativus* RefSeq Genome	NCBI, USA
Genome sequencing and assembly	Monoisolate	PRJNA827481	*Crocus sativus*	Sichuan University, China
Genome sequencing and assembly	Monoisolate	PRJNA794882	*Crocus sativus*	Shanghai Academy of Agricultural Sciences, China
Genome sequencing and assembly	Monoisolate	PRJNA739096	*Crocus sativus* cultivar:SA-2019	National Center for Biological Sciences, India
Genome sequencing and assembly	Monoisolate	PRJNA734464	*Crocus sativus* isolate:Saffron JU | cultivar:Saffron	National Center for Biological Sciences, China
Genome sequencing	Monoisolate	PRJNA609179	*Crocus sativus* isolate: MJ-001	Jawaharlal Nehru University
Raw sequence reads	Multispecies	PRJNA880437	Sugar metabolism mediates temperature-dependent flowering induction in saffron	C.S.I.R- Institute of Himalayan Bioresource Technology, India
Raw sequence reads	Monoisolate	PRJNA875658	*Crocus sativus*	Jiaxing Vocational and Technical College
Raw sequence reads	Multispecies	PRJNA757794	transcriptome of Crocus sativus L.	Zhejiang Agriculture and Forestry University
Raw sequence reads	Multispecies	PRJNA613348	Full-length transcriptome sequencing of *Crocus sativus*	Jawaharlal Nehru University, India
Raw sequence reads	Multispecies	PRJNA613151	De novo genome sequencing and assembly of *Crocus sativus*	Jawaharlal Nehru University, India
Raw sequence reads	Multiisolate	PRJNA418912	*Crocus sativus* Raw sequence reads	University of Tehran, Iran
Raw sequence reads	Monoisolate	PRJNA562470	*Crocus sativus* cultivar: *Crocus sativus*	ENEA, Italy
Raw sequence reads	Monoisolate	PRJNA528829	*Crocus sativus*	Huzhou Central Hospital, China

Contd.

Table 3 *Contd.*

Project data type	*Scope*	*Accession*	*Cultivar/Variety/Description*	*Submitter*
Raw sequence reads	Multispecies	PRJNA527934	Transcriptomic data from different periods of *crocus sativus* L.	Chengdu University of Traditional Chinese Medicine, China
Raw sequence reads	Monoisolate	PRJNA378331	*Crocus sativus* chromosome triplets 1 and 2 sequence reads	Institute of Experimental Botany, Czech Republic
Transcriptome or Gene expression	Monoisolate	PRJNA638232	Transcripts on saffron buds from corms able to flower, during the period of flower induction and development in two different environmental conditions (field and growth chambers)	Universitat Politecnica de Valencia, Spain
Transcriptome or Gene expression	Multiisolate	PRJNA524437	*Crocus sativus* Transcriptome or Gene expression	Huzhou Central Hospital, China
Transcriptome or Gene expression	Multispecies	PRJNA508499	RNA-seq analysis on *Crocus sativus* and other two Crocus species	ENEA, Italy
Transcriptome or Gene expression	Multiisolate	PRJNA400472	Dynamic transcriptomic and metabolic profiling reveal synthetic characters and regulators of apocarotenoids biosynthesis in Crocus sativus	Department of Pharmaceutical Botany, School of Pharmacy, Second Military M, China
Transcriptome or Gene expression	Monoisolate	PRJNA277895	Comprehensive transcriptome map of *Crocus sativus*	CSIR-Institute of Himalayan Bioresource Technology, India
Transcriptome or Gene expression	Multiisolate	PRJNA273002	Saffron Transcriptome and Gene Expression	Jawaharlal Nehru University, India

Contd.

Table 3 *Contd.*

Project data type	*Scope*	*Accession*	*Cultivar/Variety/Description*	*Submitter*
Transcriptome or Gene expression	Monoisolate	PRJNA222766	*Crocus sativus* Transcriptome or Gene expression	Agricultural Biotechnology Research Institute of Iran, Iran
Metagenome	Environment	PRJNA557606	Metagenome analysis of bacterial endophytes in corm and shoot tissues of *Crocus sativus* L.	CSIR-Indian Institute of Integrative Medicine, India
Metagenome	Environment	PRJNA540760	*Crocus sativus* Metagenome	CSIR-Indian Institute of Integrative Medicine, India
Variation	Multiisolate	PRJNA700580	Epigenetic variability among saffron crocus (*Crocus sativus* L.) accessions characterized by different phenotypes	NCBI (GEO), USA
Other	Monoisolate	PRJEB27327	Origin of *Crocus sativus*	Leibniz Institute of Plant Genetics and Crop Plant Research, Germany
Other	Monoisolate	PRJEB25396	Phylogeny of the saffron-crocus species group, Crocus series Crocus (Iridaceae)	Leibniz Institute of Plant Genetics and Crop Plant Research, Germany
Other	Monoisolate	PRJEB29021	*Crocus sativus* (saffron crocus) whole genome shutgun Illumina sequencing Saffron (*Crocus sativus*) is an autotriploid hybrid derived from wild *Crocus cartwrightianus* cytotypes	TU DRESDEN, Germany

Proteomics

Proteomics sheds light on saffron's biological complexity. During somatic embryogenesis, 36 differentially accumulated proteins suggest the ascorbate–glutathione cycle's role in embryo formation (Aguilar-Hernández & Loyola-Vargas, 2018). Advanced platforms like the ProteomeXchange Consortium provide detailed protein insights. iTRAQ-based proteomics with qPCR identified over 200 proteins influenced by cold stress, impacting floral initiation (Chen et al., 2021). Dormant saffron exhibit non-blooming behavior under low temperatures, linked to shifts in the ROS-antioxidant system and starch/sugar interconversion pathways (Rastegari et al., 2022). Despite progress in protein identification, predicting structures remains challenging, with only 326 saffron protein entries in GenBank. This data gap hampers functional annotation, evolutionary analysis, and network mapping. Platforms like UniProt aid in protein data compilation, while atomic-level structural refinement is crucial for drug screening and functional studies. The Protein Data Bank (PDB) offers valuable crystal structures, advancing saffron proteomics.

Metabolomics

Metabolomics offers critical insights into saffron's physiology, identifying novel metabolites like Azafrine 1 and 2, which function as phytoprotectants in corms (Moratalla-López, 2019). Advanced techniques, including 1H NMR-based metabolomics, aid in saffron quality assessment (Riswanto et al., 2022). LC-MS and UPLC-Q-TOF-MS/MS have revealed key compounds like coniferin and crocin-2 in stigmas, while tepals are enriched with flavonoids (Ho et al., 2003; Xu et al., 2019). Geographical variations influence saffron's phytochemical composition, as revealed by mass spectrometry studies. Metabolite concentrations—crocetin, crocins, and picrocrocin—rise as stigmas mature from yellow to red, offering health benefits from coronary protection to neurodegenerative and post-COVID-19 therapies (Bhuiyan et al., 2020). Saffron exhibits notable antidepressant properties, rivaling traditional medicines. Unlocking its full metabolomic potential requires robust tools and databases, such as the Saffron Metabolite Database for chemical structures, pathways, and interactions. Enhancing metabolite identification and quantification through platforms like MetaboAnalyst and KEGG could refine saffron-specific studies (Marco-Ramell et al., 2018). Advances in technology and research continue to illuminate saffron's metabolic complexities.

Integrative Omics in Saffron: A New Frontier

The integration of omics technologies—genomics, transcriptomics, proteomics, and metabolomics (Table 4)—has revolutionized biological research. Saffron, a high-value crop with therapeutic potential (Naghshineh et al., 2015), is an ideal candidate for systems biology studies. Approaches like drug-target mapping and network modeling reveal new biological insights (Vandamme et al., 2014; Harrold

Table 4 Useful system biology resources

Omics type	*Database/Tool Name*	*Description*	*URL*
Genomics	Saffron genes	EST collection from saffron stigmas	saffrongenes.org
	NCBI Plant	Database focused on plant genomes	ncbi.nlm.nih.gov/genomes/PLANTS/PlantList.html
	SAM and BCF tools	Utilities to handle and analyze alignment data in the SAM format	htslib.org
	MEGA	Tool for exploring relationships among homologous sequences using evolutionary analyses	megasoftware.net
	Trinity	Dedicated software for assembling RNA-seq data into transcripts without relying on existing annotations	https://github.com/trinityrnaseq/trinityrnaseq
	SMART 9	Online platform for analyzing protein domains and annotations within sequences	smart.embl.de
	Ensembl Plants	Database for visualizing, mining, and analyzing plant genomic data	plants.ensembl.org
	BiGGEsTS	Designed to unveil gene co-expression patterns over specific time intervals	kdbio.inesc-id.pt/software/biggests
	PlantGDB	Genomic repository focused on plant species and related sequence data	plantgdb.org
	KEGG	Comprehensive database for genome interpretation and high-throughput data analysis	https://www.genome.jp/kegg/
	TrichOME	Comparative database focused on plant trichome omics studies	planttrichome.org

Contd.

Table 4 *Contd.*

Omics type	*Database/Tool Name*	*Description*	*URL*
	PlantTFcat	Tool for identifying and categorizing plant transcription factors and regulators	zhaolab.org/PlantTFcat
	Pln TFDB	A resource tailored for studying plant transcription factors, their functions, and evolution	plntfdb.bio.uni-potsdam.de/v3.0/
	Bowtie	Alignment tool known for its speed and efficiency in aligning short DNA sequence reads to large genomes	https://bowtie-bio.sourceforge.net/index.shtml
	KaPPA-View	Web platform for the analysis of omics data in plants	kpv.kazusa.or.jp/kpv4/
	PIECE	Database dedicated to plant gene structure comparison and evolution	probes.pw.usda.gov/piece/index.php
	NIAS db	Database encompassing diverse genetic data	dna.affrc.go.jp/database/
	PlantCare	Database on cis-acting regulatory DNA elements in plants	bioinformatics.psb.ugent.be/webtools/plantcare/html/
	EXPASY	Index linking to other plant-specific databases	expasy.org/links.html
Transcriptomics	Cufflinks	Open-source tool designed for analyzing RNA-Seq data	cole-trapnell-lab.github.io/cufflinks
	GeneMarkS-T	Tool focused on identifying protein-coding regions in RNA transcripts	topaz.gatech.edu/GeneMark/license_download.cgi
	PlexDB	Data on plant expression	plexdb.org
	CaARRAY	Guides the annotation and exchange of array data	caarray.nci.nih.gov

Contd.

Table 4 *Contd.*

Omics type	*Database/Tool Name*	*Description*	*URL*
	PEDRo	Configurable data entry tool for XML	pedro.man.ac.uk/
Proteomics	UniProt	Comprehensive database of protein sequence and function data.	https://www.uniprot.org/
	Pride Archive	Repository of protein and peptide identifications.	https://www.ebi.ac.uk/pride/archive/
	Protein Data Bank (PDB)	Repository for 3D structural data of proteins.	https://www.rcsb.org/
	Plant Proteome Database (PPDB)	Focused on plants, including saffron-related proteomes.	http://ppdb.tc.cornell.edu/
	ProteomicsDB	Exploration of large collections of mass spectrometry-based proteomics data.	https://www.proteomicsdb.org/
	PeptideAtlas	Database for proteome and peptide atlas building.	https://www.peptideatlas.org/
	MaxQuant	Software for quantitative proteomics, especially for analyzing large mass-spectrometric data sets	maxquant.net/maxquant
	Perseus	Platform for interpreting protein quantification, interaction, and post-translational modification data	maxquant.net/perseus
	PRINTS	Database of protein fingerprints	bioinf.manchester.ac.uk/dbbrowser/PRINTS/index.php
	SWISS-2DPAGE	Data on proteins identified on various 2-D PAGE and SDS-PAGE reference maps	au.expasy.org/ch2d

Contd.

Table 4 *Contd.*

Omics type	*Database/Tool Name*	*Description*	*URL*
	Melanie	Two-dimensional electrophoresis (2-DE) gel analysis platform	au.expasy.org/melanie
	Flicker	An open-source stand-alone computer program for visually comparing 2D gel images	open2dprot.sourceforge.net/Flicker
	PDQuest	Software for analysis of 2D gels	proteomeworks.bio-rad.com/html/pdquest.html
	MaxQB	Stores and displays collections of large proteomics projects.	https://maxqb.biochem.mpg.de/mxdb/
	ProMEX	Mass spectral reference database for proteins and phosphorylation sites.	http://promex.pph.univie.ac.at/promex/
	GPMDB	Stores protein identifications using the GPM's X!Tandem.	http://gpmdb.thegpm.org/
	Mascot	Popular tool for protein identification using mass spectrometry data.	http://www.matrixscience.com/
	PEAKS	Suite of software tools for proteomics data analysis.	https://www.bioinfor.com/peaks-studio/
	Phenyx	Identification and characterization of proteins and peptides from mass spectrometry data.	https://phenyx.vital-it.ch/docs/pwi/
Metabolomics	CROPFORGE	Metabolomics package	cropforge.org
	SBML	Metabolomics package	sbml.org
	KNApSAcK	A species-metabolite relationship database, potentially containing data about saffron.	http://kanaya.naist.jp/KNApSAcK/
	METLIN	A metabolite and tandem mass spectrometry database.	https://metlin.scripps.edu/

Contd.

Table 4 *Contd.*

Omics type	*Database/Tool Name*	*Description*	*URL*
	HMDB	Human-centric but an extensive resource for understanding metabolic pathways.	https://hmdb.ca/
	MassBank	A high-resolution mass spectral database.	https://massbank.eu/MassBank/
	MoToDB	Database of plant secondary metabolites.	N/A (could not find a direct URL for MoToDB)
	PlantMetabolomics.org	Features plant-specific metabolomic data, including potentially that of saffron.	http://www.plantmetabolomics.org/
	MetaCyc	A comprehensive database of metabolic pathways and enzymes from all domains of life.	https://metacyc.org/
	YMDB	Yeast-centered, provides insight into metabolic processes that can be paralleled in plants.	http://www.ymdb.ca/
	Maven	A tool for interpreting and visualizing high throughput metabolomic data.	https://github.com/mavenomics/maven
	MeltDB	Software platform for analysis and integration of metabolomics experiment data.	https://meltdb.cebitec.uni-bielefeld.de/
	MetaboAnalyst	Comprehensive web-based tool for metabolomic data analysis.	https://www.metaboanalyst.ca/
	XCMS	For processing liquid chromatography-mass spectrometry (LC/MS) data.	https://xcmsonline.scripps.edu/

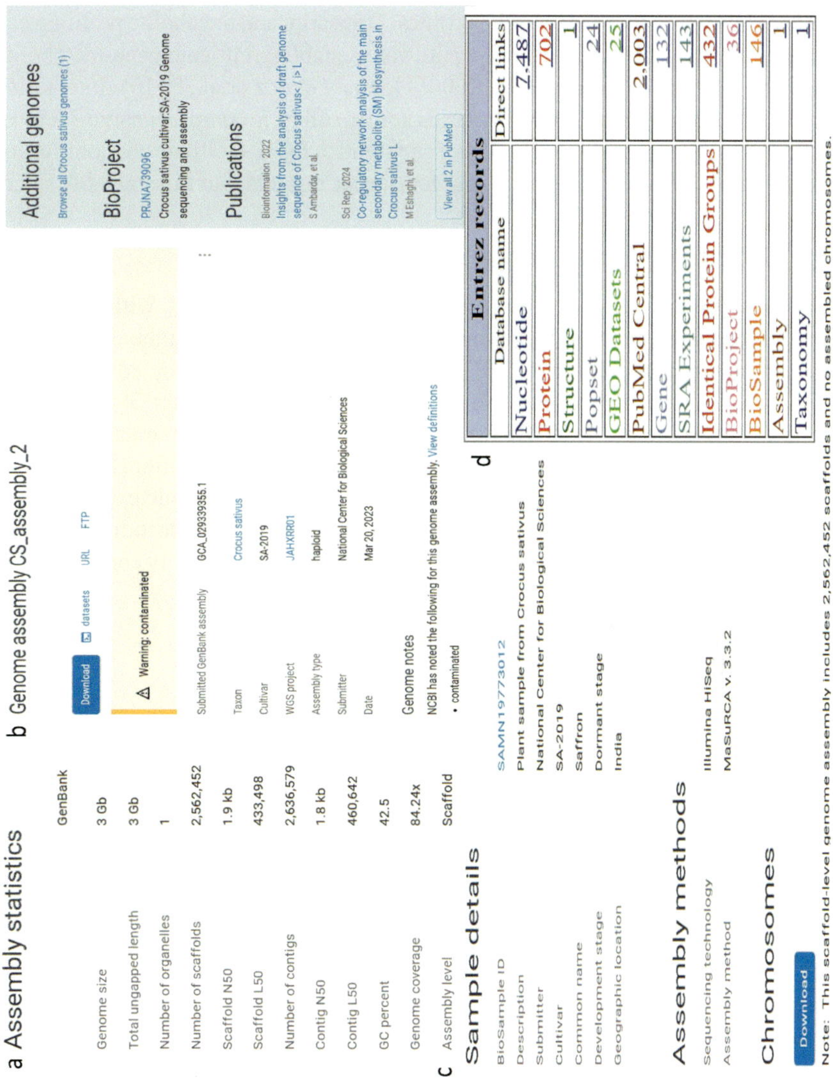

Fig. 3 Information about saffron found in the NCBI database.

et al., 2013). Techniques such as target deconvolution and molecular docking have identified functional components in saffron (Nithya & Shakthisekharan, 2015; Bhattacharjee et al., 2012). Notably, Zeraatkar et al. (2015) used reverse engineering and laser scanning to develop a 3D saffron flower model. This study not only clarifies saffron flower structure but also informs post-harvest machinery design. Its metabolic processes drive carotenoid synthesis, with each developmental stage exhibiting distinct metabolite profiles. Transcript and metabolite profiling can uncover novel genes and metabolic pathways, enabling intricate gene-metabolite network mapping (Husaini et al., 2009; Gómez-Gómez et al., 2010). Integrative omics, strengthened by bioinformatics tools, offers a comprehensive view of saffron biology. Platforms like Cytoscape and KEGG facilitate exploration of complex interactions, paving the way for deeper insights into this valuable spice.

3. Saffron's Digital Footprint on NCBI

Saffron's culinary and scientific significance is well established, with its genetic and molecular aspects gaining prominence in research. The National Center for Biotechnology Information (NCBI) serves as a key repository for saffron-related data, offering extensive resources on its genomic landscape. Figure 3(a) illustrates saffron's presence in the NCBI Entrez system, detailing available databases and research efforts. The GenBank assembly submission by the National Center for Biological Sciences further highlights the comprehensive genomic exploration of the SA-2019 cultivar, emphasizing the scientific commitment to decoding saffron's genetic blueprint. Figure 3(b) provides an overview of its assembly characteristics and metadata.

4. Challenges and Future Scope

Recent advancements have unveiled saffron's complex biology, highlighting both challenges and opportunities. Genomic engineering aims to enhance its quality and yield, with CRISPR-Cas offering potential breakthroughs in modifying genes essential for apocarotenoid biosynthesis. Whole-genome sequencing of C. sativus L. remains crucial for leveraging gene-editing technologies. As climate change intensifies, Omics tools become essential for identifying resilient traits and advancing saffron research. Omics tools help identify resistance traits and agronomic qualities in saffron, enabling targeted biotechnological improvements for environmental resilience. They also clarify genetic diversity across geographic clusters, aiding biosynthetic pathway analysis. While metabolomics has revealed key saffron characteristics, challenges remain in resolving biochemical discrepancies and linking them to phenotypes. The evolving interplay of challenges and discoveries will continue shaping saffron research.

Conclusion

Saffron, beyond its culinary prestige, holds significant therapeutic potential linked to its secondary metabolism and apocarotenoid synthesis in stigmas. Digital expression genetics (DGE) confirms stigmas as the key site for apocarotenoid biosynthesis, enriching our understanding of carotenoid metabolism. Whole-genome sequencing, integrated with omics technologies, promises deeper insights into flower development, biosynthetic pathways, and saffron's medicinal value, potentially advancing high-yield, superior-quality cultivation. The integration of omics technologies and systems biology is essential for unlocking saffron's therapeutic potential, including novel drug targets for chemotherapy. Bioinformatics plays a pivotal role in genotyping, proteomics, and systems biology, streamlining research efforts. The challenge lies in seamlessly merging advanced biotechnological methods with bioinformatics tools to drive a comprehensive exploration of saffron.

References

Aguilar-Hernández, V. and Loyola-Vargas, V.M. (2018). Advanced proteomic approaches to elucidate somatic embryogenesis. *Frontiers in Plant Science*, *9*: 1658.

Ahmed, S.A. and Husaini, A.M. (2021). Investigating binding potential of carotenoid pathway bioactive molecules for ACE2 receptor of SARS-CoV-2: Possibility of a saffron based remedy for novel coronavirus! *Journal of Horticulture and Postharvest Research*, *4*(Special Issue-Recent Advances in Saffron), 69–78.

Ahrazem, O., Argandoña, J., Fiore, A., Rujas, A., Rubio-Moraga, Á. et al. (2019). Multi-species transcriptome analyses for the regulation of crocins biosynthesis in Crocus. *BMC Genomics*, *20*: 1–15.

Ahrazem, O., López, A. J., Argandoña, J., Castillo, R., Rubio-Moraga, Á. (2020). Differential interaction of Or proteins with the PSY enzymes in saffron. *Scientific Reports*, *10*(1): 552.

Altschul, S.F., Gish, W., Miller, W., Myers, E.W. and Lipman, D.J. et al. (1990). Basic local alignment search tool. *Journal of Molecular Biology*, *215*(3): 403–410.

Ambardar, S., Vakhlu, J. and Sowdhamini, R. (2022). Insights from the analysis of draft genome sequence of Crocus sativus L. *Bioinformation*, *18*(1): 1.

Amin, J., Selwal, A. and Sabha, A. (2021). SAPS: Automatic Saffron Adulteration Prediction Systems, research issues, and prospective solutions. In *2021 Fourth International Conference on Computational Intelligence and Communication Technologies (CCICT)* (pp. 64–71). IEEE.

Amin, J., Selwal, A. and Sabha, A. (2023). SaffNet: an ensemble-based approach for saffron adulteration prediction using statistical image features. *Multimedia Tools and Applications*, 1–21.

Beiki, A.H. (2014). Classification and prediction of three and multi stigma in saffron by statistical, unsupervised machine learning tools. *Saffron Agronomy and Technology*, *2*(3): 199–204.

Bergomi, A., Comite, V., Santagostini, L., Guglielmi, V. and Fermo, P. et al. (2022). Determination of saffron quality through a multi-analytical approach. *Foods*, *11*(20): 3227.

Bhattacharjee, B., Vijayasarathy, S., Karunakar, P. and Chatterjee, J. (2012). Comparative reverse screening approach to identify potential anti-neoplastic targets of saffron functional components and binding mode. *Asian Pacific Journal of Cancer Prevention*, *13*(11): 5605–5611.

Bhuiyan, F.R., Howlader, S., Raihan, T. and Hasan, M. (2020). Plants metabolites: possibility of natural therapeutics against the COVID-19 pandemic. *Frontiers in Medicine*, *7*: 444.

Brighton, C.A. (1977). Cytology of Crocus sativus and its allies (Iridaceae). *Plant Systematics and Evolution*, *128*: 137–157.

Brighton, C.A., Mathew, B. and Marchant, C.J. (1973). Chromosome counts in the genus Crocus (Iridaceae). *Kew Bulletin*, 451–464.

Broadhead, G.K., Chang, A., Grigg, J. and McCluskey, P. (2016). Efficacy and safety of saffron supplementation: current clinical findings. *Critical Reviews in Food Science and Nutrition*, *56*(16): 2767–2776. Beiki

Busconi, M., Wischnitzki, E., Del Corvo, M., Colli, L., Soffritti, G. et al. (2021). Epigenetic variability among saffron crocus (Crocus sativus L.) accessions characterized by different phenotypes. *Frontiers in Plant Science*, *12*: 642631.

Chen, J., Zhou, G., Dong, Y., Qian, X., Li, J. et al. (2021). Screening of key proteins affecting floral initiation of saffron under cold stress using iTRAQ-based proteomics. *Frontiers in Plant Science*, *12*: 644934.

Conesa, A., Götz, S., García-Gómez, J.M., Terol, J., Talón, M. et al. (2005). Blast2GO: a universal tool for annotation, visualization and analysis in functional genomics research. *Bioinformatics*, *21*(18): 3674–3676.

Conesa, A., Madrigal, P., Tarazona, S., Gomez-Cabrero, D., Cervera, A. et al. (2016). A survey of best practices for RNA-seq data analysis. *Genome Biology*, *17*(1): 1–19.

D'Agostino, N., Pizzichini, D., Chiusano, M.L. and Giuliano, G. (2007). An EST database from saffron stigmas. *BMC Plant Biology*, *7*(1): 1–8.

De-Los-Mozos-Pascual, M., Fernández, J.A. and Roldán, M. (2009). Preserving biodiversity in saffron: The Crocusbank Project and the world saffron and crocus collection. In *III International Symposium on Saffron: Forthcoming Challenges in Cultivation, Research and Economics 850*: (pp. 23–28).

Dhar, M.K., Sharma, M., Bhat, A., Chrungoo, N.K. and Kaul, S. et al. (2017). Functional genomics of apocarotenoids in saffron: insights from chemistry, molecular biology and therapeutic applications. *Briefings in Functional Genomics*, *16*(6): 336–347.

Fernández, J.A. (2004). Biology, biotechnology and biomedicine of saffron. *Recent Research Developments in Plant Science. Vol. 2*: 127–159.

Fernández, J.A., Agayev, Y.M. and Zarifi, E. (2009). A study of karyotypes in the Crocus sativus L. aggregate and origin of cultivated saffron. In *III International Symposium on Saffron: Forthcoming Challenges in Cultivation, Research and Economics 850*: pp. 47–54.

Frusciante, S., Diretto, G., Bruno, M., Ferrante, P., Pietrella, M. et al. (2014). Novel carotenoid cleavage dioxygenase catalyzes the first dedicated step in saffron crocin biosynthesis. *Proceedings of the National Academy of Sciences*, *111*(33): 12246–12251.

Gómez-Gómez, L., Rubio-Moraga, Á. and Ahrazem, O. (2010). Understanding carotenoid metabolism in saffron stigmas: unravelling aroma and colour formation. *Func Plant Sci Biotech*, *4*, 56–63.

Guan, B., Zhou, M., Xiong, J. and Li, K. (2023). Identification of saffron origin and determination of crocin content in saffron based on machine vision and visible-near infrared spectroscopy. In *Biophysical Society of GuangDong Province Academic Forum: Precise Photons and Life Health (PPLH 2022)* (Vol. 12603, pp. 26–31). SPIE.

Harrold, J.M., Ramanathan, M. and Mager, D.E. (2013). Network-based approaches in drug discovery and early development. *Clinical Pharmacology & Therapeutics*, *94*(6): 651–658.

Hausenblas, H.A., Heekin, K., Mutchie, H.L. and Anton, S. (2015). A systematic review of randomized controlled trials examining the effectiveness of saffron (Crocus sativus L.) on psychological and behavioral outcomes. *Journal of Integrative Medicine*, *13*(4): 231–240.

Ho, C.S., Lam, C.W.K., Chan, M.H., Cheung, R.C.K., Law, L.K. et al. (2003). Electrospray ionisation mass spectrometry: principles and clinical applications. *The Clinical Biochemist Reviews*, *24*(1): 3.

Hu, J., Liu, Y., Tang, X., Rao, H., Ren, C. et al. (2020). Transcriptome profiling of the flowering transition in saffron (Crocus sativus L.). *Scientific Reports*, *10*(1): 9680.

Husaini, A.M., Haq, S.A.U. and Jiménez, A.J.L. (2022). Understanding saffron biology using omics- and bioinformatics tools: stepping towards a better Crocus phenome. *Molecular Biology Reports*, *49*(6): 5325–5340.

Husaini, A.M., Jan, K.N. and Wani, G.A. (2021). Saffron: A potential drug-supplement for severe acute respiratory syndrome coronavirus (COVID) management. *Heliyon*, *7*(5).

Husaini, A.M., Morimoto, K., Chandrasekar, B., Kelly, S., Kaschani, F. et al. (2018). Multiplex fluorescent, activity-based protein profiling identifies active α-glycosidases and other hydrolases in plants. *Plant Physiology*, *177*(1): 24–37.

Husaini, A.M., Wani, S.A., Sofi, P., Rather, A.G., Parray, G.A. et al. (2009). Bioinformatics for saffron (Crocus sativus L.) improvement. *Communications in Biometry & Crop Science, 4*(1).

Jacobsen, N. and Ørgaard, M. (2003, October). Crocus cartwrightianus on the Attica Peninsula. In *I International Symposium on Saffron Biology and Biotechnology 650*: pp. 65–69.

Jain, M., Srivastava, P.L., Verma, M., Ghangal, R. and Garg, R. et al. (2016). De novo transcriptome assembly and comprehensive expression profiling in Crocus sativus to gain insights into apocarotenoid biosynthesis. *Scientific Reports, 6*(1): 22456.

Karamplianis, T., Tsiftsis, S. and Constantinidis, T. (2013). The genus Crocus (Iridaceae) in Greece: some noteworthy floristic records and karyotypes. *Phytologia Balcanica, 19*(1): 53–66.

Kumar, A., Devi, M. and Kumar, R. (2022). Introduction of high-value Crocus sativus (saffron) cultivation in non-traditional regions of India through ecological modelling. *Sci Rep* 12, 11925.

Kumar, A., Pathak, R.K., Gupta, S.M., Gaur, V.S. and Pandey, D. et al. (2015). Systems biology for smart crops and agricultural innovation: filling the gaps between genotype and phenotype for complex traits linked with robust agricultural productivity and sustainability. *Omics: A Journal of Integrative Biology, 19*(10): 581–601.

Kyriakoudi, A., Ordoudi, S.A., Roldán-Medina, M. and Tsimidou, M.Z. (2015). Saffron, a functional spice. *Austin J Nutri Food Sci, 3*(1): 1059.

Langmead, B. (2010). Aligning short sequencing reads with Bowtie. *Current protocols in bioinformatics, 32*(1): 11–7.

López de Maturana, E., Alonso, L., Alarcón, P., Martín-Antoniano, I.A., Pineda, S. et al. (2019). Challenges in the integration of omics and non-omics data. *Genes, 10*(3): 238.

Mahmodi, P., Moeini, A., Khayam Nekoie, S.M., Mardi, M. and Hosseini Salekdeh, G. et al. (2014). Analysis of saffron stigma (Crocus sativus L.) transcriptome using SOAPdenovo and Trinity assembly software. *Crop Biotechnology, 4*(6): 35–46.

Marco-Ramell, A., Palau-Rodriguez, M., Alay, A., Tulipani, S., Urpi-Sarda, M. et al. (2018). Evaluation and comparison of bioinformatic tools for the enrichment analysis of metabolomics data. *BMC Bioinformatics, 19*: 1–11.

Mathew, B. (1977). Crocus sativus and its allies (Iridaceae). *Plant Systematics and Evolution, 128*: 89–103.

Meirmans, P.G. and Van Tienderen, P.H. (2004). GENOTYPE and GENODIVE: two programs for the analysis of genetic diversity of asexual organisms. *Molecular ecology notes, 4*(4): 792–794.

Moratalla-López, N., Bagur, M.J., Lorenzo, C., Martínez-Navarro, M.E., Salinas, M.R. et al. (2019). Bioactivity and bioavailability of the major metabolites of Crocus sativus L. flower. *Molecules, 24*(15): 2827.

Nithya, G. and Sakthisekaran, D. (2015). *In silico* docking studies on the anti-cancer effect of thymoquinone on interaction with pten-a regulator of Pi3k/Akt pathway. *Asian J Pharma Clin Res, 1*: 192–195.

Ogbe, R.J., Ochalefu, D.O. and Olaniru, O.B. (2016). Bioinformatics advances in genomics–A review. *International Journal of Current Research and Review, 8*(10): 5.

Panchangam, S.S., Vahedi, M., Megha, M.J., Kumar, A., Raithatha, K. et al. (2016). Saffron'omics': The challenges of integrating omic technologies. *Avicenna Journal of Phytomedicine, 6*(6): 604.

Ramdas, A., Zrnic, T., Wainwright, M. and Jordan, M. (2018). SAFFRON: an adaptive algorithm for online control of the false discovery rate. In *International Conference on Machine Learning* (pp. 4286–4294). PMLR.

Rastegari, S., Naser Alavi, S.M. and Mohayeji, M. (2022). Effect of salicylic acid and pre-cold treatment on flower induction in saffron. *Scientifica, 2022*.

Rhoads, A. and Au, K.F. (2015). PacBio sequencing and its applications. *Genomics, Proteomics & Bioinformatics, 13*(5): 278–289.

Riswanto, F.D.O., Windarsih, A., Lukitaningsih, E., Rafi, M., Fadzilah, N.A. et al. (2022). Metabolite Fingerprinting Based on 1H-NMR Spectroscopy and Liquid Chromatography for the Authentication of Herbal Products. *Molecules, 27*(4): 1198.

Sathasivam, R., Kim, N.S., Choi, M., Kwon, H., Nguyen, B.V. et al. (2022). Identification, in silico characterization, and differential expression profiles of carotenoid, xanthophyll, apocarotenoid biosynthetic pathways genes, and analysis of carotenoid and xanthophyll accumulation in Heracleum moellendorffii Hance. *International Journal of Molecular Sciences*, *23*(9): 4845.

Schmidt, M., Betti, G. and Hensel, A. (2007). Saffron in phytotherapy: pharmacology and clinical uses. *Wiener Medizinische Wochenschrift (1946)*, *157*: (13-14), 315–319.

Shameer, K., Naika, M.B.N., Shafi, K.M. and Sowdhamini, R. (2019). Decoding systems biology of plant stress for sustainable agriculture development and optimized food production. Progress in Biophysics and Molecular Biology, 145: 19–39,

Singla and Bhat, A., Dadras, A., Ghalandari, B., Riazi, G.H., Modaresi, S.M.S. et al. (2015). Safranal as a novel anti-tubulin binding agent with potential use in cancer therapy: An *in vitro* study. *Chemico-Biological Interactions*, *238*: 151–160.

Tan, H., Chen, X., Liang, N., Chen, R., Chen, J. et al. (2019). Transcriptome analysis reveals novel enzymes for apo-carotenoid biosynthesis in saffron and allows construction of a pathway for crocetin synthesis in yeast. *Journal of Experimental Botany*, *70*(18): 4819–4834.

Trapero, A., Ahrazem, O., Rubio-Moraga, A., Jimeno, M.L., Gómez, M.D. et al. (2012). Characterization of a glucosyltransferase enzyme involved in the formation of kaempferol and quercetin sophorosides in Crocus sativus. *Plant Physiology*, *159*(4): 1335–1354.

Vandamme, D., Minke, B.A., Fitzmaurice, W., Kholodenko, B.N. and Kolch, W. et al. (2014). Systems biology-embedded target validation: improving efficacy in drug discovery. *Wiley Interdisciplinary Reviews: Systems Biology and Medicine*, *6*(1): 1–11.

Xu, S., Ge, X., Li, S., Guo, X., Dai, D. et al. (2019). Discrimination of different parts of saffron by metabolomic-based ultra-performance liquid chromatography coupled with high-definition mass spectrometry. *Chemistry & Biodiversity*, *16*(10): e1900363.

Yadollahi, A., Azam-Ali, S., Cocking, E. and Shojaei, Z.A. (2006). Possibility of growth and development of saffron in the UK. In *II International Symposium on Saffron Biology and Technology 739*: pp. 139–149.

Yue, J., Wang, R., Ma, X., Liu, J., Lu, X. et al. (2020). Full-length transcriptome sequencing provides insights into the evolution of apocarotenoid biosynthesis in Crocus sativus. *Computational and Structural Biotechnology Journal*, *18*: 774–783.

Zeraatkar, M., Khalili, K. and Foorginejad, A. (2015). Studying and Generation of Saffron Flower's 3D Solid Model. *Procedia Technology*, *19*: 62–66

11 Molecular Markers of Saffron

Shruti Chattaraj[1*], Tulika Mishra[2], Abdel Rahman Al-Tawaha[3], Sayanti Mandal[4] and Devendra Kumar Pandey[5*]

Saffron (*Crocus sativus* L.), the most expensive spice of the Iridaceae family, is derived from dried reddish-golden stigmas. Its genetic variation is constrained by sterility and auto-triploidy, necessitating genetic diversity analysis for crop improvement. Molecular markers are crucial tools for assessing diversity, aiding breeding programs, and conserving saffron germplasm. Molecular markers such as RAPD, SSR, ISSR, AFLP, ESTs, retrotransposons, and SRAP have proven effective in analyzing genetic diversity and characterizing *C. sativus* L. Genetic barcoding and sequencing further aid phylogenetic analysis and origin detection. These techniques also authenticate saffron, distinguishing it from adulterants, making molecular markers vital for genetic diversity analysis and saffron authentication.

1. Introduction

Saffron (*Crocus sativus* L.), the most expensive spice of the Iridaceae family, is derived from dried golden stigmas, used for aroma, flavor, and dyeing (Sharaf-Eldin et al., 2015). As an autotriploid sterile plant, it propagates vegetatively via corms, limiting genetic variation in Crocus species. Meiotic abnormalities lead to defective pollen grains, low germination, and non-functional microspores and macrospores, preventing seed and fruit formation (Chichiricco, 1984; Fernandez, 2004). The sterile nature of the plant restricts introduction of genetic variation and consequently further crop improvement through the application of conventional breeding methods. Cultivated for centuries, saffron has adapted to diverse environments, undergoing spontaneous mutations—polyploidy, somatic recombination, and point mutations like deletion, duplication, and transition. Despite its sterility, these genetic changes introduce variation. However, recent land

[1] Department of Botany, Ananda Mohan College, Kolkata, India.
[2] Department of Botany, Deen Dayal Upadhyay Gorakhpur University, Gorakhpur, Uttar Pradesh, India.
[3] Department of Biological Sciences, Al-Hussein Bin Talal University, Maan, Jordan.
[4] Department of Chemistry & Biochemistry, School of Basic Sciences & Research, Sharda University, Greater Noida 201306.
[5] Department of Biotechnology, Lovely Professional University, Punjab, India.
* Corresponding author: shrutichattaraj@gmail.com; dkpandey1974@gmail.com

DOI: 10.1201/9781003403760-11

loss and biotic stress have led to genetic erosion in saffron populations. Due to high demand, low production, and cost, saffron faces adulteration by mixing *C. sativus* stigmas with other plant parts like *Carthamus tinctorius* flowers, *Bixa orellana* seeds, Hemerocallis petals, Crocus stigmas, and *Curcuma longa* rhizome powder (Marieschi et al., 2012). This underscores the need for robust conservation efforts to safeguard genetic resources and support breeding programs. Saffron taxonomy is complex due to heterogeneity in ago-morphological and cytological data, male sterility and triploidy (2n = 3x = 24) (Rubio-Moraga et al., 2009). Studies reveal significant morphological and phytochemical variations in saffron populations, including leaf count, flower number, dry stigma weight, and metabolite levels like safranal, crocins, and picrocrocin (Baghalian et al., 2010). Some researchers have reported phenotypic variants with increased stigma numbers (Ali et al., 2013). Due to saffron's sterility, morphological and physicochemical traits are non-heritable, making DNA molecular markers a more reliable alternative for identity evaluation. Genetic variation analysis across regions has employed markers like RAPD, SSR, ISSR, AFLP, microsatellite analysis, ESTs, retrotransposons, SNPs, SRAP, and IRAP, aiding in saffron characterization and diversity assessment. Early molecular marker studies classified *C. sativus* L. as monomorphic, but later analyses confirmed polymorphism. These markers aid in saffron authentication, distinguishing genuine crop from adulterants. SCAR markers have been used for commercial saffron characterization (Marieschi et al., 2012; Babaei et al., 2014; Torelli et al., 2014). Barcoding of the trnH-psbA chloroplast DNA region enables adulterant detection (Jiang et al., 2014). Some researchers combined real-time PCR assays, including DNA mini-barcodes and HRM analysis, for species authentication (Villa et al., 2016). This chapter discusses molecular markers for saffron genetic diversity, differentiation, and authentication.

2. PCR-based Molecular Markers

Molecular markers are DNA fragments that indicate genomic polymorphism but remain phenotypically neutral. They serve as crucial tools for genetic diversity analysis and DNA polymorphism exploitation in plants and animals. Since they are unaffected by environmental conditions, they can be detected at any growth stage. Various molecular markers have been utilized over time to assess saffron's genetic variation and diversity.

2.1 RAPD and ISSR Markers

RAPD uses PCR amplification with random primers (~10 nucleotides), while ISSR targets microsatellite-flanked DNA fragments. Pardo et al. (2004) analyzed *C. sativus* L. from Italy, Greece, Iran, and Spain using 38 RAPD markers, revealing limited polymorphism due to sterility. Rubio-Moraga et al. (2009) applied RAPD, ISSR, and microsatellite markers across geographic origins, finding saffron to be monomorphic, with no polymorphic bands detected—similar to findings by Grilli-

Caiola et al. (2004). RAPD and ISSR analyses revealed 59.46% polymorphism in *C. sativus* L. from Jammu and Kashmir (Mir et al., 2021). RAPD was also used to assess polymorphism between mutant and nonmutant *C. sativus* L. genotypes in Saudi Arabia (Sharaf-Eldin et al., 2018). DNA fingerprinting of saffron stigmata showed polymorphism ranging from 33.3% to 88.2% across ten RAPD primers.

2.2 AFLP Markers

AFLP a highly sensitive DNA fingerprinting method, surpasses RAPD and ISSR in genetic diversity analysis due to its superior resolution, sensitivity, and reproducibility. Germplasm characterization of 13 saffron accessions in Jammu and Kashmir revealed 98.7% polymorphism (Gaikwad et al., 2018). Torricelli et al. (2019) assessed saffron from Iran, Italy, and Spain using eight AFLP primer combinations, reporting polymorphism rates between 0-8.51%, with an average of 4.13%. Spanish saffron accessions were well distinguished from those of Iran and Italy, consistent with AFLP studies (Siracusa et al., 2012). AFLP markers revealed close genetic relationships among saffron species (Zubor et al., 2004). The MS-AFLP method, integrating genetic and epigenetic analysis, was used for saffron adulteration detection via cytosine methylation patterns (Saffritti et al., 2016), helping identify stamen adulteration from added floral parts. Busconi et al. (2015) utilized three MS-AFLP primer combinations to analyze epigenetic variability in saffron from the World Saffron and Crocus Collection (WSCC). Further, Busconi et al. (2021) examined five saffron accessions with variations in tepal pigmentation, flowering time, and yield using methylation restriction enzyme sequencing, revealing high polymorphism levels.

2.3 SRAP Markers

SRAP, a PCR-based marker targeting open reading frames (ORFs), has proven effective in genetic diversity analysis, gene tagging, and species identification (Li & Quiros, 2001; Zhoua et al., 2010). It offers superior informativeness compared to other markers (Babaei et al., 2014). In saffron, Babaei et al. (2014) applied SRAP for genetic relationship evaluation among Iranian accessions, using 19 primer combinations that amplified 147 polymorphic fragments, revealing 48.88% polymorphism with an average PIC of 0.15. SRAP markers effectively detect polymorphism in saffron genotypes by targeting functional genome regions, offering superior genetic diversity analysis compared to other PCR-based markers (Li & Quiros, 2001). The study confirms their strong potential for saffron characterization.

2.4 Long Terminal Repeat Retrotransposons (LTR-RTNs)

Retrotransposons, abundant in plant genomes, serve as molecular markers for genetic variability and phylogenetic studies in Iranian saffron species (Alavi-Kia et al., 2008). They replicate via reverse transcription, integrating new 5-10 kb cDNA copies into the genome like retroviruses (Suoniemi et al., 1996).

LTR-RTNs contain conserved sequences useful for cloning markers and flanking regions. Replicationally active RTNs introduce polymorphism via genome insertions, detectable through PCR-based methods. Barley-derived LTR-RTN primers were used for Crocus diversity analysis, with 15 of 28 primers amplifying saffron DNA, indicating moderate homology between barley and Crocus. Genetic variation was observed within and between Crocus species, suggesting *C. almehensis* and *C. michelosnii* as possible wild ancestors of cultivated saffron (Alavi-Kia et al., 2008).

2.5 Expression Sequence Tags (ESTs)

Gene expression in saffron corms across developmental stages was analyzed using expressed sequence tags (ESTs) (Orti et al., 2004). ESTs, short DNA sequences (200-500 nucleotides), were identified from two cDNA libraries, covering genes linked to cell growth, transcription factors, protein synthesis, cell wall formation, defense, and metabolism. D'Agostino et al. (2007) developed 6,603 high-quality ESTs from saffron stigma, forming the first Iridaceae genomics database for stigma biogenesis and secondary metabolism studies. Fluch et al. (2010) explored saffron genetic diversity using EST-SSR and chloroplast markers.

2.6 DNA Barcoding

DNA barcoding enables species identification via a standardized short genetic region. In plants, nuclear ribosomal DNA ITS and chloroplast matK genes serve as key barcodes (Li & Dao, 2013; Selvaraj et al., 2008). Villa et al. (2016) applied real-time PCR with HRM analysis to develop DNA mini-barcodes for distinguishing Crocus species. ITS2 was identified as *C. sativus* L.-specific, while ITS1 and matK loci were genus-specific, making mini-barcodes with HRM analysis reliable for saffron authentication. Other DNA barcodes include RUBISCO large subunit, rbcL, and trnH-psbA (Gismondi et al., 2012). Gismondi et al. (2013) applied DNA barcoding for intraspecific and population studies of *C. sativus* L. Plastidial and nuclear barcode regions (rbcL, matK, trnH-psbA, and ITS) were used to characterize *C. sativus* L., *C. carpetanus*, and *C. serotinus* from Spain, alongside Italian *C. cartwrightianus*, *C. hadriaticus*, and *C. thomasii*. Phylogenetic analysis showed *C. sativus* L. closely related to *C. cartwrightianus*. Italian and Spanish saffron species had 100% conserved rbcL and matK sequences, with ITS and trnH-psbA diverging by 10% and 1%, respectively, highlighting DNA barcoding as a reliable tool for intraspecific identification and discrimination.

Conclusion

Genetic improvement is essential for saffron conservation, with genetic diversity being key to crop enhancement. Molecular markers like ISSR, AFLP, SSR and

EST are valuable tools for diversity analysis, aiding breeding efforts. Additionally, genetic diversity studies support resource management and conservation. DNA barcodes serve as powerful tools for saffron authentication, preventing adulteration. Molecular markers also support germplasm preservation, genetic variability analysis, molecular-assisted breeding, and genetic characterization of saffron.

References

Alavi-Kia, S.S., Mohammadi, S.A., Aharizad, S. and Moghaddam, M. 2008. Analysis of Genetic Diversity and Phylogenetic Relationships in Crocus Genus of Iran Using Inter-Retrotransposon Amplified Polymorphism. *Biotechnology & Biotechnological Equipment. 22*(3): 795–800.

Ali, G., Iqbal, A.M., Nehvi, F., Samad, S.S., Nagoo, S. et al. 2013. Prospects of clonal selection for enhancing productivity in Saffron (*Crocus sativus* L.). *Afr. J. Agric. Res. 8*: 460–467.

Babaei, S., Talebi, M. and Bahar, M. 2014. Developing an SCAR and ITS reliable multiplex PCR-based assay for safflower adulterant detection in saffron samples. *Food Control. 35*(1): 323–328. http://doi.org/10.1016/j.foodcont.2013.07.019

Babaei, S., Talebi, M., Bahar, M. and Zeinali, H. 2014. Analysis of genetic diversity among saffron (*Crocus sativus*) accessions from different regions of Iran as revealed by SRAP markers. *Sci Hortic. 171*: 27–31.

Baghalian, K., Shabani, S.M. and Jamshidi, A.H. 2010. Genetic variation and heritability of agro-morphological and phytochemical traits in Iranian saffron (*Crocus sativus* L.) populations. *Ind. Crops Products. 31*: 401–406.

Busconi, M., Colli, L., Sánchez, R.A., Santaella, M., De Los Mozos Pascual et al. 2015. AFLP and MS-AFLP analysis of the variation within saffron crocus (*Crocus sativus* L.) germplasm. *PLoS One* 10:e0123434.

Busconi, M., Wischnitzki, E., Del Corvo, M., Colli, L., Soffritti, G. et al. 2021. Epigenetic variability among saffron crocus (*Crocus sativus* L.) accessions characterized by different phenotypes. *Front. Plant Sci. 12*: 642631.

Chichiricco, G. 1984. Karyotype and meiotic behavior of the triploid *Crocus sativus* L. *Caryologia. 37*(3): 233–239. https://doi.org/10.1080/00087114.1984.10797702.

D'Agostino, N., Pizzichini, D., Chiusano, M.L. and Giuliano, G. 2007. An EST database from saffron stigmas. *BMC Plant Biology. 7*(1): 53. doi:10.1186/1471-2229-7-53

Fernandez, J.A. 2004. Biology, biotechnology and biomedicine of Saffron. Recent Res. Devel. *Plant Sci. 2*: 127–159.

Fluch, S., Hohl, K., Stierschneider, M., Kopecky, D. and Kaar, B. et al. (2010). *Crocus sativus L.*- molecular evidence on its clonal origin. *Acta Horticulturae. 850*: 41–46.

Gaikwad, A.B., Saxena, S. and Archak, S. 2018. Development of AFLP markers and molecular diversity analysis in saffron (*Crocus sativus L.). Acta Horticulturae. 1200*: 1–6. doi:10.17660/actahortic.2018.1200

Gismondi, A., Fanali, F., Labarga, J.M.M., Grilli Caiola, M. and Canini, A. et al. 2013. *Crocus sativus* L. genomics and different DNA barcode applications. *Plant Syst. Evol. 299*(10): 1859–1863.

Gismondi, A., Rolfo, M.F., Leonardi, D., Rickards, O. and Canini, A. et al. 2012. Identification of ancient Olea europaea L. and Cornus mas L. seeds by DNA barcode. *C R Biologies 335*: 472–479.

Grilli-Caiola, M., Caputo, P. and Zanier, R. 2004. RAPD analysis in *Crocus sativus* L. accessions and related *Crocus* species. *Biologia Plantarum. 48*: 375–380.

Jiang, C., Cao, L., Yuan, Y., Chen, M., Jin, Y. et al. 2014. Barcoding Melting Curve Analysis for Rapid, Sensitive, and Discriminating Authentication of Saffron (*Crocus sativus* L.) from Its Adulterants. *BioMed Research International*. http://doi.org/10.1155/2014/809037

Li, G. and Quiros, C.F. 2001. Sequence-related amplified polymorphism (SRAP) a new marker system based on a simple PCR reaction: its application to mapping and gene tagging in Brassica. *Theor. Appl. Genet. 103*: 455–461.

Li, R. and Dao, Z. 2013. Identification of *Meconopsis* species by a DNA barcode sequence: the nuclear internal transcribed spacer (ITS) region of ribosomal deoxyribonucleic acid (DNA). *African Journal of Biotechnology. 10*: 15805–15807.

Marieschi, M., Torelli, A. and Bruni, R. 2012. Quality Control of Saffron (*Crocus sativus* L.): Development of SCAR Markers for the Detection of Plant Adulterants Used as Bulking Agents. *Journal of Agricultural and Food Chemistry. 60*: 10998–11004.

Mir, M.A., Mansoor, S., Sugapriya, M., Alyemeni, M.N., Wijaya, L. et al. (2021). Deciphering genetic diversity analysis of saffron (*Crocus sativus* L.) using RAPD and ISSR markers. *Saudi Journal of Biological Sciences. 28*(2): 1308–1317. doi:10.1016/j.sjbs.2020.11.063

Ortí, M.A., Gómez, L.G., Rubio, A., Escribano, J., Pardo, J., 2004. Development and gene expression in saffron corms. Acta horticulturae. *650*: 141–153. Doi:10.17660/actahortic.2004.650.

Pardo, J., Fernández, J.A. and Gómez, L.G. 2004. Development of molecular markers for origin determination in saffron. *Acta Horticulturae. 650*: 95–98. doi:10.17660/actahortic.2004.650

Rubio-Moraga, A., Castillo-López, R., Gómez-Gómez, L. and Ahrazem, O. 2009. Saffron is a monomorphic species as revealed by RAPD, ISSR and microsatellite analyses. *BMC Research Notes. 2*: 189. https://doi. org/10.1186/1756-0500-2-189.

Selvaraj, D., Sarma, R.K. and Sathishkumar, R. 2008. Phylogenetic analysis of chloroplast *matK* gene from *Zingiberaceae* for plant DNA barcoding. *Bioinformation. 3*: 24–27.

Sharaf-Eldin, M., Elkholy, S., Hosokawa, M., Yanagawa, K., Nawata, E. et al. 2015. Saffron flowers with augmented number of stigmata. *Zeitschrift für Arznei-& Gewürzpflanzen. 20*: 84–87.

Sharaf-Eldin, M.A., Alam, P. and Elkholy, S.F. 2018. Molecular and chemical characterization of mutant and nonmutant genotypes of saffron grown in Saudi Arabia. *Food Science & Nutrition.* 2018: 1–9.

Siracusa, L., Gresta, F., Avola, G., Albertini, E., Raggi, L. et al., 2012. Agronomic, chemical and genetic variability of saffron (*Crocus sativus* L.) of different origin by LC-UV–vis-DAD and AFLP analyses. *Genetic Resources and Crop Evolution. 60*(2): 711–721. doi:10.1007/s10722-012-9868-9

Soffritti, G., Busconi, M., Sánchez, R., Thiercelin, J.M., Polissiou, M. et al., (2016). Genetic and Epigenetic Approaches for the Possible Detection of Adulteration and Auto-Adulteration in Saffron *(Crocus sativus L.)* Spice. *Molecules. 21*(3): 343.

Suoniemi, A., Anamthawat-Jónsson, K., Arna, T. and Schulman, A.H. 1996. Retrotransposon BARE-1 is a major, dispersed component of the barley (Hordeum vulgare L.) genome. *Plant Mol. Biol. 30*: 1321–1329. https://doi.org/10.1007/BF00019563

Torelli, A., Marieschi, M. and Bruni, R. 2014. Authentication of saffron (*Crocus sativus* L.) in different processed, retail products by means of SCAR markers. *Food Control. 36*(1): 126–131. http://doi.org/10.1016/j.foodcont.2013.08.001

Torricelli, R., Yousefi, J., Albertini, E., Venanzoni, R. and Hosseinzadeh, Y. et al. 2019. Morphological and molecular characterization of Italian, Iranian and Spanish saffron (*Crocus sativus* L.) accessions. *Appl. Ecol. Environ. Res. 17*: 1875–1887.

Villa, C., Costa, J., Meira, L., Oliveira, M.B.P.P. and Mafra, I. et al. 2016. Exploiting DNA mini-barcodes as molecular markers to authenticate saffron (Crocus sativus L.). *Food Control. 65*: 21–31. http://doi.org/10.1016/j.foodcont.2016.01.008

Zhoua, Y., Gub, F., Zhoua, C., Yaoa, H., Duana, H. et al., 2010. Genetic diversity of Rehmannia glutinosa cultivars based on sequencerelated amplified polymorphism markers. *Sci. Hortic. 125*: 789–794.

Zubor, A.A., Suranyi, G., Gyori, Z., Borbely, G. and Prokisch, J. et al. 2004. Molecular biological approach of the systematic of Crocus sativus L. and its allies. In, Abdullaev F (ed). *First International Symposium on Saffron Biology and Biotechnology.* pp. 85–93.

12 Genome Editing in Saffron

Saba Yaseen[1] and Zulqurnain Khan[1*]

Saffron (*Crocus sativus* L.), known as "red gold," is among the world's most expensive spices, cultivated in Iran, Spain, India, and Greece. Its production is labor-intensive and costly, with yield and quality affected by drought, cold, and disease. Genome editing, particularly CRISPR-Cas9, offers precise DNA modifications to enhance stress tolerance, fragrance, and taste, potentially improving saffron's resilience to environmental challenges. Drought and cold stresshinder saffron yield. Introducing stress-tolerant genes like DREB could enhance resilience. Genome editing may improve saffron's taste and aroma by modifying genes involved in safranal, picrocrocin, and crocin biosynthesis. Additionally, altering flowering-related genes can boost bloom numbers, increasing overall yield. Modifying FLC genes can enhance saffron yield by promoting early blooming and increased flower production. Genome editing also improves stress tolerance, aroma, and taste. While still in early stages, further research is needed to evaluate risks and benefits, but genetic modification holds promise for saffron preservation and enhancement.

1. Clustered Regularly Interspaced Short Palindromic Repeats (CRISPR)

CRISPR (Clustered Regularly Interspaced Short Palindromic Repeats) revolutionized genome editing, originating from bacterial immunity against viruses. Its applications now extend across various organisms, including plants, animals, and humans (Haroon et al., 2020). CRISPR consists of guide RNA (gRNA) and CRISPR-associated (Cas) proteins. gRNA directs Cas to specific genome sites, where Cas enzymes cut DNA precisely. This system's key advantage is its specificity, allowing targeted genome modification (Prajapat et al., 2021). CRISPR enables precise genetic modifications, aiding disease correction and trait enhancement. It has advanced gene therapies for conditions like beta-thalassemia and sickle cell anemia (Hutin & Szurek, 2015). In agriculture, it enhances yield, nutrition, and pest resistance (Li et al., 2022). In biotechnology, CRISPR facilitates novel material synthesis and innovative drug development. CRISPR technology, despite its benefits, presents ethical concerns, including unintended genomic

[1] Institute of Plant Breeding and Biotechnology, MNS University of Agriculture, Multan-Pakistan.
* Corresponding author: zulqurnain.khan@mnsuam.edu.pk

DOI: 10.1201/9781003403760-12

modifications (Lee et al., 2019). Its potential for non-therapeutic applications, such as enhancing cognitive or physical traits, raises moral and socioeconomic issues. Various nations have implemented regulations to address these challenges (Jung & Altpeter, 2016). The NIH has established guidelines for genome editing in human research, while the European Court of Justice mandates CRISPR regulations align with GMO laws in the EU. China has adopted more lenient policies (Kun et al., 2020). Despite challenges, CRISPR holds promise for advancing medicine, agriculture, and biotechnology (Demorest et al., 2016). As research progresses, its impact on genetic disorders, sustainable agriculture, and material synthesis could be transformative (Haroon et al., 2020).

1.1 CRISPR/Cas9 Action Mechanism

CRISPR/Cas9, a transformative genome editing tool originates from bacterial immunity against viruses and is widely applied in plants, animals, and humans (Kohara & Akiyama, 1987). It comprises Cas9, an RNA-guided endonuclease, and guide RNA (gRNA), enabling precise DNA cleavage. Cas9's HNH and RuvC-like domains, derived from Streptococcus pyogenes, cleave complementary and non-complementary DNA strands, respectively (Marraffini & Sontheimer, 2010).

Guide RNA (gRNA) directs Cas9 to a complementary DNA sequence, activating the enzyme for precise cleavage (Fischer et al., 2012). CRISPR/Cas9 induces double-strand breaks (DSBs), repaired via homology-directed repair (HDR) or non-homologous end joining (NHEJ) (Sapranauskas et al., 2011). NHEJ, an imprecise repair process, can cause small indels, leading to gene knockouts or functional modifications (Le Rhun & Charpentier, 2019). HDR repairs DSBs using a template DNA, enabling precise genome modifications like gene insertion or mutation correction (Gupta et al., 2021). CRISPR/Cas9 surpasses earlier editing techniques due to its specificity—gRNA directs edits to exact sequences (Court et al., 2013), aiding disease correction and trait development (Bao et al., 2019). Its efficiency allows rapid, targeted genome alterations with broad applications in biotechnology, agriculture, and medicine (Prajapat et al., 2021). Despite its advantages, CRISPR/Cas9 raises ethical concerns, including unintended genomic modifications (Bao et al., 2019). Its potential use for non-therapeutic enhancements could lead to moral and socioeconomic issues (Koonin & Zhang, 2017). Various countries have implemented laws to regulate its application. The NIH has established genome editing guidelines for human research in the U.S. (Kohara & Akiyama, 1987). The EU regulates genome editing under GMO laws, while China has adopted more lenient policies. Despite challenges, CRISPR/Cas9 holds vast potential, with ongoing refinements expected to enhance its capabilities (Van der Oost & Jore, 2009).

CRISPR-Cas9 Systems and Genome Editing

CRISPR/Cas9 has revolutionized genome editing, enabling precise DNA manipulation. Cas9, an RNA-guided endonuclease, cleaves DNA at target sites dictated by guide RNA (gRNA) sequences (Shan et al., 2018). Modifying gRNA allows researchers to induce targeted mutations across the genome (Shan, 2015). CRISPR/Cas9 enables precise correction of disease-causing mutations, offering potential treatments for genetic disorders like beta-thalassemia and sickle cell anemia (Li et al., 2018). Edited cells successfully produced functional hemoglobin, highlighting the technique's future therapeutic promise. CRISPR/Cas9 has broad applications in biotechnology and agriculture beyond human gene therapy. It enables the development of pest- and disease-resistant crops with improved yields (Nekrasov et al., 2017) and facilitates microbial engineering for biofuel production and drug discovery. However, concerns remain, particularly regarding off-target effects, which can cause unintended mutations (Clasen et al., 2016). Ongoing research aims to enhance Cas9 specificity to mitigate such risks.

CRISPR/Cas9 raises concerns about its potential use in eugenics or genetic enhancement. Additionally, unintended impacts on ecosystems and biodiversity, such as GMO creation, remain a point of Academics and policymakers are developing CRISPR/Cas9 guidelines to ensure safe gene therapy and regulate GMOs in biotechnology and agriculture (Clasen et al., 2016). Despite ethical and safety concerns (Liang et al., 2018), ongoing efforts aim to establish regulations for responsible use. As the technology advances, its role in solving global challenges is expected to expand.

CRISPR Cas9 Technique Used in Genome Editing

CRISPR/Cas9 has revolutionized molecular biology, enabling precise DNA modifications in living cells for advancements in medicine, food, and industry (Hutin & Szurek, 2015). The Cas9 enzyme, guided by a single guide RNA (sgRNA), targets specific genome sites, cleaves DNA, and utilizes cellular repair mechanisms to introduce changes, such as replacing faulty genes (Jung & Altpeter, 2016). CRISPR/Cas9 is a promising tool for human gene therapy, enabling targeted correction of mutations linked to genetic diseases (Li et al., 2022). It has been used to fix mutations causing beta-thalassemia and sickle cell anemia, restoring haemoglobin production and offering potential cures for inherited disorders. CRISPR/Cas9 enhances agriculture by improving crop traits such as pest resistance, disease tolerance, and yield. It has been used to develop powdery mildew-resistant crops (Prajapat et al., 2021) and improve nutritional content, flavor, and shelf life (Demorest et al., 2016). In biotechnology, it aids biofuel production, drug discovery, and fundamental research on gene function and disease mechanisms (Prajapat et al., 2021).

Despite its potential, CRISPR/Cas9 raises safety and ethical concerns, particularly off-target effects, which may lead to unintended mutations (Hutin & Szurek, 2015). Efforts are ongoing to enhance Cas9 specificity. Ethical risks include possible misuse for eugenics and genetic enhancement, along with ecological uncertainties linked to GMOs (Lee et al., 2019). Academics and policymakers are establishing CRISPR/Cas9 guidelines to ensure safety in gene therapy and regulate GMOs in biotechnology and agriculture (Hutin & Szurek, 2015). Despite ethical concerns, efforts continue to refine regulations for responsible use. As its applications expand, CRISPR/Cas9 is expected to play a key role in addressing global challenges (Kun et al., 2020).

Table 1 Successful Reports of Gene Editing in Various Crop Plants.

Crop species	*Gene editor*	*Target gene*	*DNA repair type*	*Target trait*	*Reference*
Rice	*CRISPR/Cas9*	*OsSWEET13*	*NHEJ*	Sugar transporter	(Romero & Gatica-Arias, 2019)
Rice	*CRISPR/Cas9*	*Gn1a, GS3, DEP1*	*NHEJ*	Enhanced grain number, larger grain size and dense erect panicles	(Romero & Gatica-Arias, 2019)
Wheat	*CRISPR/Cas9*	*GW2*	*NHEJ*	Increased grain weight and protein content	(Romero & Gatica-Arias, 2019)
Camelina sativa	*CRISPR/Cas9*	*FAD2*	*NHEJ*	Decreased polyunsaturated fatty acids	(Haun et al., 2014)
Rice	*CRISPR/Cas9*	*SBEIIb*	*NHEJ*	High amylose content	(Yongwei Sun et al., 2017)
Maize	*CRISPR/Cas9*	*ZmMYB14*	*NHEJ*	Stress tolerance regulator	(Li & Ma, 2019)
Potato	*CRISPR/Cas9*	*Wx1*	*NHEJ*	High amylopectin content	(Li et al., 2019)
Wheat	*CRISPR/Cas9*	*EDR1*	*NHEJ*	Powdery mildew resistance	(Yin & Qiu, 2019)
Rice	*CRISPR/Cas9*	*OsERF922*	*NHEJ*	Enhanced rice blast resistance	(Wang et al., 2016)
Rice	*CRISPR/Cas9*	*OsSWEET13*	*NHEJ*	Bacterial blight resistance	(Zafar et al., 2020)
Tomato	*CRISPR/Cas9*	*SlMLO1*	*NHEJ*	Powdery mildew resistance	(Prajapati & Nain, 2021)

CRISPR/Cas9 has successfully enhanced disease resistance, stress tolerance, yield, and quality in various crops through targeted gene mutations. As technology progresses, gene editing is expected to further improve agricultural performance and sustainability.

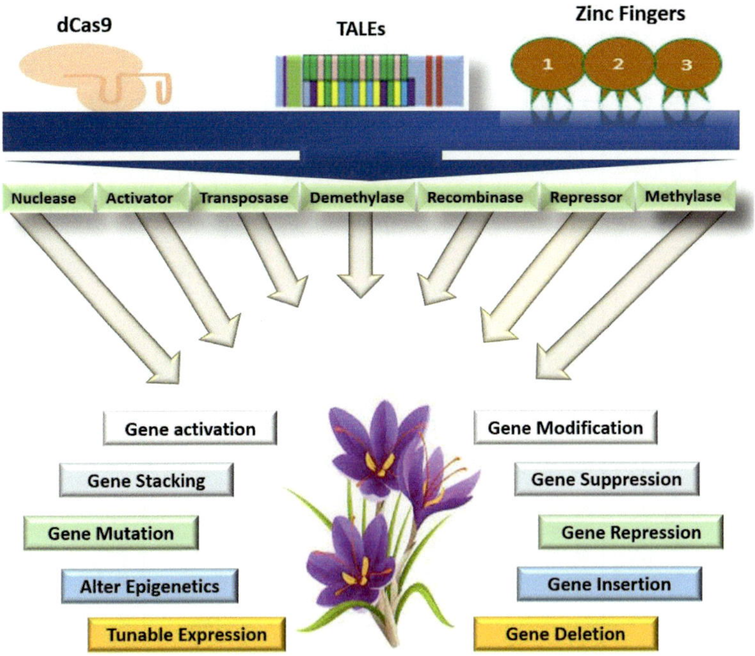

Fig. 1 Challenges in using the CRISPR-Cas system for genome editing in plants.

1.2 CRISPR for Genetic Improvements of Plants

Research for Gene Functions

Global agriculture faces challenges from biotic and abiotic factors, requiring innovations to improve yield, quality, and resilience amid population growth and environmental concerns (Popp & Nagy, 2013). CRISPR/Cas9 offers a powerful genetic tool for enhancing crop traits, with potential applications in plant breeding (Gao et al., 2018).

2. Gene Modifications for Yield Traits

Seasonal variations in day length affect crop blooming, limiting their cultivation regions. CRISPR/Cas9 has successfully modified flowering alleles, such as FLOWER Genomic and SELF-PRUNING 5G, significantly altering bloom duration in soybeans and tomatoes (Kramnik & Bloom, 1998; Soyk et al., 2017). CRISPR/Cas9 has significantly advanced rice breeding by generating 11 new TGMS lines within a year, enhancing high-yield production. It successfully modified yield-related genes—DEP1, Gn1a, IPA1, and GS3—boosting grain number and seed production per panicle (Zhou et al., 2014; Massel et al., 2002).

Table 2 Applications of CRISPR in medicinal plants.

Species	*Target gene*	*Gene description*	*Cas9/sgRNA promoter*	*Results*	*Mutation frequency*	*Reference*
Salvia miltiorrhiza	*SmCPS1*	Committed diterpene synthase gene in tanshinone biosynthetic pathway	CaMV 35S/AtU6-26	8 heterozygous and 3 homozygous hairy root mutants	11.5% and 30.8% for the homozygous and chimeric mutants	(Bai et al., 2018)
Salvia miltiorrhiza	*SmRAS*	Rosmarinic acid synthase gene in phenolic acid biosynthetic pathway	CaMV 35S/AtU6-26, OsU3	5 biallelic, 2 heterozygous and 1 homozygous hairy root mutants	50%	(Di et al., 2013)
Salvia miltiorrhiza	*SmLACs*	Laccase genes in phenolic acid and lignin biosynthetic pathway	AtUBQ/AtU6	15 single-locus crispr lines and 14 dual-locus crispr lines	90.6%	(Q. Li et al., 2019)
Salvia miltiorrhiza	*SmbZIP2*	Basic leucine zipper transcription factor; negative regulator in phenolic acid biosynthetic pathway	CaMV 35S/AtU6-26	SmbZIP2-deficient hairy roots	12%	(Shi & Kai, 2021)
Dendrobium officinale	*C3H, C4H, 4CL, CCR, and IRX*	Involved in lignocellulose biosynthetic pathway	MtHP, CVMV, MMV, PCISV, CaMV 35S / OsU3	DoLACs-deficient hairy roots	16.7%, 20%, 33.3%, 33.3% and 6.7% for C3H, C4H, 4CL, CCR and IRX	(Kui et al., 2017)
Dendrobium Chao Praya Smile	*DOTFL1*	Terminal flower 1 gene modulating flowering and inflorescence architecture	Ubi/OsU3, OsU6a	13 homozygous mutant plants	10.1%	(Y. Li, et al., 2021)
Cannabis sativa	*CsPDS1*	Phytoene desaturase gene; Marker	CaMV 35S/AtU6	CsPDS1-deficient seedlings	2.5% and 51.6% for the homozygous and chimeric mutants	(X. Zhang et al., 2021)

Gene Modifications to Improve the Quality of Products

CRISPR/Cas9 efficiently improved rice and potato quality by modifying key starch-related genes. It removed the dominant Waxy gene in rice, reducing amylose and increasing glutinosity (Sergio & Newton, 2003). It disrupted the GBSS gene in tetraploid potatoes, lowering amylose content and increasing amylopectin (Gudmundsson & Eliasson, 1990). Additionally, it reduced cadmium accumulation in rice by deleting the OsNramp5 gene, mitigating health risks (Tang et al., 2017).

Genome Modifications to Improve Abiotic Stress Resistance

CRISPR/Cas9 identified SlMAPK3, a key gene in tomato drought defense, and SAPK2, which regulates stress responses in rice (Lou et al., 2017).

CRISPR/Cas9 was used to generate cbf1, 3 dual and cbf1, 2, 3 triple mutants in Arabidopsis, revealing CBFs' role in cold stress response. Genetic markers could help develop more resilient crops, as demonstrated in drought-resistant maize (Shi et al.). Maize produces ARGOS8, an ethylene response modulator linked to stress adaptation (Shinwari & Yamaguchi-Shinozaki, 2020).

Genome Modification to Develop Biotic Stress Resistance

EDR1 reduces powdery mildew sensitivity in Arabidopsis. In wheat (Triticum aestivum), CRISPR/Cas9 removed three EDR1 homologs, increasing Taedr1 susceptibility to the disease (Zhang et al., 2017). CRISPR/Cas9 enhanced powdery mildew resistance in wheat and tomato by modifying the MLO gene (Arora & Narula, 2017). It also edited the CsLOB1 gene in grapefruit, creating canker-resistant citrus varieties (Wang et al., 2018). CRISPR/Cas9 conferred virus resistance by targeting eIF genes in cucumber and Arabidopsis (Makarova & Kalinina, 2018). It also removed eIF4G, which controls RTSV susceptibility, creating RTSV-resistant rice cultivars (Macovei et al., 2018).

3. Gene Transcription or Translation Regulation

Gene expression regulation is key to plant breeding and phenotypic diversity (Xu & Crouch, 2008). Studies link cis-regulatory elements in gene activation loops to domestication and selection in agricultural species. CRISPR/Cas9 modified three genes in tomatoes, influencing plant structure, inflorescence branching, and fruit size. Promoter changes increased trans-regulatory gene variability, resulting in shorter plants, altered leaf color, and wider tiller angles under field conditions (Yonghao Sun et al., 2020).

Mutant Libraries Construction

Evaluating gene functions in a sequenced plant genome is complex. A mutant genetic library helps address this challenge. CRISPR/Cas9 refinement, via 20-bp target-binding sgRNA modifications, enhances precise gene targeting (Tarasava, 2018). CRISPR/Cas9 enables cost-effective genome-wide modifications and genetic screening. Tomato mutant strains were generated using pooled sgRNA libraries (Lachheb et al., 2022), identifying an Arabidopsis boron transporter and an immunity-related gene (Jacobs & Martin, 2017). Additionally, rice mutant libraries created via sgRNA transformation yielded numerous loss-of-function mutations (Meng et al., 2017).

Generating Transgene-free and Genome-edited Crops

CRISPR/Cas9 enables transgene-free genome editing in crops, offering precision and efficiency. Despite challenges in distinguishing edited plants from spontaneous mutants (Wolt et al., 2016), mutant screening remains feasible (Husaini et al., 2009). Industrial applications, including waxy maize and anti-browning Agaricus bisporus, have met biosafety standards in the US (Gao et al., 2018).

4. Genome Editing in Saffron

Saffron (*Crocus sativus* L.), a rare triploid spice from the Iridaceae family, is valued for its dried red trifid stigma, widely used in cuisine (Chib et al., 2020). This perennial plant grows from underground corms, blooming in autumn, going dormant in summer, and shedding leaves in spring. Due to its triploid nature, it lacks seed production and is replanted every 3–10 years based on regional practices (Taheri-Dehkordi et al., 2020). Saffron faces environmental challenges due to its clonal structure, limiting breeding efforts due to its biological cycle, corm dependency, and high field care costs (Kashtwari et al., 2022). Mutagenesis using colchicine, x-rays, gamma rays, EMS, and ethidium bromide has produced variations, including Shalimar Saffron-1, a high-density cultivar with 14.5 kg ha^{-1} yield (Chib et al., 2020). *In-vitro* mutagenesis offers a promising alternative to vegetative propagation for saffron. Genome sequencing is crucial for advancing crop modification techniques, including TILLING and CRISPR/Cas gene editing (Basso et al., 2021). Enhancing saffron research requires specialist labs, efficient regeneration methods, and high-throughput cultivation systems (Bao et al., 2019; Taheri-Dehkordi et al., 2020). The release of the saffron transcriptome enables genetic advancements, with CRISPR expanding gene-editing possibilities for future crop development (Mirghasempour et al., 2022). Saffron (*Crocus sativus* L.), a rare triploid spice from the Iridaceae family, is prized for its dried red trifid stigma, a key culinary ingredient (Chib et al., 2020). This perennial plant, growing from underground corms, blooms in autumn, goes dormant in summer, and sheds leaves in spring. Due to its triploid nature, it lacks seed production and is replanted every 3–10 years based on regional practices (Taheri-Dehkordi et al., 2020). Saffron faces environmental stress due to its clonal structure, limiting breeding efforts given its biological cycle, corm dependency, and high field costs (Kashtwari et al., 2022). Mutagenesis using colchicine, x-rays, gamma rays, EMS, and ethidium bromide has produced variations, including Shalimar Saffron-1, a high-density cultivar yielding 14.5 kg ha^{-1} (Chib et al., 2020). *In-vitro* mutagenesis is a promising alternative to vegetative propagation in saffron. Genome sequencing is essential for advancing crop modification, enabling TILLING and CRISPR/Cas technologies (Basso et al., 2021). Genetic research requires specialist labs, efficient regeneration, and high-throughput cultivation (Bao et al., 2019; Taheri-Dehkordi et al., 2020). The saffron transcriptome supports these advancements, with CRISPR expanding possibilities for future crop improvements (Mirghasempour et al., 2022).

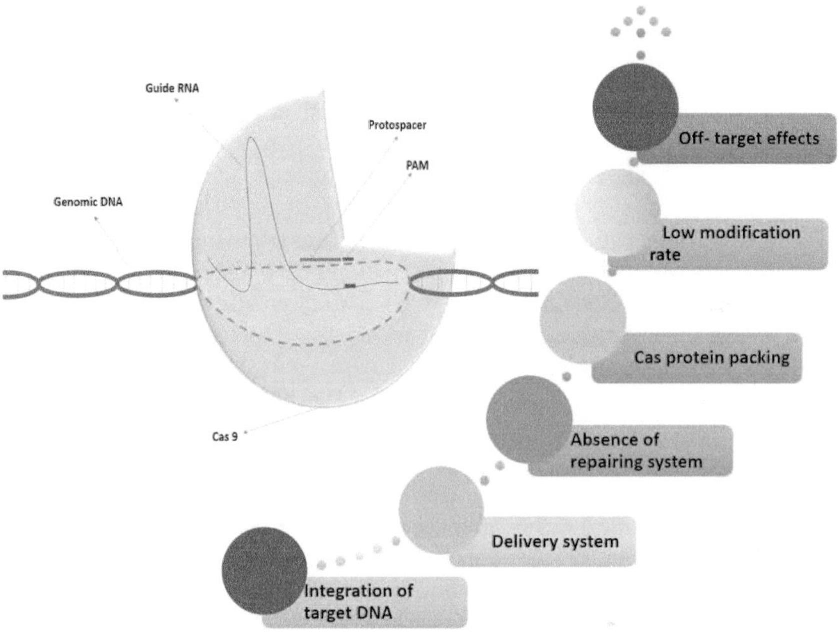

Fig. 2 Genome editing toolkit along with various effector domains for gene editing in saffron.

5. Saffron Genome Editing Through CRISPR

CRISPR enables precise genome editing, allowing targeted DNA modifications in cells. Saffron (*Crocus sativus*), a valuable spice, is widely used in cooking and traditional medicine (Kashtwari et al., 2022). CRISPR-Cas9 for genome editing in saffron begins with designing a guide RNA (gRNA) to direct Cas9 to the target DNA site for modification (Liang et al., 2017). Aviral vector or other delivery method is used to introduce gRNA and Cas9 into target saffron cells for genome editing.

Once inside the cell, gRNA guides Cas9 to the target DNA site, where Cas9 cleaves both strands (Basso et al., 2021). The cell's repair mechanisms then mend the break, enabling base pair addition, deletion, or replacement (Bao et al., 2019). Cell repair mechanisms mend double-stranded breaks, enabling base pair addition, deletion, or replacement when executed effectively (Bao et al., 2019). CRISPR-

Cas9 enables precise saffron genome editing, improving yield, pigment content, pest resistance, and quality. Altered cells undergo verification to ensure targeted modifications. Despite ethical concerns, genome editing holds immense potential for saffron enhancement (Bao et al., 2019).

5.1 Role of Genome Editing in Saffron

Saffron depends on biotic and abiotic factors for growth, including soil microbes, water, temperature, and nutrients (Keify et al., 2012). Environmental degradation alters these conditions, affecting production (Husaini et al., 2022). To thrive under climate change, saffron must develop heat tolerance, water retention, and resistance to stressors (Li et al., 2018). Synthetic biology offers solutions by reorganizing biological systems for enhanced resilience (Vakhlu et al., 2022).

Research has identified genes that enhance saffron's environmental resilience. Over 30 years, gene silencing and genome editing have refined agricultural traits (Wani et al., 2016). Genome editing (GE) enables precise DNA modifications using programmable nucleases, praised for precision and efficiency (Bhagat et al., 2022; Martignago & Caño-Delgado, 2020). DSBs, induced by specialized molecular scissors, facilitate trait enhancement (Vakhlu et al., 2022). DSB repair occurs via HDR (high accuracy, low efficiency) or NHEJ (higher efficiency, more errors) (Bhagat et al., 2022). Zinc finger nucleases, explored as an alternative to Agrobacterium-mediated genetic modification, have drawbacks like cost and preparation time, while TALEN offers an effective solution (Kohlen et al., 2012). CRISPR/Cas9, with RNA-directed targeting, surpasses protein-based methods in genome editing efficiency (Zouagui et al., 2023). Edited monocots include wheat, rice, barley, maize, and bananas, while dicots feature Arabidopsis, tomato, soybean, potato, cotton, citrus, cucumber, and grapes (Li et al., 2018).

CRISPR has been applied to Arabidopsis, tobacco, and rice, advancing stress-tolerant crop development. Efforts continue to genetically modify saffron to withstand biotic and abiotic stresses, including lethal fungal infections (Chamkhi et al., 2023). Agrobacterium rhizogenes ARqua1 enabled efficient *Crocus sativus* transformation, while Cas9 integration in Crocus calli was confirmed (Sharma et al., 2021; Chib et al., 2020). CRISPR/Cas9 is widely explored for enhancing stress resilience across crops, with potential for adaptive saffron varieties (Zouagui et al., 2023). Genome reprogramming can boost stress tolerance by targeting relevant genes (Husaini et al., 2010).

Even while greenhouse research and field trials may give enough information for the efficient regulation of stressors within the framework of GE technology, there are still certain basic issues that must be resolved before experiments are further examined (Ali et al., 2023). Problems with saffron's GE techniques are probably caused by inadequate understanding of the genes encoding important agronomic properties and the complexity of the many biochemical processes they are connected to during genome editing (Li et al., 2012). These crucial issues must be fully resolved before to starting the saffron gene-editing process.

Gene Edited by CRISPR/Cas9 in Saffron

CRISPR-Cas9 is being applied to identify and modify saffron genes for yield improvement. Key targets include genes regulating flowering time, stigma development, pigment synthesis, and stress response.

Flowering Time Genes

Saffron flowering depends on temperature and light, significantly impacting yield. Editing FT and GI genes can optimize blooming time to enhance production (Salami et al., 2022).

Stigma Development Genes

Saffron yield depends on stigma count per bloom. Editing MADS-box genes, which regulate stigma formation, can enhance flower productivity and increase yield (Hao et al., 2022).

Pigment Genes

Saffron quality and yield depend on pigment quantity and composition. Editing CHS and CHI genes, which regulate pigment biosynthesis, can enhance pigment content and yield (Mansotra et al., 2022).

Stress Response Genes

Saffron faces environmental stressors like drought, heat, and disease. Editing CBF and DREB genes can enhance stress tolerance and improve yield (Salami et al., 2022). Saffron yield can be enhanced by genome editing, with target genes selected based on project goals and varietal traits (Kohlen et al., 2012). However, factors like soil quality, water availability, and climate also influence production (Hao et al., 2022).

Applications of Genome Editing in Saffron

1. Low saffron yield is a key challenge. Genome editing can enhance stigma and flower production, boosting overall yield (Ali et al., 2023).
2. Crocin, a prized crimson pigment found in saffron, can be enhanced via genome editing, boosting stigma content, spice value, and yield (Chamkhi et al., 2023). Saffron is vulnerable to diseases and pests, affecting quality and yield. Genome editing can enhance resistance, improving sustainability and productivity (Niaz et al., 2023). Genome editing can introduce traits like enhanced environmental resistance, improved flavor, and stronger fragrance in saffron (Nemati et al., 2012).
3. Genome editing can enhance saffron quality by improving color, scent, and flavor, optimizing its desirable traits (Ramachandra et al., 2023).
4. Genome editing offers a promising approach for enhancing saffron production by introducing novel traits and increasing yields (Torricelli et al., 2019). Despite ethical challenges, its diverse applications highlight significant potential for saffron's future (Mansotra et al., 2022).

5.2 Diversity of Saffron and its Characterization

Despite advances in sequencing technology, no Crocus species of Crocus currently has a complete genome sequence. Saffron ($2n = 3x = 24$) is a triploid with eight triplets (Kumar et al., 2023). Studies using RAPD, IRAP, AFLP, and SSR markers have improved genetic insights, with barcoding confirming diversity among 86 Crocus species (Kohlen et al., 2012). RAPD and ISSR marker analysis of 43 isolates revealed high genetic variation in elite Kashmir saffron clones, while ISSR markers indicated molecular similarity between *C. sativus* and *C. cartwrightianus* (Niaz et al., 2023). DNA methylation, a key epigenetic marker, drives inheritable variation in plants. MS-AFLP analysis of 100+ saffron accessions from WSCC revealed high epigenetic variability despite low genetic diversity (Liang et al., 2022). Epigenetics likely contributes to phenotypic persistence in saffron (Ramachandra et al., 2023). AFLP and MRE-seq identified differential methylation in transcription factor genes, impacting disease susceptibility. A heatmap of 50 polymorphic GOs showed lower polymorphism in Spanish vs. Indian saffron (Liang et al., 2022).

C. sativus shows strong genetic links to *C. cartwrightianus*, per phylogenetic studies on nuclear loci, chloroplast genomes, and DNA polymorphism (Uranga et al., 2021). Genome sequencing and FISH suggest C. cartwrightianus may be saffron's ancestor, though C. pallasii subsp. Pallasii is also a potential progenitor (Kohlen et al., 2012).

5.3 The Medicinal Properties of Saffron

Antidepressant Properties

Saffron, used for centuries in depression treatment, has shown antidepressant effects comparable to prescription drugs in clinical studies (Rizzo et al., 2023). Further research is needed to optimize its use and understand underlying mechanisms (Kumar et al., 2023).

Anti-inflammatory Properties

Saffron shows promise in treating inflammatory conditions like osteoporosis, cardiovascular disease, and arthritis. Animal studies suggest it enhances joint and heart function while reducing inflammation and oxidative stress (Taheri-Dehkordi et al., 2020).

Antioxidant Properties

Saffron, rich in antioxidants, helps counter free radical damage, linked to cancer, heart disease, and Alzheimer's (Kumar et al., 2023). Its properties may support disease prevention and overall well-being.

Cognitive Benefits

Saffron has demonstrated cognitive benefits, potentially aiding Alzheimer's and age-related decline. Animal studies suggest it enhances memory, learning, and reduces oxidative stress and brain inflammation (Chib et al., 2020).

Other Potential Benefits

Saffron may offer benefits for respiratory issues, skin disorders, and menstrual cramps (Li et al., 2023). Further research is needed to refine its applications and understand its mechanisms. Current data highlights its potential as a natural medicine for various health conditions and overall well-being (Zhou et al., 2023).

5.4 Future Prospects and Conclusion

CRISPR-Cas9 genome editing is gaining importance in saffron production, addressing low yield, quality issues, and stress susceptibility. With high demand for premium saffron due to its taste, color, and medicinal value, genome editing offers a sustainable, ethical solution to enhance yield and resilience (Torricelli et al., 2019; Mansotra et al., 2022).

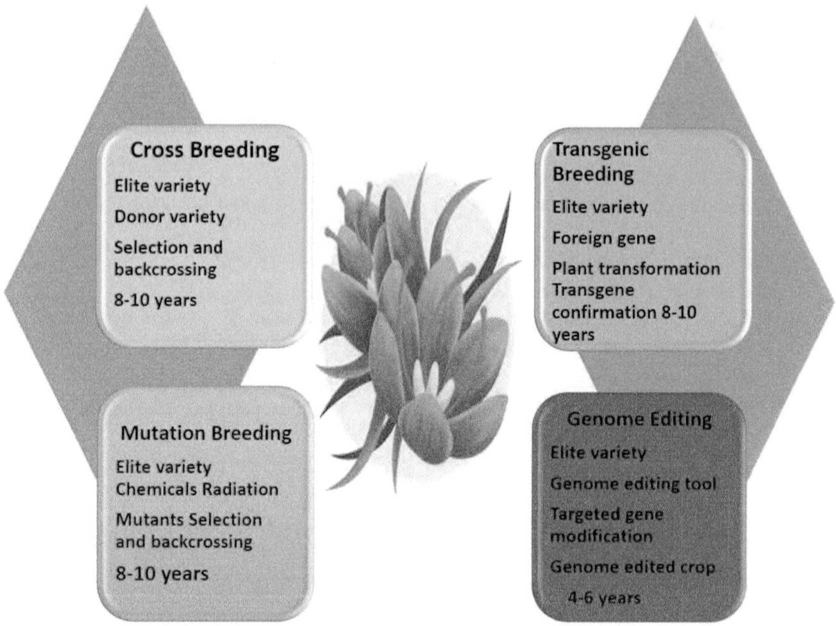

Fig. 3 Advantages of gene editing over other breeding techniques.

CRISPR-Cas9 genome editing enhances saffron yield by modifying blooming genes (FT, GI) for optimized flowering and MADS-box genes for increased stigma production (Salami et al., 2022; Hao et al., 2022). It also improves quality by altering CHS, CHI genes for higher pigment concentration (Mansotra et al., 2022). Stress tolerance can be strengthened by editing CBF, DREB genes, aiding resistance to drought, heat, and soil salinity (Salami et al., 2022). Genome editing can reduce saffron's environmental impact of saffron production by enhancing stress resistance, lowering reliance on water, fertilizers, pesticides, and fungicides.

Improved biotic stress tolerance aids sustainability, enabling high-yield, resilient saffron varieties suited to local conditions (Torricelli et al., 2019; Mansotra et al., 2022). Genome editing, especially CRISPR-Cas9, enhances saffron yield and quality while reducing environmental impact, ensuring long-term sustainability (Torricelli et al., 2019; Mansotra et al., 2022). Responsible implementation through collaboration among researchers, farmers, and communities can maximize benefits while addressing risks.

References

Abudayyeh, O.O., Gootenberg, J.S., Konermann, S., Joung, J., Slaymaker, I.M. et al. (2016). C2c2 is a single-component programmable RNA-guided RNA-targeting CRISPR effector. *Science* 353, aaf5573.

Abudayyeh, O.O., Gootenberg, J.S., Konermann, S., Joung, J., Slaymaker, I.M. et al. (2016). C2c2 is a single-component programmable RNA-guided RNA-targeting CRISPR effects.

Ali, N., Swarnkar, M.K., Veer, R., Kaushal, P. and Pati, A.M. et al. (2023). Temperature-induced modulation of stress-tolerant PGP genes bioprospected from Bacillus sp. IHBT-705 associated with saffron (Crocus sativus) rhizosphere: A natural-treasure trove of microbial biostimulants. *Frontiers in Plant Science, 14.*

Arora, L. and Narula, A. (2017). Gene editing and crop improvement using CRISPR-Cas9 system. *8*: 1932.

Bai, Z., Li, W., Jia, Y., Yue, Z., Jiao, J. et al. (2018). The ethylene response factor SmERF6 co-regulates the transcription of SmCPS1 and SmKSL1 and is involved in tanshinone biosynthesis in Salvia miltiorrhiza hairy roots. *Planta, 248*(1): 243–255.

Bao, A., Burritt, D.J., Chen, H., Zhou, X., Cao, D. et al. (2019). The CRISPR/Cas9 system and its applications in crop genome editing. *39*(3): 321–336.

Basso, M.F., Duarte, K.E., Santiago, T.R., de Souza, W.R., de Oliveira Garcia, B. et al. (2021). Efficient genome editing and gene knockout in Setaria viridis with CRISPR/Cas9 directed gene editing by the non-homologous end-joining pathway. *38*(2): 227–238.

Bhagat, N., Mansotra, R. and Patel, K. (2022). Saffron-Fusarium oxysporum R1 dual transcriptomics unravels, defense mechanism of saffron and robust pathogenicity of Fusarium oxysporum R1.

Chamkhi, I., Sbabou, L. and Aurag, J. (2023). Improved growth and quality of saffron (Crocus sativus L.) in the field conditions through inoculation with selected native plant growth-promoting rhizobacteria (PGPR). *Industrial Crops and Products, 197*: 116606.

Chib, S., Thangaraj, A., Kaul, S., Dhar, M.K. and Kaul, T. et al. (2020). Development of a system for efficient callus production, somatic embryogenesis and gene editing using CRISPR/Cas9 in Saffron (Crocus sativus L.). *Plant Methods, 16*(1): 1–10.

Chib, S., Thangaraj, A., Kaul, S., Dhar, M.K. et al. (2020). Development of a system for efficient callus production, somatic embryogenesis and gene editing using CRISPR/Cas9 in Saffron (Crocus sativus L.). *Plant Methods, 16*(1): 1–10.

Clasen, B.M., Stoddard, T.J., Luo, S., Demorest, Z., Cedrone, F. et al. (2016). Improving cold storage and processing traits in potato through targeted gene knockout. *Plant Biotechnology Journal, 14*(1): 169–176.

Clasen, B.M., Stoddard, T.J., Luo, S., Demorest, Z.L. et al. (2016). Improving cold storage and processing traits in potato through

Court, D.L., Gan, J., Liang, Y.-H., Shaw, G.X., Tropea, J.E. et al. (2013). RNase III: genetics and function; structure and mechanism. *47*: 405–431.

Demorest, Z.L., Coffman, A., Baltes, N.J., Stoddard, T.J., Clasen, B.M., Luo, S. et al. (2016). Direct stacking of sequence-specific nuclease-induced mutations to produce high oleic and low linolenic soybean oil. *BMC Plant Biology, 16*(1): 1–8.

Doudna, J.A. and Charpentier, E. (2014). The new frontier of genome engineering with CRISPR-Cas9. Science 346, 1258096.

Fischer, S., Maier, L.-K., Stoll, B., Brendel, J., Fischer, E. et al. (2012). An archaeal immune system can detect multiple protospacer adjacent motifs (PAMs) to target invader DNA. *287*(40): 33351–33363.

Gao, C. (2018). *The future of CRISPR technologies in agriculture. 19*(5): 275–276.

Gao, W., Xu, W.-T., Huang, K.-L., Guo, M.-z. and Luo et al. (2018). *Risk analysis for genome editing-derived food safety in China. 84*: 128–137.

Gudmundsson, M. and Eliasson, A. (1990). Retrogradation of amylopectin and the effects of amylose and added surfactants/emulsifiers. *13*(3): 295–315.

Gupta, R., Gupta, D., Ahmed, K.T., Dey, D., Singh, R. et al. (2021). Modification of Cas9, gRNA and PAM: key to further regulate genome editing and its applications. *178*: 85–98.

Hao, L., Xu, W., Qi, G., Xin, T., Xu, Z. et al. (2022). GAGE is a method for identification of plant species based on whole genome analysis and genome editing. *Communications Biology*, *5*(1): 947.

Haroon, M., Zafar, M.M., Farooq, M.A., Afzal, R., Batool, M. et al. (2020). Conventional Breeding, Molecular Breeding and Speed Breeding; Brave Approaches to Revamp the Production of Cereal Crops.

Haun, W., Coffman, A., Clasen, B.M., Demorest, Z.L., Lowy, A. et al. (2014). Improved soybean oil quality by targeted mutagenesis of the fatty acid desaturase 2 gene family. *Plant Biotechnology Journal, 12*(7): 934–940.

Husaini, A.M. and Ashraf, N. (2010). Understanding Saffron biology using bioinformatics tools. *Saffron. Functional Plant Science and Biotechnology*, *4*: 31–37.

Husaini, A.M., Haq, S.A.U. and Jiménez, A.J.L. (2022). Understanding saffron biology using omics- and bioinformatics tools: stepping towards a better Crocus phenome. *Molecular Biology Reports*, *49*(6): 5325–5340.

Husaini, A.M., Wani, S.A., Sofi, P., Rather, A.G., Parray, G.A. et al. (2009). Bioinformatics for saffron (Crocus sativus L.) improvement. *Communications in Biometry & Crop Science*, *4*(1):

Hutin, M., Sabot, F., Ghesquière, A., Koebnik, R. and Szurek, B. et al. (2015). A knowledge-based molecular screen uncovers a broad-spectrum Os SWEET 14 resistance allele to bacterial blight from wild rice. *The Plant Journal, 84*(4): 694–703.

Jacobs, T.B., Zhang, N., Patel, D. and Martin, G. (2017). Generation of a collection of mutant tomato lines using pooled CRISPR libraries. *174*(4): 2023–2037.

Jung, J.H. and Altpeter, F. (2016). TALEN mediated targeted mutagenesis of the caffeic acid O-methyltransferase in highly polyploid sugarcane improves cell wall composition for production of bioethanol. *Plant Molecular Biology, 92*(1): 131–142.

Kashtwari, M., Mansoor, S., Wani, A.A., Najar, M.A., Deshmukh, R.K. et al. (2022). Random mutagenesis in vegetatively propagated crops: opportunities, challenges and genome editing prospects. *Molecular Biology Reports*, *49*(6): 5729–5749.

Keify, F. and Beiki, A.H. (2012). Exploitation of random amplified polymorphic DNA (RAPD) and sequence-related amplified polymorphism (SRAP) markers for genetic diversity of saffron collection. *Journal of Medicinal Plants Research, 6*(14): 2761–2768.

Khan, Z., Khan, S.H., Mubarik, M.S. and Ahmad, A. (2018). Targeted genome editing for cotton improvement. *Past, Present and Future Trends in Cotton Breeding, 11*.

Kohara, Y., Akiyama, K. and Isono, K. (1987). The physical map of the whole E. coli chromosome: application of a new strategy for rapid analysis and sorting of a large genomic library. *50*(3): 495–508.

Kohlen, W., Charnikhova, T., Lammers, M., Pollina, T., Tóth, P. et al. (2012). The Tomato Carotenoid Cleavage Dioxygenase 8 (S L Ccd 8) Regulates Rhizosphere Signaling, Plant Architecture and Affects Reproductive Development Through Strigolactone Biosynthesis. *196*(2): 535–547.

Koonin, E.V., Makarova, K.S. and Zhang, F. (2017). Diversity, classification and evolution of CRISPR-Cas systems. *37*: 67–78.

Kramnik, I., Demant, P. and Bloom, B.B. (1998). *Susceptibility to tuberculosis as a complex genetic trait: analysis using recombinant congenic strains of mice.* Paper presented at the Novartis Foundation symposium.

Kumar, M., Kumar, R., Singh, V.P., Pathak, S., Kamboj, A.D. et al. (2023). Breeding Approaches for Quality Improvement in Fruit Crops: Strategies and Achievements. *International Journal of Environment and Climate Change*, *13*(5): 75–93.

Kun, R.S., Meng, J., Salazar-Cerezo, S., Mäkelä, M.R., De Vries, R.P. et al. (2020). CRISPR/Cas9 facilitates rapid generation of constitutive forms of transcription factors in Aspergillus niger through specific on-site genomic mutations resulting in increased saccharification of plant biomass. *Enzyme and Microbial Technology, 136*: 109508.

Lachheb, M., El Merzougui, S., Boudadi, I., El Caid, M.B., El Mousadik, A. et al. (2022). Assessing genetic diversity using the first polymorphic set of EST-SSRs markers and barcoding of Moroccan saffron. *Journal of Applied Research on Medicinal and Aromatic Plants*, *29*: 100376.

Le Rhun, A., Escalera-Maurer, A., Bratovič, M. and Charpentier, E. (2019). CRISPR-Cas in Streptococcus pyogenes. *16*(4): 380–389.

Lee, K., Zhang, Y., Kleinstiver, B.P., Guo, J.A., Aryee, M. et al. (2019). Activities and specificities of CRISPR/Cas9 and Cas12a nucleases for targeted mutagenesis in maize. *Plant Biotechnology journal, 17*(2): 362–372.

Li, C., Liu, C., Qi, X., Wu, Y., Fei, X. et al. (2017). RNA-guided Cas9 as an in *vivo* desired-target mutator in maize. *15*(12): 1566–1576.

Li, J., Li, Y. and Ma, L. (2019). CRISPR/Cas9-Based Genome Editing and its Applications for Functional Genomic Analyses in Plants. *Small Methods, 3*(3): 1800473.

Li, M., Foster, C., Kelkar, S., Pu, Y., Holmes, D. et al. (2012). Structural characterization of alkaline hydrogen peroxide pretreated grasses exhibiting diverse lignin phenotypes. *Biotechnology for Biofuels*, *5*(1): 1–15.

Li, Q., Feng, J., Chen, L., Xu, Z., Zhu, Y. et al. (2019). Genome-wide identification and characterization of Salvia miltiorrhiza laccases reveal potential targets for salvianolic acid B biosynthesis. *Frontiers in Plant Science, 10*: 435.

Li, S., Lin, D., Zhang, Y., Deng, M., Chen, Y. et al. (2022). Genome-edited powdery mildew resistance in wheat without growth penalties. *Nature, 602*(7897): 455–460.

Li, X., Liu, M. and Dong, C. (2023). Hydrophobin Gene Cmhyd4 Negatively Regulates Fruiting Body Development in Edible Fungi Cordyceps militaris. *International Journal of Molecular Sciences*, *24*(5): 4586.

Li, X., Wang, Y., Chen, S., Tian, H., Fu, D. et al. (2018). Lycopene is enriched in tomato fruit by CRISPR/Cas9-mediated multiplex genome editing. *9*: 559.

Li, X., Wang, Y., Chen, S., Tian, H., Fu, D. et al. (2018). Lycopene is Enriched in Tomato Fruit by CRISPR/Cas9-Mediated Multiplex Genome Editing. *Frontiers in Plant Science 9*: 559.

Li, Y., Zhang, B., Wang, Y., Gong, X. and Yu, H. et al. (2021). DOTFL1 affects the floral transition in orchid Dendrobium Chao Praya Smile. *Plant Physiology, 186*(4): 2021–2036.

Liang, G., Zhang, H., Lou, D. and Yu, D. (2016). Selection of highly efficient sgRNAs for CRISPR/Cas9-based plant genome editing. *6*(1): 1–8.

Liang, Y., Xu, X., Xu, L., Iqbal, Z., Ouyang, K. et al. (2022). Chondrocyte-specific genomic editing enabled by hybrid exosomes for osteoarthritis treatment. *Theranostics*, *12*(11): 4866.

Liang, Z., Chen, K., Li, T., Zhang, Y., Wang, Y. et al. (2017). Efficient DNA-free genome editing of bread wheat using CRISPR/Cas9 ribonucleoprotein complexes. *8*(1): 1–5.

Liang, Z., Chen, K., Zhang, Y., Liu, J., Yin, K. et al. (2018). Genome Editing of Bread Wheat Using Biolistic Delivery of CRISPR/Cas9 in Embryogenic Calli. *Frontiers in Plant Science 9*: 1594.

Lou, D., Wang, H., Liang, G. and Yu, D. (2017). OsSAPK2 confers abscisic acid sensitivity and tolerance to drought stress in rice. *8*: 993.

Macovei, A., Sevilla, N.R., Cantos, C., Jonson, G.B., Slamet-Loedin et al. (2018). Novel alleles of rice eIF4G generated by CRISPR/Cas9-targeted mutagenesis confer resistance to Rice tungro spherical virus. *16*(11): 1918–1927.

Makarova, K.S., Zhang, F. and Koonin, E.V. (2017). SnapShot: Class 1 CRISPR-Cas Systems. Cell, *168*(5): 946–946.e1. https://doi.org/10.1016/j.cell.2017.02.018

Mansotra, R. and Vakhlu, J. (2022). Crocus Sativus Saffron: A 360-Degree Overview. In *The Saffron Genome* (pp. 3-25). Cham: Springer International Publishing.

Marraffini, L.A. and Sontheimer, E.J. (2010). CRISPR interference: RNA-directed adaptive immunity in bacteria and archaea. *11*(3): 181–190.

Martignago, D., Rico-Medina, A., Blasco-Escámez, D., Fontanet-Manzaneque, J.B. and Caño-Delgado, A. et al. (2020). Drought resistance by engineering plant tissue-specific responses. *10*, 1676.

Massel, K., Lam, Y., Wong, A., Hickey, L.T., Borrell, A.K. et al. (2021). Hotter, drier, CRISPR: The latest edit on climate change. *134*(6): 1691–1709.

Meng, X., Yu, H., Zhang, Y., Zhuang, F., Song, X. et al. (2017). Construction of a genome-wide mutant library in rice using CRISPR/Cas9. *10*(9): 1238–1241.

Mir, J.I., Ahmed, N., Khan, M.H., Mokhdomi, T.A., Wani, S.H. et al. (2015). Molecular characterization of saffron-potential candidates for crop improvement. *Notulae Scientia Biologicae*, *7*(1): 81–89.

Mirghasempour, S.A., Studholme, D.J., Chen, W., Zhu, W. and Mao, B. et al. (2022). Molecular and Pathogenic Characterization of Fusarium Species Associated with Corm Rot Disease in Saffron from China. *Journal of Fungi*, *8*(5): 515.

Nekrasov, V., Wang, C., Win, J., Lanz, C., Weigel, D. et al. (2017). Rapid generation of a transgene-free powdery mildew resistant tomato by genome deletion. Scientific Reports *7*: 482.

Nemati, Z., Zeinalabedini, M., Mardi, M., Pirseyediand, S.M., Marashi, S.H. et al. (2012). Isolation and characterization of a first set of polymorphic microsatellite markers in saffron, Crocus sativus (Iridaceae). *American Journal of Botany*, *99*(9): e340–e343.

Niaz, M., Zhang, B., Zhang, Y., Yan, X., Yuan, M. et al. (2023). Genetic and molecular basis of carotenoid metabolism in cereals. *Theoretical and Applied Genetics*, *136*(3): 63.

Popp, J., Pető, K. and Nagy, J. (2013). Pesticide productivity and food security. A review. *33*(1): 243–255.

Prajapat, R.K., Mathur, M., Upadhyay, T.K., Lal, D., Maloo, S. et al. (2021). Genome Editing for Crop Improvement *Crop Improvement* (pp. 111–123): CRC Press.

Prajapati, A. and Nain, V. (2021). Screening of CRISPR/Cas9 gRNA for mimicking Powdery Mildew resistant MLO ol-2 mutant. *Bioinformation, 17*(6): 637.

Ramachandra, S. and Vidya, B. (2023). Agricultural revolutions which moved india from food deficit to food surplus since independence. *Delta National Journal of Multidisciplinary Research*, *10*(spl), 110–112.

Rizzo, P., Chavez, B.G., Dias, S.L. and D'Auria, J.C. (2023). Plant synthetic biology: from inspiration to augmentation. *Current Opinion in Biotechnology*, *79*: 102857.

Romero, F.M. and Gatica-Arias, A. (2019). CRISPR/Cas9: development and application in rice breeding. *Rice science, 26*(5): 265–281.

Salami, S.A. (2022). Tissue Culture Techniques for Saffron Improvement. In *The Saffron Genome* (pp. 195–220). Cham: Springer International Publishing.

Salami, S.A. and Husaini, A.M. (2022). SaffronOMICS: Novel Approaches Toward Putting Saffron Data at Work. In *The Saffron Genome* (pp. 43-62). Cham: Springer International Publishing.

Sapranauskas, R., Gasiunas, G., Fremaux, C., Barrangou, R., Horvath, P. et al. (2011). The Streptococcus thermophilus CRISPR/Cas system provides immunity in Escherichia coli. *39*(21): 9275–9282.

Sergio, F. and Newton, I. (2003). Occupancy as a measure of territory quality. *72*(5): 857–865.

Shan, Q., Zhang, Y., Chen, K., Zhang, K. and Gao, C. et al. (2015). Creation of fragrant rice by targeted knockout of the Os BADH 2 gene using TALEN technology. *Plant biotechnology journal, 13*(6): 791–800.

Shan, Q., Zhang, Y., Chen, K., Zhang, K., Gao, C. et al. (2018). Creation of fragrant rice by targeted knockout of the OsBADH2 gene using TALEN technology. *Plant Biotechnology Journal 16*: 1468–1477.

Shen, C., Que, Z., Xia, Y., Tang, N., Li, D. et al. (2017). Genome editing using CRISPR/Cas9 reveals that the zinc finger protein ZmZF3 is required for drought tolerance in maize. *Plant Biotechnology Journal 15*: 117–119.

Shi, M., Du, Z., Hua, Q. and Kai, G. (2021). CRISPR/Cas9-mediated targeted mutagenesis of bZIP2 in Salvia miltiorrhiza leads to promoted phenolic acid biosynthesis. *Industrial Crops and Products, 167*: 113560.

Shinwari, Z.K., Jan, S.A., Nakashima, K. and Yamaguchi-Shinozaki, K. (2020). Genetic engineering approaches to understanding drought tolerance in plants. *14*(2): 151–162.

Soyk, S., Müller, N.A., Park, S.J., Schmalenbach, I., Jiang, K. et al. (2017). Variation in the flowering gene SELF PRUNING 5G promotes day-neutrality and early yield in tomato. *49*(1): 162–168.

Sun, Y., Jiao, G., Liu, Z., Zhang, X., Li, J. et al. (2017). Generation of high-amylose rice through CRISPR/Cas9-mediated targeted mutagenesis of starch branching enzymes. *Frontiers in Plant Science, 8*: 298.

Taheri-Dehkordi, A., Naderi, R., Martinelli, F. and Salami, S.A. (2020). A robust workflow for indirect somatic embryogenesis and cormlet production in saffron (Crocus sativus L.) and its wild allies; C. caspius and C. speciosus. *Heliyon, 6*(12): e05841.

Tang, L., Mao, B., Li, Y., Lv, Q., Zhang, L. et al. (2017). Knockout of OsNramp5 using the CRISPR/Cas9 system produces low Cd-accumulating indica rice without compromising yield. *7*(1): 1–12.

Tarasava, K., Oh, E.J., Eckert, C.A. and Gill, R. (2018). CRISPR-enabled tools for engineering microbial genomes and phenotypes. *13*(9): 1700586.

Torricelli, R., Yousefi, J., Albertini, E., Venanzoni, R. and Hosseinzadeh, Y. et al. (2019). Morphological and molecular characterization of Italian, Iranian and Spanish saffron (Crocus sativus L.) accessions. *Appl. Ecol. Environ. Res, 17*: 1875–1887.

Uranga, M., Vazquez-Vilar, M., Orzáez, D. and Daròs, J.A. (2021). CRISPR-Cas12a genome editing at the whole-plant level using two compatible RNA virus vectors. *The CRISPR Journal, 4*(5): 761–769.

Vakhlu, J., Ambardar, S., Salami, S.A. and Kole, C. (Eds.). (2022). *The Saffron Genome*. Springer Nature.

Van der Oost, J., Jore, M.M., Westra, E.R., Lundgren, M. et al. (2009). CRISPR-based adaptive and heritable immunity in prokaryotes. *34*(8): 401–407.

Wang, F., Wang, C., Liu, P., Lei, C., Hao, W. et al. (2016). Enhanced rice blast resistance by CRISPR/Cas9-targeted mutagenesis of the ERF transcription factor gene OsERF922. *PloS one, 11*(4): e0154027.

Wani, Z.A., Mirza, D.N., Arora, P. and Riyaz-Ul-Hassan, S. (2016). Molecular phylogeny, diversity, community structure, and plant growth promoting properties of fungal endophytes associated with the corms of saffron plant: An insight into the microbiome of Crocus sativus Linn. *Fungal biology, 120*(12): 1509–1524.

Wolt, J.D., Wang, K. and Yang, B. (2016). The regulatory status of genome-edited crops. *14*(2): 510–518.

Xu, Y. and Crouch, J. (2008). Marker-assisted selection in plant breeding: From publications to practice. *48*(2): 391–407.

Yin, K. and Qiu, J. (2019). Genome editing for plant disease resistance: applications and perspectives. *Philosophical Transactions of the Royal Society B, 374*(1767): 20180322.

Zafar, K., Khan, M.Z., Amin, I., Mukhtar, Z., Yasmin, S. et al. (2020). Precise CRISPR-Cas9 mediated genome editing in super basmati rice for resistance against bacterial blight by targeting the major susceptibility gene. *Frontiers in plant science, 11*: 575.

Zetsche, B., Gootenberg, J.S., Abudayyeh, O.O., Slaymaker, I.M., Makarova, K.S. et al. (2015). Cpf1 Is a Single RNA-Guided Endonuclease of a Class 2 CRISPR-Cas System. Cell *163*: 759–771.

Zhang, X., Xu, G., Cheng, C., Lei, L., Sun, J. et al. (2021). Establishment of an Agrobacterium-mediated genetic transformation and CRISPR/Cas9-mediated targeted mutagenesis in Hemp (Cannabis Sativa L.). *Plant biotechnology journal, 19*(10): 1979–1987.

Zhang, Y., Bai, Y., Wu, G., Zou, S., Chen, Y. et al. (2017). Simultaneous modification of three homoeologs of Ta EDR 1 by genome editing enhances powdery mildew resistance in wheat. *91*(4): 714–724.

Zhou, H., Zhou, M., Yang, Y., Li, J., Zhu, L. et al. (2014). RNase ZS1 processes UbL40 mRNAs and controls thermosensitive genic male sterility in rice. *5*(1): 1–9.

Zhou, X.R., Liu, Q. and Singh, S. (2023). Engineering Nutritionally Improved Edible Plant Oils. *Annual Review of Food Science and Technology, 14*: 247–269.

Zouagui, R., Zouagui, H., Chemao-Elfihri, M.W., Chamkhi, I., Aurag, J. et al. (2023). Draft Genome Sequences of Four Plant Growth-Promoting Rhizobacteria Isolated from Saffron (Crocus sativus L.) Rhizosphere in Morocco. *Microbiology Resource Announcements*, e01046-22.

13 Microbiome and Endophytic Interactions of Saffron

Saif Khan[1], Atharva Naik[2] and Priya Sundarrajan[2,3*]

Saffron (*Crocus sativus* L.), a perennial stemless herb, is grown in Greece, Iran, and India. It is one of the most costly and ancient spices. Primarily used for medicinal purposes, the plant's stigmas. Saffron cultivation requires skilled laborers for processing due to its labor-intensive cultivation and low yield. With a primary focus on diversity and the isolation of microbes that promote plant growth, the microbiome of saffron has been studied using both cultivation- dependent and -independent methods. This review article summarizes what is currently known about the saffron microbiome, taking into account various plant parts like the roots, corm, leaves, stamen, petal, and stigma. The article also covers the biology of saffron, including its classification, distribution, pharmacology, and endophytic relationships as well as its botanical description. This information is important for understanding the interaction between saffron and its microbiome, which could have significant implications for its growth, yield, bioactivity, microbial interaction and medicinal properties.

1. Introduction

Saffron (*Crocus sativus* L., Iridaceae), a perennial stemless herb, is cultivated in Greece and India. Commercial saffron consists of dried red stigmas with traces of yellow style (Zargari, 1990).

Saffron, among the oldest spices, dates back thousands of years. A 1600–1700 BC fresco in Crete's Minos Palace depicts its harvest (Algrech, 2001). Around 2300 BC, the Akkadian empire's founder, Sargon, was born in Azupirano, possibly meaning "Saffron town" (Gadd, 1971). While Vavilov linked saffron's origin to the Middle East, others argue for Central Asia or Greek islands (Ammaro, 1989). Arabs later spread saffron across the Mediterranean (Ait-Oubahou et al., 2001). The wild ancestor of domesticated saffron was *Crocus cartwrightianus*, with its first recorded mention by Ashurbanipal in the 7th century BC. Evidence suggests saffron has treated 90 diseases for over 4,000 years (Nakarni, 2000). Today, it is

[1] Department of Botany, St. Xavier's College, Mumbai.
[2] Department of Life Science and Biochemistry, St. Xavier's College, Mumbai.
[3] Caius Research Laboratory, St. Xavier's College, Mumbai.
* Corresponding author: priya.s@xaviers.edu

DOI: 10.1201/9781003403760-13

cultivated in Tibet, Southern Europe, and India, with Kashmir and Uttaranchal as key growing regions. Saffron stigmas are primarily used medicinally (Bhargava, 2011). Its labor-intensive cultivation requires manual stigma extraction and trained workers (Melnyk et al., 2010). Due to high labor costs, low yield, and careful handling, saffron remains expensive. Major producers include Iran, India, Greece, Morocco, Italy, Spain, Turkey, Switzerland, France, Pakistan, Japan, China, and Australia, with Iran supplying 80% of global demand (Menia et al., 2018). India contributes 5% to global saffron production, with Jammu and Kashmir supplying 90%, mainly from Pulwama and Budgam (Tanra et al., 2017). Saffron's microbiome has been explored using cultivation- dependent (culturomic) and -independent (metabarcodic) methods, though studies on leaves, stamens, petals, and stigmas remain limited. Researchers focusing on Plant Growth-Promoting Microbes via culturomics emphasize microbial diversity (Bhagat et al., 2022).

2. Biology of Saffron

2.1 Botanical Description

Crocus sativus L., a grass-like tuber,produces large lilac flowers with a long, white stalk emerging from a bulb. Its dark green leaves, enclosed in a membranous sheath, persist through winter. The corolla, divided into two parts, features long styles and three thick, bright orange stigmas with rolled edges. Medicinally valuable, saffron stigmas offer a warm, slightly bitter taste, containing volatile oil and extractive matter, though age and exposure degrade their quality (Bhargava, 2011).

2.2 Classification

The taxonomic classification of *C. sativus* as per the Bentham and Hooker system of classification (1862-1863) is as follows:

Division:	Spermatophyta
Sub-division:	Angiospermae
Class:	Monocotyledonae
Sub-class:	Liliidae
Order:	Liliales
Family:	Iridaceae
Genus:	*Crocus*
species:	*sativus* L.

per – perianth
lvs – foliage leaves
scl. lf – scale leaves
crm – corm
adv.rt – adventitious root
sti – stigma
(Original illustration by Saif Khan)

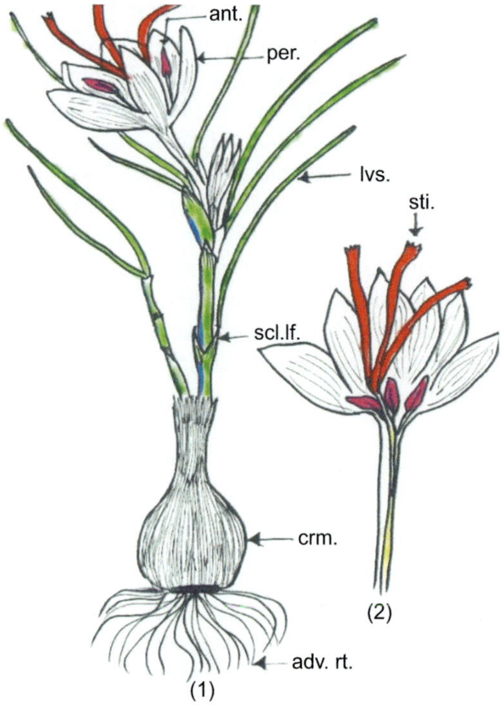

Fig. 1 Label 1 – Habit of Saffron; Label 2 – L.S of flower of Saffron showing stigmas in red; ant - anther; per- perianth; lvs - foliage leaves; scl. lf - scale leaves; crm - corm; adv.rt - adventitous root; sti - stigma (Original Illustration made by Saif Khan)

2.3 Geographical Distribution

Saffron is cultivated in Iran, Algeria, Italy, India, France, Russia, Morocco, Persia, Turkey, and Spain, with its earliest cultivation traced to Greece, Asia Minor, and Persia (Khan et al., 2011). In India, it is primarily grown in Pulwama, Baramulla, Badgam, Anantnag, and Kishtwar (Dhar and Mir, 1997). Pulwama district in Kashmir remains the hub, with Pampore Tehsil hosting key saffron villages like Khunmoh, Zewan, Balhama, Sampora, Ladhoo, Chandhara, and others. Over time, saffron cultivation expanded to nearby districts (Ganaie and Singh, 2019).

3. Endophytes – The Microbial Inhabitants of Plants

Plant-associated microorganisms play a crucial role in growth and health (Mendes et al., 2013; Philippot et al., 2013). These include epiphytes, residing in the phyllosphere and rhizosphere, and endophytes, inhabiting stems, roots, and leaves (Oldroyd et al., 2011; Turner et al., 2013; Andreote et al., 2014). While most are neutral or beneficial, only a small fraction are pathogenic (Andreote et al., 2014). Endophytes are categorized as obligate (entirely dependent on plants) or

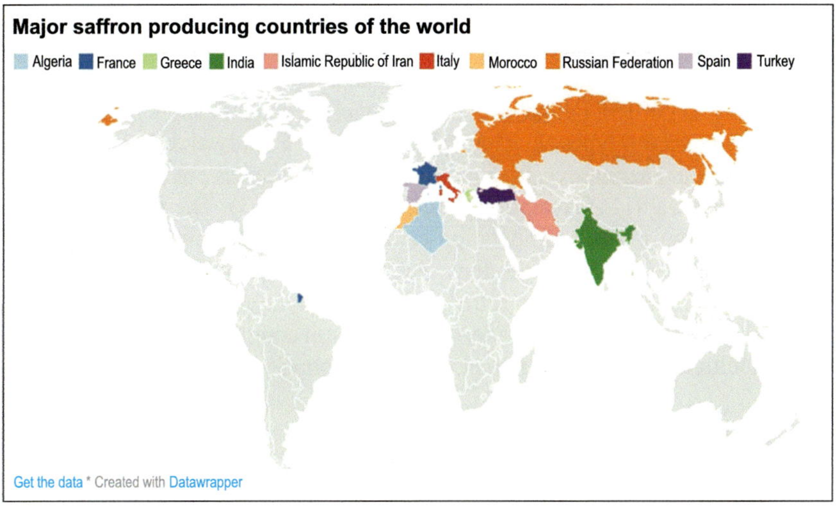

Fig. 2 World's top saffron-producing nations (Map created and customized by author Saif Khan using datawrapper)

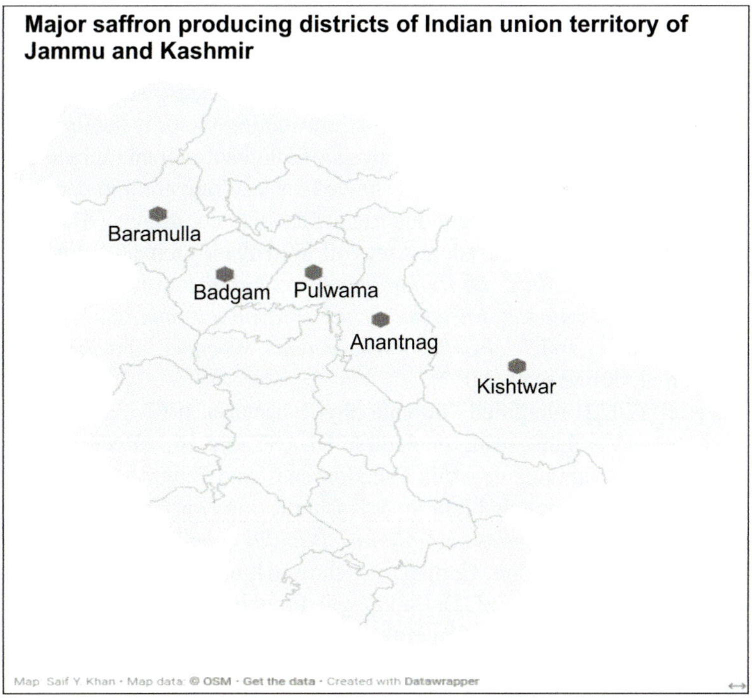

Fig. 3 Map highlighting Pulwama district, Jammu and Kashmir, India, as the leading saffron-producing region, along with other significant saffron-producing districts. (Map created and customized by author Saif Khan using datawrapper)

facultative (partially plant-associated) (Hardoim et al., 2008; Abreu-Tarazi et al., 2010). Predominantly bacteria and fungi, over 200 bacterial genera have been identified, including *Bacillus*, *Pseudomonas*, and *Agrobacterium* (Golinska et al., 2015; Sun et al., 2013). Fungal endophytes are classified into clavicipitaceous and non-clavicipitaceous types (Jalgaonwala et al., 2011; Bhardwaj & Agrawal, 2014). Mycoplasma and actinomycetes have also been identified as potential endophytes (Hollants et al., 2011; Gayathri & Muralikrishnan, 2013). Their isolation relies on fluorescence in situ hybridization-confocal laser scanning microscopy and cultivation-dependent/-independent techniques (Menpara & Chanda, 2013; Mitter et al., 2013; Berg et al., 2014). The endophytic profile depends on plant physiology, genotype, growth stage, tissue type, soil composition, season, sterility, and agricultural practices (Gaiero et al., 2013; Golinska et al., 2015).

3.1 Endophytes Associated with Saffron

Endophytes (fungi and bacteria) in saffron have been identified in leaves, roots, stigmas, anthers, and corms through culture-based methods (Ambardar & Vakhlu, 2013; Gupta & Vakhlu, 2015; Kour et al., 2018; Magotra et al., 2021; Hu et al., 2021; Rasool et al., 2021). These microorganisms interact positively or negatively with the host, influencing plant growth, disease resistance, stress tolerance, phenotypic traits, and structural integrity (Adeleke & Babalola, 2021; Mukherjee et al., 2021). Saffron production has declined by 72% in Kashmir, India over a decade, largely due to poor agricultural practices, disease susceptibility, and sterility (Wani et al., 2016, 2017). A major contributing factor is the understudied endophytic relationships in saffron. Analysis of 306 bacterial endophyte cultures from Pulwama, Srinagar, and Budgam showed 55% colonization in corm tissues, compared to 31% in shoots, likely due to nutrient-rich conditions (Ahmad et al., 2022). Proteobacteria were most dominant, followed by Firmicutes, Actinobacteria, and Bacteroidetes. *Bacillus* and *Burkholderia* were abundant in both corms and shoots, while *Xanthomonas*, *Microbacterium*, *Flavobacterium*, and *Alcaligenes* were corm-specific, and *Klebsiella*, *Pseudomonas*, *Kocuria*, and *Erwinia* were shoot-specific (Ahmad et al., 2022).

Jan et al. (2022) identified 700 endophytic bacteria in *Crocus sativus* from corms, roots, and anthers, using nucleotide BLAST and 16S rRNA sequencing. Bacillus was most abundant, while Pseudomonas and Enterobacter were least dominant. Study sites included Jammu & Kashmir, Baroosa, Letpur, and Konibal. *Bacillus proteolyticus, Bacillus halotolerans, Bacillus cereus, and Staphylococcus felis* were found across all sites. Corm tissues showed higher endophyte colonization than shoot tissues (Jan et al., 2022). Endophytic fungi in saffron vary across tissues based on substrate specificity (Wani et al., 2016; Jan et al., 2022). Wani et al. (2016, 2018) identified 36 OTUs in saffron corms, including *F. oxysporum, Cadophora malorum, Talaromyces pinophilus, Aspergillus flavipes, Epicoccum nigrum, and Botrytis cinerea*. A South Italian study by Belifiori et al. (2021) found 34 OTUs, with Ascomycota dominating, though only *F. oxysporum*, *C. malorum*, and *T.*

pinophilus were common across studies. Jan et al. (2022) identified T. pinophilus and *Aspergillus ustus* as the most abundant species.

3.2 Endophytic Interactions of Saffron

Endophytes influence key plant functions, including growth, pathogen defense, metabolism, abiotic stress adaptation, and gene expression. While 90% of interactions are beneficial, some studies highlight negative effects on host plants.

3.3 Positive Interactions

Bacterial endophytes in *Crocus sativus* enhance plant growth by synthesizing Indole-3-acetic acid (IAA) and regulating ethylene levels via acid deaminase (Sharma et al., 2015; Ahmad et al., 2022; Jan et al., 2022). Nine strains, including Bacillus stratosphericus, *Pantoea eucalypti*, and *Paraburkholderia phenazinium*, produce >100 mg/L IAA, while *Ba. megaterium, Bu. gladioli*, and *Erwinia persicina* generate high acid deaminase levels (Ahmad et al., 2022). Additionally, these bacteria aid in nutrient supply via phosphate solubilization, nitrogen fixation, and siderophore production (Adeleke et al., 2021; Husseiny et al., 2021; Patel et al., 2021). *O. proteus*, *Bu. gladioli* and *R. aquatilis* efficiently solubilize organic phosphate, while Paraburkholderia soli, Phyllobacterium ifriqiyense, Bu. gladioli, and P. conspicua are top siderophore producers (Ahmad et al., 2022). Staphylococcus, Enterobacter, Paenibacillus, Brevibacterium, Bacillus, and Pseudomonas, found in corm and leaf tissues, produce lipase, cellulase, protease, phytochrome, and chitinase, enhancing plant survival and phosphate solubilization (Sharma et al., 2015). Additionally, bacterial endophytes help *C. sativus* withstand environmental stresses, including drought, temperature fluctuations, and poor soil conditions (Tufail et al., 2021). Endophytic bacteria, including *Streptomyces achromogenes, Ba. mojavensis, Ba. siamensis, Ba. amyloliquefaciens, Bu. gladioli* and *Ba. halotolerans,* exhibit antifungal activity via chitinase, targeting F. oxysporum, a major saffron pathogen. As biocontrol agents, they mitigate chemical side effects (Sharaf-Eldin et al., 2008). *Bacillus subtilis* spore solution applied to saffron corms in Italy boosted flower count, leaf length, early sprouting, and levels of Crocetin, Safranal, and Picrocrocin (Sharaf-Eldin et al., 2008). Additionally, B. subtilis enhanced auxin and gibberellin production, leading to longer flowers and stigmas, competing with pathogens for resources while secreting allo-chemicals that inhibit their growth (Gutierrez-Manero et al., 2001; Choudhary & Varma, 2016). *Bacillus* sp. strain D5 enhances saffron growth by increasing roots, shoots, daughter cormlets, and reducing corm root disease. Its bioformulation significantly improves flower count and stigma length (Magotra et al., 2021). Additionally, S. achromogenes, Ba. mojavensis, and Bu. gladioli boost adventitious root formation, flower production, chlorophyll content, and Crocin levels in saffron (Ahmad et al., 2022). Endophyte treated plants regulate key carotenoid and apocarotenoid biosynthesis genes (LYC, UGT, CCD2, PSY, ALDH, etc.), supporting metabolite production (Ahmad et al.,

2022). Rhizobacteria aid nitrogen fixation in *Crocus sativus* (Ambardar et al., 2016). Endophytic fungi, as mutualists, protect against abiotic and biotic stress while deriving nutrients from the host (Adeleke & Babalola, 2021; Suryanarayanan & Shaanker, 2021). They are rich reservoirs of secondary metabolites, benefiting industry, medicine, and agriculture (Adeleke & Babalola, 2021; Bian et al., 2021; Ghasemnezhad et al., 2021). *Talaromyces assiutensis,* known for nematicidal and antimicrobial metabolites, naturally occurs in saffron, protecting it from nematodes and harmful microbes (Hamza et al., 2017; Deka & Jha, 2020; Jan et al., 2022). Mycorrhizal fungi, with their hyphal structures, enhance photosynthesis, moisture retention, and nutrient absorption (copper, zinc, nitrogen, etc.), serving as biofertilizers and stress protectors (Harrier & Watson, 2004).

P. mustea and *C. malorum*, long-associated with saffron, possess melanized hyphae, aiding oxygen radical elimination and Indole-3-acetic acid production, crucial for plant survival (Wani et al., 2016). *T. viride, T. harzianum*, and *Pseudomonas fluorescens* act as bio-pesticides against Rhizoctonia solani and *F. oxysporum. Penicillium, Saccharomycetales, Aspergillus*, and Nectriaceae support corm germination and blooming (Ambardar et al., 2016). Mortierella alpina enhances abiotic stress resistance by synthesizing tetraterpenoid-linked phytohormones (Wani et al., 2017).

3.4 Negative Interactions

While endophytes generally benefit *Crocus sativus*, certain fungi—*Penicillium, Aspergillus, Epicoccum, Sclerotium, Cochliobolus, Phoma, Macrophomina, Rhizopus, Rhizoctonia, Fusarium, Porostereum, Stromatinia, Talaromyces, and Colletotrichum*—are associated with saffron disorders (Ahrazem et al., 2010; Wani et al., 2016). (Refer to Table 2 for details.) Corm rot, caused by *Fusarium oxysporum*, is the most severe saffron disease, significantly impacting global production (Husaini et al., 2010). Symptoms include tissue desiccation, pigmentation, reduced flower, corm, and stigma count, leading to plant death. First reported in Japan (Yamamoto et al., 1954), the pathogen persists in soil as chlamydospores, mycelium, macroconidia, or microconidia, germinating and infecting corms and roots. Wounds caused by insects, rodents or nematodes enable pathogen entry, triggering mechanical structure formation and microbial activity, leading to tissue decomposition (Katan, 2017). Pathogens alter and inhibit essential plant functions, causing poor growth. Infection spreads, manifesting as leaf yellowing, shoot wilting, and vegetative decay, ultimately resulting in plant death (Cappelli & Di Minco, 1999; Rubio-Moraga et al., 2009). Extensive research has examined *Crocus sativus* pathogens globally (Cappelli & Di Minco, 1999; Palmero et al., 2014; Gupta & Vakhlu, 2015; Wani et al., 2016). Disease-associated saffron corms commonly harbor *Talaromyces, Penicillium, Trichoderma, Fusarium*, and *Paecilomyces*. Pathogenicity assays identified latent pathogens, such as *Epicoccum nigrum, Alternaria alternata, Acremonium sp., F. oxysporum, Talaromyces cellulolyticus*, and *Penicillium pinophilum*, with moderate to high virulence,

Table 1 Positive interactions of endophytes with different parts of saffron.

Micro-organisms	*Interactions*	*Reference*
Bacillus stratosphericus	Produces large amounts of indole-3-acetic acid, helps in promotion of plant growth.	Sharma et al., 2015; Ahmad et al., 2022; Jan et al., 2022
Pantoea eucalypti	Produces large amounts of indole-3-acetic acid, helps in promotion of plant growth.	Sharma et al., 2015; Ahmad et al., 2022; Jan et al., 2022
Bacillus megaterium, Klebsiella oxytoca, Pantoea conspicua, Obesumbacterium proteus, Paraburkholderia phenazinium, Rahnella aquatilis, Achromobacter xylosoxidans	Produces large amounts of indole-3-acetic acid, helps in promotion of plant growth, produces large quantities of deaminase, which inhibits ethylene, provides nutrients.	Sharma et al., 2015; Ahmad et al., 2022; Jan et al., 2022
Burkholderia gladioli, Erwinia persicina	Produces large quantities of deaminase, which inhibit ethylene, provide nutrients, act as organic phosphate solubilizers, produce siderophores, help in increasing the number and length of adventitious roots, increased flower count, chlorophyll and crocin content.	Ahmad et al., 2022
Phyllobacterium ifriqiyense, Paraburkholderia soli	Produces siderophores.	Ahmad et al., 2022
Staphylococcus sp., Enterobacter sp., Paenibacillus sp., Brevibacterium sp., Pseudomonas sp.	Produce lipase, cellulase and protease enzymes which are essential for many vital activities of plant survival, help in production of phytochrome and chitinase that positively influences plant growth.	Sharma et al., 2015
Streptomyces achromogenes, Bacillus mojavenis, Bacillus siamensis, Bacillus amoliquefaciens, Bacillus haloterans	Possess antifungal properties, helps in increase number and length of adventitious roots, increased flower count, chlorophyll and crocin content.	Rasool et al., 2021; Ahmad et al., 2022

Contd.

Table 1 *Contd.*

Micro-organisms	*Interactions*	*Reference*
Bacillus subtilis	Increase in the number of flowers, length of leaves, early sprouting of corms with potential rise in crocetin, safranal and picrocrocin, enhances production of auxins and gibberellins.	Gutierrez-Manero et al., 2001; Sharaf-Eldin et al., 2008
Bacillus sp. strain D5	Promotes plant growth, possesses antifungal properties, helps in decreasing the incidence of corm rot disease.	Magotra et al., 2021
Talaromyces assiutensis	Protects the plant from various nematodes and deleterious microbes.	Hamza et al., 2017; Deka and Jha, 2020; Jan et al., 2022
Phialophora mustea,	Helps in oxygen radical eradication, production of Indole acetic acid.	Wani et al., 2016
Mortierella alpina	Helps in protection against abiotic stresses.	Wani et al., 2017

while *Botrytis fabiopsis, Talaromyces verruculosus, Aspergillus pseudodeflectus, Porostereum sp., Talaromyces pinophilus, Paecilomyces marquandii, and Penicillium canescens* exhibited low virulence, classifying them as low-risk pathogens (Wani et al., 2016). A study using PCR and Miseq techniques identified pathogenic endophytes in saffron corms across three groups: infected (Group A), non-germinating (Group B), and germinating but non-blooming (Group C). Group A contained Trichocomaceae, Leotiomycetes, and Helitiales, Group B included Eurotiomycetes, Aspergillaceae, and Eurotiales, while Group C had Capnodiales, Cladosporiaceae, and Pseudeurotiaceae (Shuwen et al., 2021).Several fungi cause diseases in saffron: Uromyces croci induces rust disease (Ahrazem et al., 2010), *Aspergillus* niger leads to black mould during rooting, while Penicillium species cause blue mould rot, resulting in brown corm patches, shoot wilting, and dark lesions (Cappelli et al., 1991; Ahrazem et al., 2010). Rhizoctonia triggers corm chlorosis and rot (Madan et al., 1966), Talaromyces pinophilus is an opportunistic pathogen in rotten corms (Wani et al., 2016), and Stemphylium vesicarium is a latent pathogen in saffron leaves (Belifiori et al., 2021). Among bacterial endophytes, Burkholderia gladioli causes bacterial soft rot, *Xanthomonas campestris* results in leaf spots and blights, and *Microbacterium oxydans* contributes to corm rot disease (Ahmad et al., 2022).

Table 2 Negative interactions of endophytes with different parts of saffron.

Micro-organisms	*Interactions*	*Reference*
Fusarium oxysporum	Causes corm rot disease, desiccation of the tissue, Pigmentation, reduction in number of flowers, corm and stigma and ultimately death of the plant.	Husaini et al., 2010
Uromyces croci	Causes “rust disease” where large destruction to leaves occur.	Ahrazem et al., 2010
Aspergillus niger	Infects and causes ‘black mould ’at the rooting stages.	Cappelli et al., 1991, Ahrazem et al., 2010
Penicillium sp.	Cause “blue mould rot” characterized by brown patches on the corm, shoot wilting and dark lesions.	Cappelli et al., 1991, Ahrazem et al.., 2010
Rhizoctonia sp.	C Causes chlorosis when it infects corm which ultimately leads to rotting of the entire corm.	Madan et al., 1966
Talaromyces pinophilus	O Opportunistic or low risk pathogen found in rotten corms.	Wani et al., 2016
Stemphylium vesicarium	Latent pathogen found in saffron leaves.	Belifiori et al., 2021
Burkholderia gladioli	C Cause bacterial soft rot.	Ahmad et al., 2022

Contd.

Table 2 *Contd.*

Micro-organisms	*Interactions*	*Reference*
Xanthomonas campestris	Causes spots and blights of leaves	Ahmad et al.,2022
Microbacterium oxydans	Causes corm rot diseases	Ahmad et al., 2022

4. Bioactive Secondary Metabolites of Endophytes In Saffron

Endophytes benefit both plants and humans by producing bioactive secondary metabolites, widely utilized in agriculture, medicine, and the food industry (Tariq et al., 2022). They can be isolated and cultured in vitro, enabling controlled extraction of valuable compounds. Techniques like UHPLC/MS help analyze endophytic secondary metabolites and quantify differences between endophyte-rich and endophyte-deprived saffron plants (Jiang et al., 2022). Saffron, known as "red gold", is prized for its medicinal applications. Its key constituents—Safranal, Picrocrocin, Crocin, and Crocetin—offer antioxidant, anti-arthritic, and anti-inflammatory benefits (Poma et al., 2012). Crocin has shown potential in Parkinson's treatment (Mohammadzadeh et al., 2018), while saffron petals, rich in flavonol and Crocin, serve as anti-depressants, antioxidants, and antinociceptive agents in animal feed (Hosseinzadeh & Talebzadeh, 2005; Moshiri et al., 2006; Sánchez-Vioque et al., 2012, 2016). Safranal aids in seizure management and convulsion delay (Hosseinzadeh & Talebzadeh, 2005). Endophytic organisms play a vital role in enhancing compound concentration by providing essential precursors, substrates, and enzymes (Kosar & Baser, 2020; Salehi et al., 2022). Endophyte extracts from *C. sativus* exhibit strong antimicrobial properties. Phialomustin, derived from *P. mustea*, shows anti-cancer effects by inhibiting telomerase, regulating cell proliferation, and inducing apoptosis, alongside antimicrobial activity (Nalli et al., 2015; Colapietro et al., 2019). *Penicillium vinaceum* produces quinazoline alkaloid, known for antifungal and cytotoxic properties (Zheng et al., 2012). With environmental concerns surrounding synthetic fertilizers, the focus is shifting towards bio-fertilizers for sustainable crop yield.

Endophytic organisms function as natural bio-fertilizers by producing secondary metabolites that support plant growth, nutrient absorption, and hormone production. They enhance saffron yield and quality by supplying essential nutrients like nitrogen, potassium, and phosphorus (Salehi et al., 2022). Endophytes exhibit strong biocontrol potential by inhibiting pathogenic fungi in saffron. M. alpina produces arachidonic acid, protecting against corm rot disease (Wani et al., 2017), while Talaromyces spp. acts as an insecticidal, nematicidal, antimicrobial, and antifungal agent (Hamza et al., 2017; Li et al., 2017; Vinale et al., 2017). Fungal endophytes in *Crocus sativus* synthesize anti-mycotic volatile organic compounds (VOCs), serving as natural pesticides and info-chemicals vital for ecosystem dynamics (Wheatley, 2002). Endophytes contribute to plant hormone production, with Bacillus subtilis enhancing gibberellin, auxin, and cytokinin synthesis in *Crocus sativus*, promoting cell growth and division (Salehi et al., 2022). Additionally, saffron-associated endophytes produce enzymes like cellulase, amylase, protease, lipase, chitinase, pectinase, and endoglucanase, valuable for industrial applications (Sharma et al., 2015; Li et al., 2017). ACC deaminase, a key enzyme from saffron-associated endophytic bacteria, supports bioremediation, root nodule formation, and ethylene regulation, enhancing plant growth (Ma et al., 2004; Glick et al., 2007). *M. alpina* produces polyunsaturated fatty acids and arachidonic acid, influencing saffron metabolism by activating apocarotenoid pathway genes, making its interaction with *Crocus sativus* a valuable apocarotenoid source (Wani et al., 2017). Saffron contains ~200 aromatic compounds and carotenoids (lycopene, aldehydes, carotenes, zeaxanthin glycosides, monoterpenes, anthocyanins, flavonoids, amino acids, and vitamins), synthesized with endophyte assistance (Gupta et al., 2021). Porostereum sp. produces chlorinated aromatic compounds, useful in various applications (Wani et al., 2018). Endophytes also enhance phenolic content and anthocyanins in saffron petals, improving their use as food additives (Lotfi et al., 2015). These endophytes hold potential for bioactive product extraction, industrial and pharmacological applications, and sustainable saffron cultivation.

Conclusion

Saffron's microbiome and endophytes play a key role in growth, quality, and pathogen defense, supporting nutrient absorption and secondary metabolite production. Endophytic fungi show promise in enhancing saffron yield, though further research is needed to unravel their complex interactions and explore applications in agriculture and medicine. A deeper understanding could lead to novel therapeutics and sustainable cultivation strategies.

References

Abreu-Tarazi, M.F., Navarrete, A.A., Andreote, F.D., Almeida, C.V., Tsai, S.M. et al. (2010). Endophytic bacteria in long-term in vitro cultivated "axenic" pineapple microplants revealed by PCR–DGGE. *World Journal of Microbiology and Biotechnology*, *26*(3): 555–560. https://doi.org/10.1007/s11274-009-0191-3

Adeleke, B.S. and Babalola, O.O. (2021b). The plant endosphere-hidden treasures: A review of fungal endophytes. *Biotechnology and Genetic Engineering Reviews*, *37*(2): 154–177. https://doi.org/10.1080/02648725.2021.1991714

Adeleke, B.S., Babalola, O.O. and Glick, B.R. (2021a). Plant growth-promoting root-colonizing bacterial endophytes. *Rhizosphere*, *20*, 100433. https://doi.org/10.1016/j.rhisph.2021.100433

Ahmad, T., Farooq, S., Mirza, D.N., Kumar, A., Mir, R.A. et al. (2022). Insights into the endophytic bacterial microbiome of *Crocus sativus*: Functional characterization leads to potential agents that enhance the plant growth, productivity, and key metabolite content. *Microbial Ecology*, *83*(3): 669–688. https://doi.org/10.1007/s00248-021-01810-y

Ahrazem, O., Rubio-Moraga, A., Castillo-Lopez, R., Trapero-Mozos, A. and Gómez et al. (2010). *Crocus sativus* pathogens and defence responses. Gómez L. In A. M. Husaini (Ed.), *Functional Plant Science and Biotechnology Saffron* (pp. 81–90). Global Science Publications Ltd.

Aït Hamza, M., Lakhtar, H., Tazi, H., Moukhli, A., Fossati-Gaschignard, O. et al. (2017). Diversity of nematophagous fungi in Moroccan olive nurseries: Highlighting prey–predator interactions and efficient strains against root-knot nematodes. *Biological Control*, *114*: 14–23. https://doi.org/10.1016/j.biocontrol.2017.07.011

Ait-Oubahou, A. and El-Otmani, M. (2001). The saffron of Quercy. Saffron cultivation in Morocco. In *Saffron: Crocus sativus L. Quercy Recherche* M. Negbi & C. Algrech (Eds.), *97*: 20–27.

Ambardar, S. and Vakhlu, J. (2013). Plant growth promoting bacteria from *Crocus sativus* rhizosphere. *World Journal of Microbiology and Biotechnology*, *29*(12): 2271–2279. https://doi.org/10.1007/s11274-013-1393-2

Ambardar, S., Singh, H.R., Gowda, M. and Vakhlu, J. (2016). Comparative metagenomics reveal phylum level temporal and spatial changes in mycobiome of belowground parts of *Crocus sativus*. *PLOS ONE*, *11*(9): e0163300. https://doi.org/10.1371/journal.pone.0163300

Ammaro, F. (October 27–29, 1989). *Crocus sativus* L.—Cv. Piano di Navelli (L'Aquila saffron): Environment, cultivation, morphometric characteristics, active principles, uses. In *Proceedings of the International Conference on Saffron* (*Crocus sativus* L.), *L'Àquilla*, Italy (pp. 47–57).

Andreote, F.D., Gumiere, T. and Durrer, A. (2014). Exploring interactions of plant microbiomes. *Scientia Agricola*, *71*(6): 528–539. https://doi.org/10.1590/0103-9016-2014-0195

Belfiori, B., Rubini, A. and Riccioni, C. (2021). Diversity of endophytic and pathogenic fungi of saffron (*Crocus sativus*) plants from cultivation sites in Italy. *Diversity*, *13*(11): 535. https://doi.org/10.3390/d13110535

Berg, G., Grube, M., Schloter, M. and Smalla, K. (2014). Unraveling the plant microbiome: Looking back and future perspectives. *Frontiers in Microbiology*, *5*: 148. https://doi.org/10.3389/fmicb.2014.00148

Bhagat, N., Mansotra, R., Ambardar, S. and Vakhlu, J. (2022). Cultromic and Metabarcodic insights into saffron-microbiome associations. In *The Saffron Genome* (pp. 113–135). Springer International Publishing. https://doi.org/10.1007/978-3-031-10000-0_7

Bhardwaj, A. and Agrawal, P. (2014). A review of fungal endophytes: As a store house of bioactive compound. *World Journal of Pharmacy and Pharmaceutical Sciences*, *3*: 228–237.

Bhargava, V. (2011). Medicinal uses and pharmacological properties of Crocus sativus Linn (Saffron). *International Journal of Pharmacy and Pharmaceutical Sciences*, *3*(3): 22–26.

Bian, J.Y., Fang, Y.L., Song, Q., Sun, M.L., Yang, J.Y et al. (2021). The Fungal Endophyte *Epicoccum dendrobii* as a Potential Biocontrol Agent against *Colletotrichum gloeosporioides*. *Phytopathology*, *111*(2): 293–303. https://doi.org/10.1094/PHYTO-05-20-0170-R

Cappelli, C. and Di Minco, G. (1999). Three-years of trials on saffron diseases in Abruzzo (Central Italy) *Crocus sativus* L. Inf Fitopatol, *49*: 27–32.

Cappelli, C., Buonaurio, R. and Polverari, A. (1991). Occurrence of *Penicillium corymbiferum* on saffron in Italy. *Plant Pathology*, *40*(1): 148–149. https://doi.org/10.1111/j.1365-3059.1991.tb02304.x

Choudhary, D. and Varma, A. (2016). *Microbial-mediated Induced Systemic Resistance in Plants* p. 226. Springer.

Colapietro, A., Mancini, A., D'Alessandro, A.M. and Festuccia, C. (2019). Crocetin and crocin from saffron in cancer chemotherapy and chemoprevention. *Anti-Cancer Agents in Medicinal Chemistry*, *19*(1): 38–47. https://doi.org/10.2174/1871520619666181231112453

Deka, D. and Jha, D.K. (2020). Bioactivity assessment of endophytic fungi associated with *Citrus macroptera* Montr.: An endangered ethnomedicinal plant used in folk medicines in North-East India. *Indian Phytopathology*, *73*(1): 21–33. https://doi.org/10.1007/s42360-019-00179-w

Dhar, A.K. and Mir, G.M. (1997). Saffron in Kashmir-VI: A review of distribution and production. *Journal of Herbs, Spices and Medicinal Plants*, *4*(4): 83–90. https://doi.org/10.1300/J044v04n04_09

Gadd, C.J. (1971). In I.E.S. Edwards, C.J. Gadd & N.G.L. Hammand (Eds.). *The dynasty of Agade and the Guitan Invasion* (pp. 417–463). Cambridge University Press.

Gaiero, J.R., McCall, C.A., Thompson, K.A., Day, N.J., Best, A.S. et al. (2013). Inside the root microbiome: Bacterial root endophytes and plant growth promotion. *American Journal of Botany, 100*(9): 1738–1750. https://doi.org/10.3732/ajb.1200572

Ganaie, D.B. and Singh, Y. (2019). Saffron in Jammu and Kashmir. *Int. J. Res. Geogr.*, *5*: 1–12.

Gayathri, P. and Muralikrishnan, V. (2013). Isolation and characterization of endophytic actinomycetes from mangrove plants for antimicrobial activity. *International Journal of Current Microbiology and Applied Sciences*, *2*: 78–79.

Ghasemnezhad, A., Frouzy, A., Ghorbanpour, M. and Sohrabi, O. (2021). Microbial endophytes: New direction to natural sources. In *Endophytes, 3*: (pp. 123–155). https://doi.org/10.1007/978-3-030-65447-4_6

Glick, B.R., Cheng, Z., Czarny, J. and Duan, J. (2007). Promotion of plant growth by ACC deaminase-producing soil bacteria. *European Journal of Plant Pathology*, *119*(3): 329–339. https://doi.org/10.1007/s10658-007-9162-4

Golinska, P., Wypij, M., Agarkar, G., Rathod, D., Dahm, H. et al. (2015). Endophytic Actinobacteria of medicinal plants: Diversity and bioactivity. *Antonie van Leeuwenhoek*, *108*(2): 267–289. https://doi.org/10.1007/s10482-015-0502-7

Gupta, R. and Vakhlu, J. (2015). Native *Bacillus amyloliquefaciens* W2 as a potential biocontrol for *Fusarium oxysporum* R1 causing corm rot of *Crocus sativus*. *European Journal of Plant Pathology*, *143*(1): 123–131. https://doi.org/10.1007/s10658-015-0670-3

Gupta, R., Anand, G., Gaur, R. and Yadav, D. (2021). Plant–microbiome interactions for sustainable agriculture: A review. *Physiology and Molecular Biology of Plants*, *27*(1): 165–179. https://doi.org/10.1007/s12298-021-00927-1

Gutiérrez-Mañero, F.J., Ramos-Solano, B., Probanza, A., Mehouachi, J., Tadeo, F.R. et al. (2001). The plant growth-promoting rhizobacteria *Bacillus pumilus* and *Bacillus licheniformis* produce high amounts of physiologically active gibberellins. *Physiologia Plantarum*, *111*(2): 206–211. https://doi.org/10.1034/j.1399-3054.2001.1110211.x

Hardoim, P.R., van-van Overbeek, L.S. and van-Elsas, J.D. (2008). Properties of bacterial endophytes and their proposed role in plant growth. *Trends in Microbiology*, *16*(10): 463–471. https://doi.org/10.1016/j.tim.2008.07.008

Harrier, L.A. and Watson, C.A. (2004). The potential role of arbuscular mycorrhizal (AM) fungi in the bioprotection of plants against soil-borne pathogens in organic and/or other sustainable farming systems. *Pest Management Science*, *60*(2): 149–157. https://doi.org/10.1002/ps.820

Hollants, J., Leroux, O., Leliaert, F., Decleyre, H., De Clerck, O. et al. (2011). Who is in there? Exploration of endophytic bacteria within the siphonous green seaweed Bryopsis (Bryopsidales, Chlorophyta). *PLOS ONE*, *6*(10): e26458. https://doi.org/10.1371/journal.pone.0026458

Hosseinzadeh, H. and Talebzadeh, F. (2005). Anticonvulsant evaluation of safranal and crocin from *Crocus sativus* in mice. *Fitoterapia*, *76*(7–8): 722–724. https://doi.org/10.1016/j.fitote.2005.07.008

Hu, S., Wang, X., Sun, W., Wang, L. and Li, W. et al. (2021). *In vitro* study of biocontrol potential of rhizospheric Pseudomonas aeruginosa against pathogenic fungi of saffron (*Crocus sativus L.*). *Pathogens*, *10*(11): 1423. https://doi.org/10.3390/pathogens10111423

Hui, S., Yan, H., Qing, X., Renyuan, Y. and Yongqiang, T. et al. (2013). Isolation, characterization, and antimicrobial activity of endophytic bacteria from *Polygonum cuspidatum*. *African Journal of Microbiology Research*, *7*(16): 1496–1504. https://doi.org/10.5897/AJMR12.899

Husaini, A.M., Hassan, B., Ghani, M.Y., Teixeira da Silva, J.A. and Kirmani, N.A. et al. (2010). Saffron (*Crocus sativus* L. Kashmirianus) cultivation in Kashmir: Practices and problems. *Funct. Plant Sci. Biotechnol.*, *4*: 108–115.

Husseiny, S., Dishisha, T., Soliman, H.A., Adeleke, R. and Raslan, M. et al. (2021). Characterization of growth promoting bacterial endophytes isolated from *Artemisia annua L*. *South African Journal of Botany*, *143*: 238–247. https://doi.org/10.1016/j.sajb.2021.07.042

Jalgaonwala, R.E., Mohite, B.V. and Mahajan, R.T. (2011). Natural products from plant associated endophytic fungi. *Journal of Microbiology and Biotechnology Research, 1*: 21–32.

Jan, B., Reshi, Z.A. and Mohiddin, F.A. (2022). Site and organ-specific culture-dependent endophytic diversity of Crocus sativus L. (Saffron) in Kashmir Himalaya, India. *Microbial Ecology, 83*(4): 989–1006. https://doi.org/10.1007/s00248-021-01817-5

Jiang, Y., Wu, J., Kawagishi, H., Jiang, C., Zhou, Q. et al. (2022). Study on secondary metabolites of endophytic fungus, *Aspergillus fumigatus, from Crocus sativus* L. Guided by UHPLC-HRMS/MS-based molecular network. *International Journal of Analytical Chemistry, 2022*, 7067665. https://doi.org/10.1155/2022/7067665

Katan, J. (2017). Diseases caused by soil borne pathogens: Biology managements and challenges. *Journal of Plant Pathology, 99*: 305–315.

Khan, M.A., Naseer, S., Nagoo, S. and Nehvi, F.A. (2011). Behaviour of saffron (*Crocus sativus* L.) corms for daughter corm production. *Journal of Phytology, 3*: 47–49.

Koşar, M. and Baser, K.H. (2020). Beneficial effects of saffron (*Crocus sativus* L.) in ocular diseases. In M. Sarwat & S. Sumaiya (Eds.), *Saffron: The Age-old Panacea in a New Light* (pp. 155–161). Academic Press.

Kour, R., Ambardar, S. and Vakhlu, J. (2018). Plant growth promoting bacteria associated with corm of *Crocus sativus* during three growth stages. *Letters in Applied Microbiology, 67*(5): 458–464. https://doi.org/10.1111/lam.13042

Li, C.X., Zhao, S., Zhang, T., Xian, L., Liao, L.S. et al. (2017). Genome sequencing and analysis of *Talaromyces pinophilus* provide insights into biotechnological applications. *Scientific Reports, 7*(1): 490. https://doi.org/10.1038/s41598-017-00567-0

Lotfi, L., Kalbasi-Ashtari, A., Hamedi, M. and Ghorbani, F. (2015). Effects of enzymatic extraction on anthocyanins yield of saffron tepals (*Crocos sativus*) along with its color properties and structural stability. *Journal of Food and Drug Analysis, 23*(2): 210–218. https://doi.org/10.1016/j.jfda.2014.10.011

Ma, W., Charles, T.C. and Glick, B.R. (2004). Expression of an Exogenous 1-aminocyclopropane-1-carboxylate deaminase Gene in *Sinorhizobium meliloti* Increases Its Ability to Nodulate alfalfa. *Applied and Environmental Microbiology, 70*(10): 5891–5897. https://doi.org/10.1128/AEM.70.10.5891-5897.2004

Madan, C.L., Kapur, B.M. and Gupta, U.S. (1966). Saffron. *Economic Botany, 20*(4): 377–385. https://doi.org/10.1007/BF02904059

Magotra, S., Bhagat, N., Ambardar, S., Ali, T., Hurek, B.R. et al. (2021). Field evaluation of PGP *Bacillus* sp. strain D5 native to Crocus sativus, in traditional and non traditional areas, and mining of PGP genes from its genome. *Scientific Reports, 11*(1): 5454. https://doi.org/10.1038/s41598-021-84585-z

Melnyk, J.P., Wang, S. and Marcone, M.F. (2010). Chemical and biological properties of the world's most expensive spice: Saffron. *Food Research International, 43*(8): 1981–1989. https://doi.org/10.1016/j.foodres.2010.07.033

Mendes, R., Garbeva, P. and Raaijmakers, J.M. (2013). The rhizosphere microbiome: Significance of plant beneficial, plant pathogenic, and human pathogenic microorganisms. *FEMS Microbiology Reviews, 37*(5): 634–663. https://doi.org/10.1111/1574-6976.12028

Menia, M., Iqbal, S., Zahida, R., Tahir, S., Kanth, R.H. et al. (2018). Production technology of saffron for enhancing productivity. *Journal of Pharmacognosy and Phytochemistry, 7*(1): 1033–1039.

Menpara, D. and Chanda, S. (2013). Endophytic bacteria – Unexplored reservoir of antimicrobials for combating microbial pathogens. In A. Méndez-Vilas (Ed.), *Microbial Pathogens and Strategies for Combating them: Science, Technology and Education* (pp. 1095–1103). Formatex Research Center.

Mitter, B., Petric, A., Shin, M.W., Chain, P.S., Hauberg-Lotte, L. et al. (2013). Comparative genome analysis of Burkholderia phytofirmans PsJN reveals a wide spectrum of endophytic lifestyles based on interaction strategies with host plants. *Frontiers in Plant Science*, *4*: 120. https://doi.org/10.3389/fpls.2013.00120

Mohammadzadeh, L., Hosseinzadeh, H., Abnous, K. and Razavi, B.M. (2018). Neuroprotective potential of crocin against malathion-induced motor deficit and neurochemical alterations in rats. *Environmental Science and Pollution Research International*, *25*(5): 4904–4914. https://doi.org/10.1007/s11356-017-0842-0

Moraga, A.R., Nohales, P.F., Pérez, J.A. and Gómez-Gómez, L. (2004). Glucosylation of the saffron apocarotenoid crocetin by a glucosyltransferase isolated from *Crocus sativus* stigmas. *Planta*, *219*(6): 955–966. https://doi.org/10.1007/s00425-004-1299-1

Moshiri, E., Basti, A.A., Noorbala, A.A., Jamshidi, A.H., Hesameddin Abbasi, S. and Akhondzadeh, S. (2006). *Crocus sativus* L. (petal) in the treatment of mild-to-moderate depression: A double-blind, randomized and placebo-controlled trial. *Phytomedicine: International Journal of Phytotherapy and Phytopharmacology*, *13*(9–10): 607–611. https://doi.org/10.1016/j.phymed.2006.08.006

Mukherjee, A., Bhowmick, S., Yadav, S., Rashid, M.M., Chouhan, G.K. et al. (2021). Re-vitalizing of endophytic microbes for soil health management and plant protection. *3 Biotech*, *11*(9): 399. https://doi.org/10.1007/s13205-021-02931-4

Nadkarni, K.M. (2000). *Indian materia medica*, *1*: (pp. 390–391). Popular Press Prakashan.

Nalli, Y., Mirza, D.N., Wani, Z.A., Wadhwa, B., Mallik, F.A. et al. (2015). Phialomustin A–D, new antimicrobial and cytotoxic metabolites from an endophytic fungus, *Phialophora mustea*. *RSC Advances*, *5*(115): 95307–95312. https://doi.org/10.1039/C5RA18121F

Oldroyd, G.E.D., Murray, J.D., Poole, P.S. and Downie, J.A. (2011). The rules of engagement in the legume-rhizobial symbiosis. *Annual Review of Genetics*, *45*: 119–144. https://doi.org/10.1146/annurev-genet-110410-132549

Palmero, D., Rubio-Moraga, A., Galvez-Patón, L., Nogueras, J., Abato, C. et al. (2014). Pathogenicity and genetic diversity of *Fusarium oxysporum* isolates from corms of *Crocus sativus*. *Industrial Crops and Products*, *61*: 186–192. https://doi.org/10.1016/j.indcrop.2014.06.051

Patel, J.K., Gohel, K., Patel, H. and Solanki, T. (2021). Wheat growth dependent succession of culturable endophytic bacteria and their plant growth promoting traits. *Current Microbiology*, *78*(12): 4103–4114. https://doi.org/10.1007/s00284-021-02668-6

Philippot, L., Raaijmakers, J.M., Lemanceau, P. and van der Putten, W.H. (2013). Going back to the roots: The microbial ecology of the rhizosphere. *Nature Reviews. Microbiology*, *11*(11): 789–799. https://doi.org/10.1038/nrmicro3109

Poma, A., Fontecchio, G., Carlucci, G. and Chichiriccò, G. (2012). Anti-inflammatory properties of drugs from saffron crocus. *Anti-Inflammatory and Anti-Allergy Agents in Medicinal Chemistry*, *11*(1): 37–51. https://doi.org/10.2174/187152312803476282

Rasool, A., Imran Mir, M., Zulfajri, M., Hanafiah, M.M., Azeem Unnisa, S. et al. (2021). Plant growth promoting and antifungal asset of indigenous rhizobacteria secluded from saffron (*Crocus sativus L.*) rhizosphere. *Microbial Pathogenesis*, *150*: 104734. https://doi.org/10.1016/j.micpath.2021.104734

Salehi, F., Aelaei, M., Mortazavi, S.N., Salami, S.A. and Rezadoost Chahardeh, H. et al. (2022). Study of morphophysiological attributes of two saffron ecotypes treated with *Bacillus subtilis* under Zanjan climatic conditions. *Iran J. Hortic. Sci.*, *53*(1): 15–30.

Sánchez-Vioque, R., Rodríguez-Conde, M.F., Reina-Ureña, J.V., Escolano-Tercero, M.A., Herráiz-Peñalver, D. et al. (2012). *In vitro* antioxidant and metal chelating properties of corm, tepal and leaf from saffron (*Crocus sativus* L.). *Industrial Crops and Products*, *39*: 149–153. https://doi.org/10.1016/j.indcrop.2012.02.028

Sánchez-Vioque, R., Santana-Méridas, O., Polissiou, M.G., Vioque, J., Astraka, K. et al. (2016). Polyphenol composition and in vitro antiproliferative effect of corm, tepal and leaf from *Crocus sativus* L. on human colon adenocarcinoma cells (Caco-2). *Journal of Functional Foods*, *24*: 18–25. https://doi.org/10.1016/j.jff.2016.03.032

Sharaf-Eldin, M., Elkholy, S., Fernández, J.A., Junge, H., Cheetham, R. et al. (2008). *Bacillus subtilis* FZB24 affects flower quantity and quality of saffron (*Crocus sativus*). *Planta Medica*, *74*(10): 1316–1320. https://doi.org/10.1055/s-2008-1081293

Sharma, T., Kaul, S. and Dhar, M.K. (2015). Diversity of culturable bacterial endophytes of saffron in Kashmir, India. *SpringerPlus*, *4*: 661. https://doi.org/10.1186/s40064-015-1435-3

Shuwen, H., Qing, Z., Jin, L. and Miao, D. (2021). Fungal diversity on the surface of saffron corms with different growth characteristics. Plant Biosyst int J dealing aspects. *Plant Biology*, *155*(2): 302–309.

Suryanarayanan, T.S. and Shaanker, R.U. (2021). Can fungal endophytes fast-track plant adaptations to climate change? *Fungal Ecology*, *50*: 101039. https://doi.org/10.1016/j.funeco.2021.101039

Tanra, M.A., Dar, B.A. and Sing, S. (2017). An economic analysis of production and marketing of saffron in Jammu and Kashmir. *J. Social Relevance Concern*, *5*(10): 12–19.

Tariq, M., Jameel, F., Ijaz, U., Abdullah, M. and Rashid, K. et al. (2022). Biofertilizer microorganisms accompanying pathogenic attributes: A potential threat. *Physiology and Molecular Biology of Plants*, *28*(1): 77–90. https://doi.org/10.1007/s12298-022-01138-y

Tufail, M.A., Bejarano, A., Shakoor, A., Naeem, A., Arif, M.S. et al. (2021). Can bacterial endophytes be used as a promising bio-inoculant for the mitigation of salinity stress in crop plants?–A Global Meta–Analysis of the Last Decade (2011–2020) (2011–20). *Microorganisms*, *9*(9): 1861. https://doi.org/10.3390/microorganisms9091861

Turner, T.R., James, E.K. and Poole, P.S. (2013). The plant microbiome. *Genome Biology*, *14*(6): 209. https://doi.org/10.1186/gb-2013-14-6-209

Vinale, F., Nicoletti, R., Lacatena, F., Marra, R., Sacco, A. et al. (2017). Secondary metabolites from the endophytic fungus *Talaromyces pinophilus*. *Natural Product Research*, *31*(15): 1778–1785. https://doi.org/10.1080/14786419.2017.1290624

Wani, Z.A., Ahmad, T., Nalli, Y., Ali, A., Singh, A.P. et al. (2018). *Porostereum* sp., Associated with Saffron (*Crocus sativus L.*), is a Latent Pathogen Capable of Producing Phytotoxic Chlorinated aromatic compounds. *Current Microbiology*, *75*(7): 880–887. https://doi.org/10.1007/s00284-018-1461-9

Wani, Z.A., Kumar, A., Sultan, P., Bindu, K., Riyaz-ul-Hassan, S. et al. (2017). *Mortierella alpina* CS10E4, an oleaginous fungal endophyte of *Crocus sativus L.* enhances apocarotenoid biosynthesis and stress tolerance in the host plant. *Scientific Reports*, *7*(1): 8598. https://doi.org/10.1038/s41598-017-08974-z

Wani, Z.A., Mirza, D.N., Arora, P. and Riyaz-ul-Hassan, S. (2016). Molecular phylogeny, diversity, community structure, and plant growth promoting properties of fungal endophytes associated with the corms of saffron plant: An insight into the microbiome of *Crocus sativus* Linn. *Fungal Biology*, *120*(12): 1509–1524. https://doi.org/10.1016/j.funbio.2016.07.011

Wheatley, R.E. (2002). The consequences of volatile organic compound mediated bacterial and fungal interactions. *Antonie van Leeuwenhoek*, *81*(1–4): 357–364. https://doi.org/10.1023/a:1020592802234

Yamamoto, W., Omatsu, T. and Takami, K. (1954). Studies on the corm roots of *Crocus sativus* and saprophytic propagation of *Sclerotinia gladioli* and *Fusarium oxysporum* f. sp. gladioli on various plants and soils. *Sci Rep Hyogo Univ Agric*, *1*: 64–70.

Zargari, A. (1990). *Medicinal Plant* p. 574. Tehran University Press.

Zheng, C.J., Li, L., Zou, J.P., Han, T. and Qin, L.P. et al. (2012). Identification of a quinazoline alkaloid produced by *Penicillium vinaceum*, an endophytic fungus from *Crocus sativus*. *Pharmaceutical Biology*, *50*(2): 129–133. https://doi.org/10.3109/13880209.2011.569726

14 Micropropagation as an Aid to Saffron Conservation

Shruti Chattaraj[1*], Vijaykumar Shivaji Gadekar[2], Abdel Rahman Al-Tawaha[3], Sayanti Mandal[4] and Devendra Kumar Pandey[5*]

Saffron, the world's most expensive spice, comes from the stigmas of *Crocus sativus* L., an autotriploid sterile plant that propagates vegetatively via daughter corms. Conventional propagation faces challenges like slow growth and fungal infections, limiting high-quality corm production. Micropropagation offers a solution, producing disease-free, high-quality planting material rapidly. Recent advancements include direct and indirect shoot induction, *in vitro* microcorm development, callus-induced plantlet regeneration, and somatic embryogenesis, with studies exploring plant growth regulators (PGRs), explant types, salt concentrations, and media optimization. This chapter explores advances in saffron microcorm regeneration and micropropagation, highlighting challenges in commercializing these techniques for corm production and plantlet regeneration.

1. Introduction

Saffron (*Crocus sativus*. L), a fall-flowering perennial from the Iridaceae family, has been cultivated for its golden-red stigmas, used in seasoning, coloring, and medicine (Fallahi & Mahmoodi, 2018; Mzabri et al., 2019). This sterile triploid thrives in tropical and subtropical northern regions, producing ~150 volatile compounds (e.g., safranal, dihydrooxophorone, ketoisophorone) and non-volatile carotenoids (lycopene, zeaxanthin, crocetin, crocein) (Lozano et al., 2000). *C. sativus* L. propagates vegetatively via corms, producing ~10 cormlets per cycle, with each corm lasting one growing season (Deo, 2003). Its triploid (2n=3x=24) sterile nature prevents sexual reproduction, limiting genetic variation and traditional breeding. The slow daughter corm development restricts

[1] Department of Botany, Ananda Mohan College, Kolkata, India.
[2] Zoology Department, Sangola College (Affiliated to Punyashlok Ahilyadevi Holkar Solapur University), Solapur, India.
[3] Department of Biological Sciences, Al-Hussein Bin Talal University, Maan, Jordan.
[4] Department of Chemistry & Biochemistry, School of Basic Sciences & Research, Sharda University, Greater Noida, India.
[5] Department of Biotechnology, Lovely Professional University Punjab, India.
* Corresponding author: shrutichattaraj@gmail.com; dkpandey1974@gmail.com

DOI: 10.1201/9781003403760-14

large-scale cultivation, making saffron labor-intensive and costly. Developing faster, improved propagation techniques is crucial for saffron conservation and sustainable production. Advancements in plant tissue culture have enabled efficient propagation and conservation of vegetatively propagated plants. Compared to traditional saffron cultivation, micropropagation, organogenesis, and somatic embryogenesis show promising results (Devi et al., 2014). For large-scale multiplication, micropropagation is a highly efficient technique for producing disease-free, high-quality planting material. This chapter explores recent research, highlighting tissue culture's role in germplasm conservation and commercial saffron propagation.

2. Regeneration of Saffron Plantlet from Protoplast

Protoplast culture and regeneration, crucial for somatic hybridization and genetic manipulation, has been explored in saffron using alginate entrapment and nurse culture. Isa et al. (1990) successfully regenerated saffron plantlets from callus-derived protoplasts, isolated with cellulase, driselase, pectinase, and pectolyase Y23 in 0.3M mannitol. Immobilized in Ca-alginate beads, protoplasts formed cell colonies within two months, later cultured in MS medium with NAA and BA, leading to shoot and root regeneration. A study reported 80% calli regeneration, while another optimized protoplast isolation and culture using embryogenic saffron calli (Kramian & Ranjbar, 2010). Shoot meristems were induced to form embryogenic callus in LS medium with 4 mg/L kinetin and 1–4 mg/L 2,4-D. Protoplasts were isolated with an enzyme cocktail, cultured in MS medium with nurse cells, and immobilized in Ca-alginate beads. After 4–5 weeks, microcalli developed, forming somatic embryos in half-strength MS medium with 1 mg/L ABA. Mature embryos emerged in 25 mg/L GA3, with plantlets achieving 88% conversion efficiency on MS medium with 0.1 mg/L NAA and 1 mg/L BA.

3. *In Vitro* Microcorm Development

Saffron propagates via cormlets,with each mother corm producing ~10 per cycle. However, the slow development rate and fungal infections reduce corm production and quality, limiting availability and driving up costs. Micropropagation offers an efficient alternative, rapidly multiplying corms. *In vitro* saffron shoots swell at the base, forming microcorms, which are ideal for large-scale production, germplasm storage, and transport due to their small size and rapid formation. Plessner et al. (1990) first reported microcorm induction in saffron from directly generated shoots. Subsequent studies found microcorm formation from various explants like corm segments, ovaries, floral parts, leaves, and apical buds (Bhagyalakshmi et al., 1999; Raja et al., 2007; Sharma et al., 2008). *In vitro* cormogenesis is influenced by factors such as explant type, age, culture medium, ethylene treatment, PGRs, salt concentration, and growth retardants (Sharma et al., 2008; Sharma & Piqueras, 2010). Bunch shoots exhibit better microcorm development than individual shoots

(Sharma et al., 2008), and Mir et al. (2014) reported microcorm formation from apical bud explants cultured in various medium compositions. Optimal microcorm development (up to 10 corms) was observed in MS medium with BAP (2 mg/L), NAA (0.5 mg/L), and paclobutrazol (1.5 mg/L), while Gamborg (G-5) medium with the same additives yielded 8.8 microcorms on average. Cavusoglu et al. (2013) reported 76.7% corm formation from corm segments with meristematic regions using MS medium supplemented with 1 mg/L IAA. Sucrose concentration (6–9%) significantly enhances microcorm formation, development, and mass (Raja et al., 2007; Sharma et al., 2008).

Low salt concentration favors microcorm induction, with half-strength MS medium being optimal for daughter corm development when supplemented with BA, NAA, IBA, IAA, and ABA (Sharma et al., 2008; Sharma & Piqueras, 2010; Cavusoglu et al., 2013; Mir et al., 2014). BA combined with other growth regulators enhances cormogenesis (Sharma et al., 2008; Parray et al., 2012; Mir et al., 2014; Tahiri et al., 2022). GA3 and ABA promote in vitro corm formation (Sharma et al., 2008). Dewir et al. (2022) found mother corm explants produced more corms with greater weight and diameter than apical bud explants, with liquid culture proving more effective than solid media. Salicylic acid and glutamine enhance cormlet diameter and fresh weight, with 20°C being optimal for cormlet development in saffron. However, the role of light remains unclear, requiring further research. (See Figure 1 for microcorm regeneration.)

4. Somatic Embryogenesis

Somatic embryogenesis, a high-frequency micropropagation technique, has been reported in saffron, primarily indirectly via callus formation, leading to cormlet development from somatic embryo-derived shoots (Karamian, 2004; Raja et al., 2007; Sheibani et al., 2007). Ding et al. (1979, 1981) successfully induced callus formation and plantlet regeneration from corm explants using IAA and 2,4-D, with auxin-to-cytokinin ratio influencing callus induction (Amini et al., 2013). Somatic embryogenesis has been observed in vegetative apex, shoot meristem, leaf primordia, apical buds, central meristem sections, and leaf base (Blázquez et al., 2004; Karamian, 2004; Sharifi et al., 2010; Devi et al., 2014). Plant growth regulators (PGRs) are crucial for somatic embryogenic callus induction in saffron. 2,4-D is essential, as its absence leads to non-embryogenic callus formation (Karamian, 2004; Raja et al., 2007). Kinetin + 2,4-D or 2,4-D + BA effectively induce embryogenic callus from shoot and leaf explants (Karamian, 2004; Raja et al., 2007). Jasmonic acid, thidiazuron (TDZ), and picloram also support direct somatic embryogenesis (Devi et al., 2014). TDZ (2.27 μM) promotes callus formation and somatic embryo induction (Sheibani et al., 2007). Sucrose (6%) aids embryo maturation in hormone-free MS medium, while 1 mg/L ABA further enhances development (Karamian, 2004). GA3-supplemented half strength MS medium with ABA, and BAP enhances embryo maturation and germination, while picloram regulates embryogenic stages (Ahmed et al., 2011). Sucrose

Fig. 1 *In vitro* microcorm regeneration. a) Corm explants on medium on first day. b) Two-week-old culture showing callus induction on MS medium with 2,4-D (0.5mg/L), BAP (1mg/L), IAA (1mg/L). c) four-week-old culture showing embryogenic callus on MS medium supplemented with TDZ (0.5mg/L) + IAA (1mg/L) + Activated charcoal (0.1g/L). d) Multiplied somatic embryos on MS medium supplemented with TDZ (0.5mg/L) + IAA (1mg/L) + Activated charcoal (0.1g/L). e) Shoot initiation in four-week-old culture. f) Eight-week-old culture showing cormlet regeneration (Chib et al.,2020); it is an open access journal available under the creative commons attribution (CC-BY) license.

concentration influences germination and biomass, with 4.5% sucrose + 2.38 μM jasmonic acid aiding embryo maturation (Blázquez et al., 2004). Saffron somatic embryos matured and germinated in TDZ-supplemented medium, then transferred to PGR-free MS medium with 6% sucrose for development (Sheibani et al., 2007). Auxin type and concentration influence somatic embryo conversion into plantlets. Devi et al. (2014) reported direct somatic embryogenesis in saffron using MS medium with TDZ (2.5 μM) and picloram (1.0 μM), with embryo maturation favored in half-strength EPM. Somatic embryos retained embryogenic potential for over three years, developing shoots and plantlets, with cormlet performance tested under greenhouse conditions.

5. Shoot Regeneration from Saffron Explants

Direct shoot regeneration in saffron, bypassing callus formation, ensures genetic uniformity and requires less time than indirect methods. It has been achieved from apical/lateral buds, small corms, and ovaries, with efficiency influenced by explant type, age, PGRs, and incubation temperature (Sharma et al., 2008; Ascough et al.,

2009; Demeter et al., 2010). Plessner et al. (1990) reported shoot regeneration from apical buds using cytokinins at 15–20°C. Direct shoot regeneration was reported from corm meristematic node explants cultured in half-strength MS medium with 6mg/L BAP, achieving 96.7% shoot regeneration and 93.3% foliation in 90 days (Cavusoglu et al., 2013). Zeybek et al. (2012) identified MS medium with 100 mg/L ascorbic acid, 1 mg/L BAP, and 3% sucrose as optimal for shoot growth in direct organogenesis. Bhagyalakshmi (1999) found 2,4-D necessary in early shoot induction, while its absence later enhanced growth, with BAP essential throughout (Zeybek et al., 2012). Multiple shoot initiation in saffron was achieved using NAA (21.6 μM) and BAP (22.2 μM) in MS medium, with shoots reaching 11.4 cm (Mir et al., 2014). While BAP was generally the most effective cytokinin for shoot induction, Sharifi et al. (2010) found TDZ more effective for apical bud explants at concentrations below 10 μM. Sucrose concentration also influences shoot regeneration, with 50 g/L yielding better results than 30 g/L (Sharma & Piqueras, 2010; Sharma et al., 2008). Light and temperature influence direct shoot induction in saffron, with 20–25°C being optimal for shoot development, while dark conditions enhance shoot bud induction from ovaries (Bhagyalakshmi, 1999). Sharifi & Ebrahimzedah (2010) noted the role of antioxidant enzymes like catalase, peroxidase, esterase, superoxide dismutase, and polyphenoloxidase in shoot induction.

Conclusion

Micropropagation is crucial for saffron conservation, overcoming limitations of traditional cultivation. Significant progress has been made in direct and indirect organogenesis, somatic embryogenesis, microcorm generation, and stigma-like structure (SLS) formation. Among micropropagation methods, direct shoot induction followed by microcorm formation holds commercial potential, enabling rapid production of genetically superior corms. Additionally, tissue culture offers a platform for bioactive compound production (crocin, picrocrocin, safranal, crocetin). Further research is needed to optimize growth conditions, refine media composition, and assess light and temperature effects for commercial applications. Advancing somaclonal variation techniques could aid saffron breeding programs and genetic diversification.

References

Ahmed, A.B.A., Rao, A.S., Rao, M.V. and Taha, R.M. 2011. Effect of picloram, additives and plant growth regulators on somatic embryogenesis of Phyla nodiflora (L.) *Greene. Braz. Arch. Biol. Technol. 54*: 7–13.

Amini, F., Ghanbarzadeh, Z. and Askary Mehrabadi, M. 2013. Optimization of callus production and plant regeneration in *Salsola arbuscular* pall. *Journal of Cell and Tissue. 4*(2): 129–137.

Ascough, G.D., Erwin, J.E. and Staden, J. 2009. *Micropropagation of Iridaceae-a review. PCTOC. 97*: 1–19.

Bhagyalakshmi, N. 1999. Factors influencing direct shoot regeneration from ovary explants of saffron. *PCTOC. 58*: 205–211.

Blázquez, S., Piqueras, A., Serna, M.D., Casas, J.L. and Fernández, J.A. et al. 2004. Somatic embryogenesis in Saffron: optimization through temporary immersion and polyamine metabolism. Acta Horticult. (ISHS) *650*: 269–276.

Cavusoglu, A., Sulusoglu, M. and Erkal, S. 2013. Plant regeneration and corm formation of saffron (*Crocus sativus* L.) *in vitro*. *Res. J. Biotechnol.* *8*(12): 128–133.

Chib, S., Thangaraj, A., Kaul, S., Dhar, M.K. and Kaul, T. et al. 2020. Development of a system for efficient callus production, somatic embryogenesis and gene editing using CRISPR/Cas9 in Saffron (*Crocus sativus* L.). *Plant Methods.* *16*(1).

Demeter, Z., Gyula Suranyi, V., Molnar, A., Sramko, G., Beyer, D. et al. 2010. Somatic embryogenesis and regeneration from shoot primordial of *Crocus heuffelianus*. *PCTOC.* *100*: 349–352.

Deo, B. 2003. Growing Saffron–The World's Most Expensive Spice, Crop & Food Research (New Zealand Institute for Crop & Food Research), no. 20. http://www.crop.cri.nz/home/productsservices/ publications/broadsheets/020Saffron.pdf

Devi, K., Sharma, M. and Ahuja, P. 2014. Direct somatic embryogenesis with high frequency plantlet regeneration and successive cormlet production in saffron (*Crocus sativus* L.). *S. Afr. J. Bot.* *93*: 207–216.

Dewir, Y.H., Alsadon, A., Al-Aizari, A.A. and Al-Mohidib, M. 2022. *In Vitro* Floral Emergence and Improved Formation of Saffron Daughter Corms. *Horticulturae.* *8*: 973.

Ding, B., Bai, S.H., Wu, Y. and Fan, X.P. 1981. Induction of callus and regeneration of plantlets from corm of *Crocus sativus* L. *Acta. Bot. Sin.* *23*: 119–120.

Ding, B., Bai, S.H., Wu, Y. and Wang, B.K. 1979. Preliminary report on tissue culture of corms of *Crocus sativus*. *Acta Bot. Sin.* *21*: 387.

Fallahi, H.R. and Mahmoodi, S. 2018. Impact of water availability and fertilization management on saffron (*Crocus sativus* L.) biomass allocation. *J. Hortic. Postharvest. Res.* *1*(2): 131–146.

Isa, T., Ogasawara, T. and Kaneko, H. 1990. Regeneration from the protoplast of saffron (*Crocus sativus* L.). *Japanese Journal of Breeding.* *40*: 153–158.

Karamian, R., 2004. Plantlet regeneration via somatic embryogenesis in four species of Crocus. *Acta Horticult. (ISHS) 650*: 253–259.

Karamian, R., and Ranjbar, M. 2010. Efficient somatic embryogenesis and plantlet regeneration from protoplast culture of *Crocus sativus* L. Acta horticult. *850*: 113–116.

Lozano, P., Delgado, D., Gomez, D., Rubio, M. and Iborra, J. et al. 2000. A non-destructive method to determine the safranal content of saffron (*Crocus sativus* L.) by supercritical carbon dioxide extraction combined with high performance liquid chromatography and gas chromatography. *J. Biochem. Biophys. Methods.* *43*(1-3): 367–378.

Mir, J.I., Ahmed, N., Wajida, S., Rizwan, R., Mudasir, H.K. et al. 2014. *In vitro* development and regeneration of microcorms in saffron (*Crocus sativus* L). *Afri. J. Biotechnol.* *13*(26): 2637–2640.

Mzabri, I., Addi, M. and Berrichi, A. 2019. Traditional and Modern Uses of Saffron *(Crocus sativus).* Cosmetics. *6*(63):1–11.

Parray, J.A., Kamili, A.N., Hamid, R. and Husaini, A.M. 2012. *In vitro* cormlet production of saffron (*Crocus sativus* L. Kashmirianus) and their flowering response under greenhouse. *GM Crops Food.* *3*(4): 289–95.

Plessner, O., Ziv, M. and Negbi, M. 1990. *In vitro* corm production in the saffron crocus (*Crocus sativus* L.). *PCTOC.* *20*: 89–94.

Raja, W., Zaffer, G. and Wani, S.A. 2007. *In vitro* microcorm formation in saffron (*Crocus sativus* L.) Acta Horti. *739*: 291–296.

Sharifi, G. and Ebrahimzadeh, H. 2010. Interaction of IBA and NAA with enzymes in root induction of *Crocus sativus* (L). *Afr. J. Biotechnol.* *9*(2): 217–225.

Sharifi, G., Ebrahimzadeh, H., Ghareyazie, B. and Karimi, M. 2010. Globular embryo-like structures and highly efficient thidiazuron-induced multiple shoot formation in saffron (*Crocus sativus* L.). *In Vitro* Cell. Dev. *Biol. Plant.* *46*: 274–280.

Sharma, K.D. and Piqueras, A. 2010. Saffron (*Crocus sativus* L.) tissue culture: micropropagation and secondary metabolite production. *Functional Plant Science and Biotechnology. 4*(2): 15–24.

Sharma, K.D., Rathour, R., Sharma, R., Goel, S., Sharma, T.R. et al. 2008. *In vitro* cormlet development in *Crocus sativus. Biol. Plant. 52*: 709–712.

Sheibani, M., Nemati, S.H., Davarinejad, G.H., Azghandi, A.V. and Habashi, A.A. et al. 2007. Induction of somatic embryogenesis in saffron using thidiazuron (TDZ). *Acta Horticulturae* (ISHA). *739*: 259–267.

Tahiri, A., Mazri, M.A., Karra, Y., Aabd, N.A., Bouharroud, R. et al. 2022. Propagation of saffron (*Crocus sativus* L.) through tissue culture: a review, *The J. Hortic. Sci. Biotechnol. 98*(1): 10–30. DOI: 10.1080/14620316.2022.2078233

Zeybek, E., Onde, S. and Kaya, Z. 2012. Improved in vitro micropropagation method with adventitious corms and roots for endangered saffron. *Cent. Eur. J. Biol. 7*(1): 138–145. DOI: 10.2478/s11535-011-0102-0

15 Saffron: Phytochemistry and uses in Human Health and Diseases

Bisma Hassan Bhat[1], Tabasum Ashraf[1] and Ashraf Dar[1*]

Saffron, derived from *Crocus sativus*, is a valuable spice used in culinary, pharmaceuticals, cosmetics, perfumes, and textiles. Its crimson-red stigmas and styles, dried for seasoning, impart distinct aroma, color, and taste. Used for over 3600 years in cooking and medicine, saffron is rich in bioactive compounds like picrocrocin, kaempferol, safranal, phenol, delphinidin, flavonoids, and crocetin, offering antioxidant, anti-inflammatory, and antinociceptive properties. It benefits gastrointestinal health, respiratory conditions, dysmenorrhea, and cardiovascular diseases. Saffron aids cardiovascular health by reducing vasoconstriction, oxidative stress, and dyslipidemia. It may help prevent peptic ulcers and inhibit cancer cell growth. Its antidepressant effects stem from safranal and crocin, which suppress dopamine, norepinephrine, and serotonin absorption.

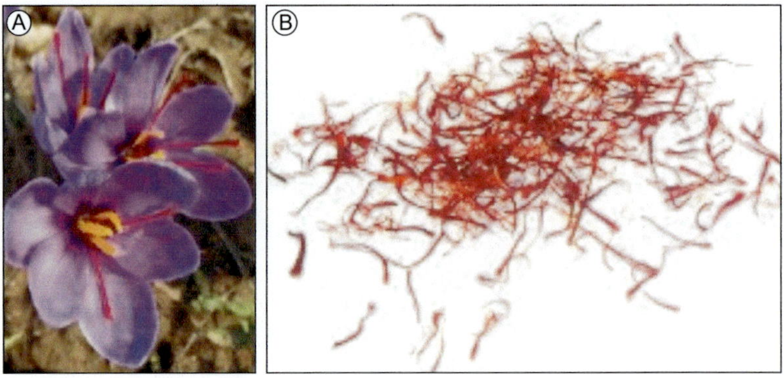

Fig. 1 Saffron flower and Saffron spice. (A) A saffron flower with crimson red stigmas and styles. (B) The harvested stigmas and styles, sold as saffron spice. [Source: Creative Commons (Openverse)]

1. Introduction

Saffron threads are harvested from *Crocus sativus* blooms (Fig. 1A). The term "Saffron" may derive from the French "Safran" or Arabic "Zaffaran" meaning

[1] Department of Biochemistry, University of Kashmir, Srinagar, India.
* Corresponding author: ashrafdar@kashmiruniversity.ac.in

DOI: 10.1201/9781003403760-15

"Yellow" (Trease & Evans, 1997). Major producers include Iran, Spain, Greece, Italy, Pakistan, and India (Kashmir), with Iran contributing ~88% of global production. Due to labour-intensive cultivation, saffron remains one of the world's costliest spices. Saffron, processed from the dried stigmas of Crocus sativus (Iridaceae), is widely used in medicine, cosmetics, perfumes, textiles, and cooking to enhance aroma (Fig. 1B). Its culinary and medicinal applications date back 3600 years (Siddiqui et al., 2018). Saffron stigmas contain primary and secondary metabolites, making them rich in bioactive compounds with strong antioxidant properties (Popović-Djordjević et al., 2021). The crimson-red color results from carotenoid degradation, primarily crocin and crocetin.

Saffron is widely used in traditional medicine, including Ayurveda, as a sedative, expectorant, anti-asthma, emmenagogue, and adaptogen. It exhibits antioxidant effects, preventing lipid peroxidation in human platelets (Gohari et al., 2013). Its antidepressant properties are well recognized. Recent research explores its role in cancer, heart disease, and Alzheimer's prevention (Popović-Djordjević et al., 2021).

Scientific Classification

Kingdom:	Plantae
Division:	Magnoliophyta
Class:	Liliopsida
Order:	Asparagales
Family:	Iridaceae
Genus:	*Croccus*
Species:	*sativus*

1.1 Botanical Description

Saffron spice is derived from the 25–30 mm long crimson-red stigmas of *Crocus sativus*, with ~150,000–200,000 flowers needed per kilogram (Gohari et al., 2013; Maliha Sami et al., 2022). Despite its light purple flowers, the stigmas hold high economic value for their aroma, color, and taste (Gohari et al., 2013). Saffron propagates via corms, an underground stem enabling vegetative reproduction (Pandita, 2021). It blooms in autumn (late September), featuring three internal and three external tepals that encase the ovary (Maliha Sami et al., 2022).

Uses of Saffron in Historical Perspective

Saffron, valued for its low side effects, has been used for centuries across civilizations for health interventions (Mir et al., 2022). Known for its vivid color and therapeutic properties, it served as a spice, herbal remedy, food coloring, and flavoring for over 3000 years. In Minoan society, saffron appeared in artefacts and sacred rituals (Wenger, 2022). In ancient India, its stigmas produced a golden-hued dye, later becoming the official color of Buddhist robes. Various cultures incorporated saffron-dyed fabrics in royal attire, and it is referenced in Song of

Solomon 4:14 as a fragrant herb. Saffron was widely used in Greek and Roman baths, theatres, and courts as a fragrance, notably linked to Greek courtesans (hetairai). Nero's arrival in Rome saw streets strewn with saffron (Britannica, 2024). In Kashmir, India, it remains integral to Wazwan cuisine and Kehva, prized for its color and aroma. Frequently used in fragrances and cosmetics, its key constituents—safranal, picrocrocin, and crocin—define its color, flavor, and scent (Mzabri et al., 2019).

2. Phytochemical Composition of Saffron and its Constituents

Phytochemicals, present in various plant parts, exhibit physiological benefits (Maqbool et al., 2022). Extensive research on saffron highlights its medicinal properties, linked to bioactive compounds like picrocrocin, kaempferol, safranal, phenol, delphinidin, flavonoids, and crocetin. Secondary metabolites include anthocyanins, carotenoids, minerals, carbohydrates, and vitamins. The three key components—crocin (color), picrocrocin (bitter taste), and safranal (aroma)—are crucial to saffron's applications (Maqbool et al., 2022; Srivastava et al., 2010). Widely used in food processing, saffron functions as a flavoring and coloring agent. Saffron essential oils contain safranal, terpene derivatives, ketones, and aldehydes, contributing to anti-inflammatory, antinociceptive, and blood pressure-lowering effects (Popović-Djordjević et al., 2021). (See Table 1 for key phytochemicals.)

Table 1 Phytochemicals in saffron and their benefits in human health.

Phytochemical	*Active substances*	*Health benefits*
Carotenoids	β-Carotene, α-carotene, crocetin, and crocins	Antioxidant, anti-inflammatory and immunomodulatory
Monoterpene	Monoterpenoid Derivatives, Amino Acids, and Alkaloids	Antioxidants, improves cognitive functions, support weight loss, and elevate mood.
Flavanoids	Isophorone, Isomer of Isophorones	Improve heart health, bone health, and balance hormones.
Phenolic acids	Hydroxycinnamic acids and Hydroxybenzoic acids	Antioxidant, anti-inflammatory, antimicrobial, and antihyperglycemic
Phytosterols	–	Manage cholesterol levels and reduce the risk of heart disease.

2.1 Carotenoids

Carotenoids, naturally occurring yellow to red pigments in fungi, algae, plants, and animals, are governed by a conjugated double-bond system, influencing their chemical, biological, and physical properties. In saffron, carotenoids exhibit antioxidant, anti-inflammatory, and immunomodulatory activities. Provitamin A (carotenoid) breaks down into retinol, essential for vision, skin maintenance, immunity, and reproductive health (Maqbool et al., 2022). These compounds enhance immune response and reduce risks of degenerative diseases like cataracts, cardiovascular disease, cancer, and muscle degeneration. Saffron's carotenoids include crocetin di-(β-D-glucosyl)-ester (dicrocin), crocetin-(β-D-gentiobiosyl)-(β-D-glucosyl)-ester (tricrocin), and crocetin di-(β-D-gentiobiosyl)-ester (crocin), with α-crocin imparting golden-yellow coloration. The key crocin compound is trans-crocetin di-(β-D-gentiobiosyl) ester, with an IUPAC name of 8,8-diapo-8,8-carotenoic acid (Christodoulou et al., 2015; Srivastava et al., 2010).

Table 2 Various carotenoids found in saffron.

Name	*Other Name*	*Additional Information*
Crocetin	crocetin II or 13-cis-crocetin isomer α-crocetin or trans-crocetin isomer or crocetin I β-crocetin or monomethyl ester of crocetin	–
Methyl-crocetin	crocin-1 or α-crocin or digentiobioside crocetin γ-crocetin or dimethyl ester of crocetin	–
Dimethyl-crocetin	crocin-3 or gentiobioside crocetin crocin-2 or tricrocin or gentioglucoside crocetin	–
Crocins	crocin-4 or glucoside crocetin crocin-5 or dicrocin or diglucoside crocetin	Major active constituent of saffron and water-soluble carotenoid
α-carotene	–	–
β-carotene	–	–
Lycopene	–	–
Zeaxanthin	–	–
Mangicrocin	–	–
xanthone-carotenoid	–	–
Glycosidic conjugate	-	–

2.1.1 *Crocins, Picrocrocin and Safranal*

Crocins, hydrophilic crocetin mono- or di-glycosylated polyene esters, provide stable, water-soluble red coloration for food applications (Maqbool et al., 2022). Crocetin, an oil-soluble, hydrophobic polyene dicarboxylic acid, esterifies with two gentiobioses, yielding α-crocin, constituting ~10% of saffron's dry weight. Upon drying, protocrocin degrades into crocin and picrocrocin, with picrocrocin hydrolyzing into glucose and safranal, while crocin generates gentiobiose and crocetin (Srivastava et al., 2010). Picrocrocin, a bitter glycoside, imparts saffron's flavor. Crocin and saffron extract may improve memory and spatial learning under persistent stress. Crocetin and its derivatives, used as colorants in food and pharmaceuticals, offer therapeutic benefits. Saffron's scent comes from volatile compounds like isophorone and safranal, which exhibit antiseizure and anticonvulsant effects. Crocetin, crocin, and picrocrocin provide cancer-preventive, memory-enhancing, and heart-protective properties. Crocin levels serve as a key saffron quality indicator, while picrocrocin and safranal arise from zeaxanthin degradation. Picrocrocin (C16H26O7, 330.37 g/mol) is a monoterpene glycoside in saffron essential oil, with safranal (150.21 g/mol) forming through glucosidase action during post-harvest dehydration (Maqbool et al., 2022; Srivastava et al., 2010). Crocin and crocetin exhibit anti-hyperlipidemic, anti-atherosclerotic, and cancer-preventive effects (Kosar et al., 2017). (See Fig. 2 for chemical structures.)

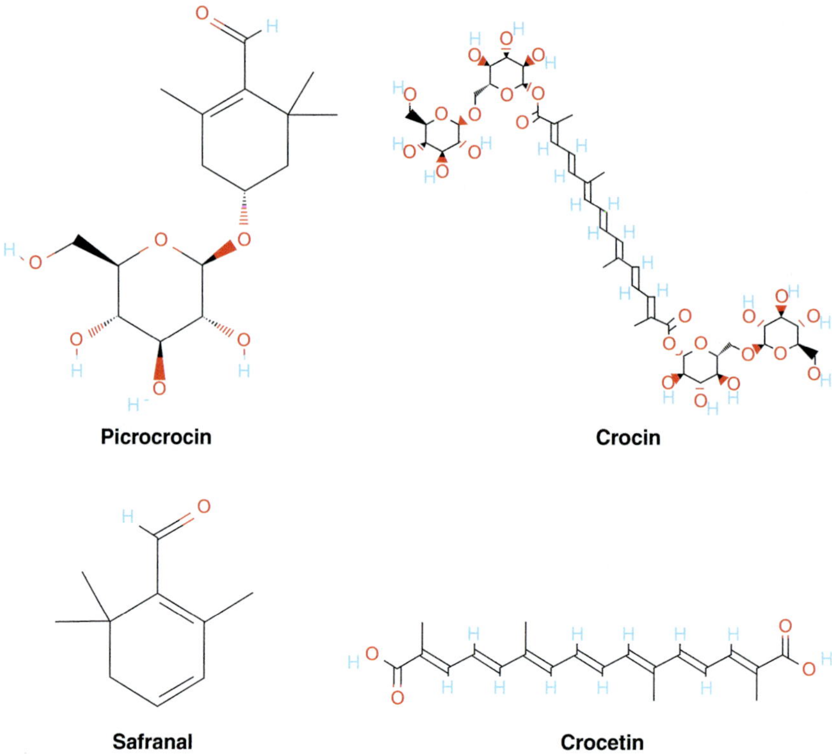

Fig. 2 The chemical structures of four principal bioactive carotenoids found in saffron. (Adapted from PubChem)

2.2 *Flavonoids*

Flavonoids, secondary polyphenolic metabolites, exhibit cardioprotective, anti-inflammatory, pro-apoptotic, and anti-tumor properties. Saffron stigmas contain three primary kaempferol glucosides. In plants, flavonoids aid UV defense, floral pigmentation, allelopathy, oxidative stress regulation, and auxin transport suppression. They also support pollen viability and influence signal transduction, transcriptional regulation, and intercellular communication. In animals, flavonoids provide therapeutic benefits and play a key role in human nutrition (Maqbool et al., 2022).

Table 3 Flavinoids in saffron

Name of the flavinoid	*Alternate name*	*Importance*
Isophorone	3,5,5-trimethyl-2-cyclohexen-1-one	Important for saffron differentiation
Isomer of isophorone	3,5,5-trimethyl-3-cyclohexen-1-one	–

2.3 *Phenolic Acids*

Phenolic acids, key bioactive compounds in plants, include hydroxycinnamic and hydroxybenzoic acids, which shape food's organoleptic properties. Saffron contains gallic acid, pyrogallol, caffeic acid, methylparaben, and chlorogenic acid, while hydroxybenzoic acids serve as flavonoid precursors. P-hydroxybenzoic acid, h-coumaric acid, sinapic acid, and vanillic acid are found in saffron petals, with benzoic and p-hydroxybenzoic acids present in pollen (Maqbool et al., 2022).

2.4 *Monoterpenes, Monoterpenoid Derivatives, Amino Acids, and Alkaloids*

Terpenoids (monoterpenes), formed from two isoprene molecules, include 38 monoterpene glycosides and monoterpenoids isolated from saffron. Alkaloids and amino acids, present in saffron stigmas, exhibit antibacterial, anti-inflammatory, and antiviral properties, while amino acids enhance food flavor (Maqbool et al., 2022; Środa-Pomianek et al., 2020). (See Table 4 for key monoterpenes and monoterpenoids.)

Table 4 List of monoterpenes and monoterpenoids found in saffron.

Name	*Group*	*Further Information*
Crocusatin D	Monoterpene	
Crocusatin F	Monoterpene	
Crocusatin G	Monoterpene	Colourless, optically active oil
Crocusatin H	Monoterpenoid	
Crocusatin E	Monoterpenoid	
Crocusatin I	Monoterpenoid	

2.5 *Phytosterols*

Phytosterols, natural steroid alcohols, help reduce blood cholesterol and LDL levels, promoting nutritional wellness (Britannica, 2024). Saffron contains stigmasterol and β-sitosterol, while fecosterol, stigmasterol, and β-sitosterol have been isolated from flower petals (Ajam et al., n.d.).

3. Therapeutic Properties of Saffron and its Constituents

Saffron, widely studied for its pharmacological applications, has long been central to traditional medicine. Its stigmas, rich in crocetin, crocin, and flavonoids, offer therapeutic benefits (Maqbool et al., 2022). It supports gastrointestinal health by stimulating appetite and preventing atonia (Mir et al., 2022). Safranal, a key compound, aids chronic bronchitis treatment by numbing vagal nerves in the alveolar sac (Christodoulou et al., 2015; Giaccio, 2004). Additionally, crocin may help reduce uterine contractions in dysmenorrhea (Christodoulou et al., 2015). Picrocrocin may help relieve spasms and lumbar discomfort, while crocetin, known for its oxygen transport capacity, is a key therapeutic agent for arthritis, tumors, cell regeneration, bleeding, and coronary artery disease (Ichimiya et al., 2007).

3.1 Saffron Promotes Cardiovascular Health

Cardiovascular disease (CVD) is a leading cause of mortality worldwide. Saffron supports cardiovascular health by maintaining steady-state function (Maqbool et al., 2022). Its antioxidants, including lycopene, help reduce CVD risk. Crocetin aids cholesterol reduction, mitigating atherosclerosis and heart attack risk. Crocin combats dyslipidemia and oxidative stress, potentially preventing diabetes-linked heart issues. Saffron's anti-vasoconstriction properties may also lower blood pressure (Maqbool et al., 2022). Additionally, crocin protects endothelial cells by upregulating Bcl2/Bax expression (Assimopoulou et al., 2005). An *in-vivo* study found aqueous saffron extract significantly reduced aortic atheromatic plaques in ApoE transgenic mice, improving plaque composition. Saffron, especially crocin, enhances plaque stability, reducing rupture risk by influencing collagen and elastin concentrations in atheromatic plaques (Christodoulou et al., 2015). Safranal improves myocardial hemodynamics, left ventricular function, and infarct size in in-vivo studies. Frequent saffron consumption may contribute to Spain's lower CVD incidence (Christodoulou et al., 2015).

3.2 Impact of Saffron on Gastric Acid and Pepsin Secretion

A study by Nabavzadeh et.al (2009) investigated the effects of saffron extract on gastric acid secretion and pepsin in Wistar rats. Rats were divided into three groups: saffron (100 mg/kg orally), LNAME + saffron (40 mg/kg LNAME injection before saffron), and control (normal saline). Results showed significantly increased acid and pepsin secretion in the saffron group, while the LNAME + saffron group exhibited no significant difference compared to controls. The findings suggest saffron enhances gastric secretions via increased NO levels (Nabavizadeh et al., n.d.). Additionally, saffron may protect against ulcer formation (Al-Mofleh et al., 2006).

3.3 Chemopreventive and Anticancer Properties of Saffron

Cancer, the second leading global cause of mortality, claims 6 million lives annually (Abdullaev et al., 2000). Saffron extract has shown efficacy in inhibiting cancer cell growth and tumor progression. Its crocetin and crocin suppress colorectal cancer cell division and tumor size, with crocin encapsulation aiding prevention. Carotenoids exhibit antioxidant activity, regulate cell growth, and influence gene expression and immune response. Research highlights saffron's potential in colon, stomach, pancreatic, ovarian, cervical, lung, skin, and breast cancer treatment (Bhandari, 2015). *In-vivo* studies confirm saffron extract delays or prevents neoplasia (Bathaie et al., n.d.). Its anticarcinogenic effects stem from DNA/RNA synthesis inhibition (Nair et al., 1991) and free radical scavenging by carotenoids (Negbi, 1999). Additionally, saffron's cytotoxicity may be linked to carotenoid interaction with topoisomerase II, a key enzyme in DNA and protein production (Shakeri et al., 2020).

3.4 Anti-depressant Properties of Saffron

Saffron petals have been studied for antidepressant effects in a 6-week placebo-controlled trial, confirming its efficacy (Akhondzadeh Basti et al., 2007). Crocin enhances its action by boosting brain-derived neurotrophic factor, nerve growth factor, and cAMP response element binding protein in rat hippocampal tissues. Additionally, kaempferol demonstrated antidepressant effects in mice and rats using the forced swimming test. Saffron's antidepressant properties stem from its serotonergic, antioxidant, anti-inflammatory, neuro-endocrine, and neuroprotective effects, primarily through dopamine, norepinephrine, and serotonin absorption suppression by safranal and crocin (Lopresti & Drummond, 2014). It has also been linked to decreased plasma corticosterone in response to stress, suggesting potential HPA axis modulation. However, no human study has assessed saffron's impact on acute stress response. Saffron fragrance reduced salivary cortisol levels and anxiety in healthy females (Jackson et al., 2021). Clinical trials compared fluoxetine (20 mg/day) and imipramine (100 mg/day) with saffron extract (20–30 mg/day) for mild to moderate depression, finding saffron equally effective, with anticonvulsant benefits and fewer side effects (Akhondzadeh et al., 2004). C. sativus corm extracts showed antidepressant-like effects in tail-suspending and forced swimming tests at 150, 300, and 600 mg/kg (Wang et al., 2010).

3.5 Saffron in Managing Hyperglycemia and Diabetes

Hyperglycemia, linked to diabetes, increases reactive oxygen species (ROS) production, causing oxidative stress, a key factor in diabetes progression (Bonnefont-Rousselot, 2002; Rajaei et al., 2013). Crocin's antioxidant properties may enhance beta-pancreatic cell insulin secretion and lower blood sugar by scavenging ROS (Kianbakht & Hajiaghaee, 2011; Shirali et al., 2013). In C2C12 skeletal muscle cells, saffron improved glucose uptake, increasing AMPK/ACC and MAPK phosphorylation, but not PI 3-kinase/Akt (Kang et al., 2012). Crocetin (40 mg/kg/d) countered dexamethasone-induced insulin resistance in rats, reducing serum insulin, free fatty acids, triglycerides, and TNF-α, while boosting liver glycogen levels (Xi et al., 2005). In fructose-fed rats, crocetin normalized adiponectin expression, enhanced TNF-α and leptin, and improved insulin sensitivity by restoring adipocyte function (Xi et al., 2007).

3.6 Immunomodulatory Properties of Saffron

Studies exhibits immunomodulatory effects, stimulating Tlymphocyte proliferation, potentially mediating antitumor activity (Nair et al., 1995). Crocetin alters cytokine expression in hemorrhagic shock-induced rats (Zeinali et al., 2019). Aqueous extracts from eight herbs, including saffron, inhibited histamine release in rats and alleviated allergic asthma clinically. While crocin derivatives were ineffective against lymphoma multidrug resistance, crocin-1 inhibited Epstein-Barr virus early antigen expression, with varied suppression of adenovirus-infected cell tumor antigens (Sotoodeh Jahromi et al., 2021). Saffron corn proteoglycan induces plasma membrane evaginations in HeLa cells, leading to fluid uptake. It shows cytotoxic effects against human cervical carcinoma cells and activates macrophages, evidenced by nitric oxide release at lower concentrations. It rapidly stimulates NF-κB and protein kinase C, akin to other macrophage activators (Escribano et al., 1996).

3.7 Effect of Saffron on Apoptosis or Programmed Cell Death

Saffron and its carotenoids influence apoptotic and signal transduction pathways, benefiting cancer and neurodegenerative diseases. Crocin prevents oxidant-induced endothelial apoptosis, while crocetin suppresses TNF-α-induced cell death in PC-12 cells and inhibits TPA-induced PKC activity in NIH/3T3 fibroblasts. Crocetin selectively inhibits cyclic AMP-dependent protein kinase (PKA) and prevents hemorrhagic shock-induced cellular damage in rats. Saffron extract and crocetin bind NMDA receptors, potentially explaining their central nervous system and anticancer effects. Dimethyl crocetin interacts with retinoic acid receptors, triggering signal transduction (Bao et al., 2023). Crocin and crocetin suppress microglial activation, reducing neuroinflammation and neurodegeneration. They inhibit the release of neurotoxic compounds like nitric oxide, TNF-α, IL-1β, and ROS, while also downregulating NF-κB, a key inflammatory transcription factor. Crocin lowers nitric oxide release in interferon-γ and amyloid-β stimulated microglia. In hippocampal slice cultures, both compounds prevent LPS-induced cell death, suggesting neuroprotective potential (Nam et al., 2010).

4. Saftey and Toxixity (by Saffron)

Saffron's safety was confirmed in clinical and experimental studies, showing no toxicity at therapeutic levels (Bostan et al., 2017). A double-blind study found no adverse effects with 200–400 mg/day saffron tablets for 1 week (Ayatollahi et al., 2014). Another study on schizophrenia patients reported no toxicity in the kidney, liver, thyroid, or hematologic system at 15 mg twice daily (Mousavi et al., 2015). Saffron extract exhibited teratogenic effects in albino mice, causing morphological abnormalities and increased mortality at higher doses (Hosseini et al., 2009). Crocin and safranal disrupted skeleton formation in mice, while crocetin impaired embryo development in frog embryos. High saffron exposure in pregnant females was linked to increased abortion rates, likely due to uterine contractions and bleeding (Ajam et al., 2014). Drugs.com advises against saffron supplementation during pregnancy, citing toxicity risks at ≥5 g (uterine stimulant, thrombocytopenia, bleeding, abortifacient effects), with 10 g inducing abortion and 20 g being fatal. However, culinary use poses no toxicity concerns. Crocetin's median lethal dose is 20.7 g/kg, while saffron extract is lethal at 600 mg/kg in mice (Gainer & Chisolm, 1974). Human ingestion (1200–2000 mg) may cause acute nausea, vomiting, diarrhea, and bleeding, while extended consumption (60 mg/day for 26 weeks) led to reduced blood cell counts, lower blood pressure, sedation, hypomania, and appetite changes (Gainer & Chisolm, 1974).

Conclusion

Crocus sativus, valued for food and economic uses, holds medicinal significance, showing potential therapeutic effects for diabetes, cancer, cardiovascular disease, immune disorders, and depression (Mir et al., 2022). Its bioactive components—flavonoids, kaempferol, safranal, crocin, crocetin, phenolics, monoterpenes, and picrocrocin—exhibit antioxidant benefits. Despite promising preclinical findings, comprehensive clinical trials remain scarce, and standardized methods for assessing saffron's bioavailability and efficacy in humans are needed. Due to toxicity risks, saffron supplements should be taken under medical supervision. Future research could improve nutritional applications and enhance disease prevention strategies.

References

Abdullaev, F.I., Rivera-Luna, R., Roitenburd-Belacortu, V. and Espinosa-Aguirre, J. (2000). Pattern of Childhood Cancer Mortality in Mexico. *Archives of Medical Research*, *31*(5): 526–531. https://doi.org/10.1016/S0188-4409(00)00094-1

Ajam, M., Reyhani, T., Roshanravan, V. and Zare, Z. (n.d.). *Increased Miscarriage Rate in Female Farmers Working in Saffron Fields: A Possible Effect of Saffron Toxicity.*

Akhondzadeh Basti, A., Moshiri, E., Noorbala, A.-A., Jamshidi, A.-H., Abbasi, S.H. et al. (2007). Comparison of petal of Crocus sativus L. and fluoxetine in the treatment of depressed outpatients: A pilot double-blind randomized trial. *Progress in Neuro-Psychopharmacology and Biological Psychiatry*, *31*(2): 439–442. https://doi.org/10.1016/j.pnpbp.2006.11.010

Akhondzadeh, S., Fallah-Pour, H., Afkham, K., Jamshidi, A.-H. and Khalighi-Cigaroudi, F. et al. (2004). Comparison of Crocus sativus L. and imipramine in the treatment of mild to moderate depression: A pilot double-blind randomized trial [ISRCTN45683816]. *BMC Complementary and Alternative Medicine*, *4*(1): 12. https://doi.org/10.1186/1472-6882-4-12

Al-Mofleh, I.A., Alhaider, A.A., Mossa, J.S., Al-Sohaiba, M.O., Qureshi, S. et al. (2006). Antigastric Ulcer Studies on 'Saffron' Crocus sativus L. in Rats. *Pakistan Journal of Biological Sciences*, *9*(6): 1009–1013. https://doi.org/10.3923/pjbs.2006.1009.1013

Assimopoulou, A.N., Sinakos, Z. and Papageorgiou, V.P. (2005). Radical scavenging activity ofCrocus sativus L. extract and its bioactive constituents. *Phytotherapy Research*, *19*(11): 997–1000. https://doi.org/10.1002/ptr.1749

Bao, X., Hu, J., Zhao, Y., Jia, R., Zhang, H. et al. (2023). Advances on the anti-tumor mechanisms of the carotenoid Crocin. *PeerJ*, *11*, e15535. https://doi.org/10.7717/peerj.15535

Bathaie, S.Z., Miri, H., Mohagheghi, M.-A., Mokhtari, M., Shahbazfar, A.-A. et al. (n.d.). *Saffron Aqueous Extract Inhibits the Chemically-induced Gastric Cancer Progression in the Wistar Albino Rat.*

Bhandari, P.R. (2015). Crocus sativus L. (saffron) for cancer chemoprevention: A mini review. *Journal of Traditional and Complementary Medicine*, *5*(2): 81–87. https://doi.org/10.1016/j.jtcme.2014.10.009

Bonnefont-Rousselot, D. (2002). Glucose and reactive oxygen species: *Current Opinion in Clinical Nutrition and Metabolic Care*, *5*(5): 561–568. https://doi.org/10.1097/00075197-200209000

Bostan, H.B., Mehri, S., Hosseinzadeh, H. (2017). Toxicology effects of saffron and its constituents: a review. *Iran J Basic Med Sci.;20*: 110–121. doi: 10.22038/ijbms.2017.8230. 00016

Christodoulou, E., Kadoglou, N.P., Kostomitsopoulos, N. and Valsami, G. (2015). Saffron: A natural product with potential pharmaceutical applications. *Journal of Pharmacy and Pharmacology*, *67*(12): 1634–1649. https://doi.org/10.1111/jphp.12456

Escribano, J., Alonso, G.-L., Coca-Prados, M. and Fernández, J.-A. (1996). Crocin, safranal and picrocrocin from saffron (Crocus sativus L.) inhibit the growth of human cancer cells in vitro. *Cancer Letters*, *100*(1–2): 23–30. https://doi.org/10.1016/0304-3835(95)04067-6

Gainer, J.L. and Chisolm, G.M. (1974). Oxygen diffusion and atherosclerosis. *Atherosclerosis*, *19*(1): 135–138. https://doi.org/10.1016/0021-9150(74)90049-5

Giaccio, M. (2004). Crocetin from Saffron: An Active Component of an Ancient Spice. *Critical Reviews in Food Science and Nutrition*, *44*(3): 155–172. https://doi.org/10.1080/10408690490441433

Gohari, A. Saeidnia, S. and Mahmoodabadi, M. (2013). An overview on saffron, phytochemicals, and medicinal properties. *Pharmacognosy Reviews*, *7*(1): 61. https://doi.org/10.4103/0973-7847.112850

Ichimiya, M., Takita, Y., Yamaguchi, M. and Muto, M. (2007). Case of purpura fulminans due to septicemia after artificial abortion. *The Journal of Dermatology*, *34*(11): 786–789. https://doi.org/10.1111/j.1346-8138.2007.00385.x

Jackson, P.A., Forster, J., Khan, J., Pouchieu, C., Dubreuil, S. et al. (2021). Effects of Saffron Extract Supplementation on Mood, Well-Being, and Response to a Psychosocial Stressor in Healthy Adults: A Randomized, Double-Blind, Parallel Group, Clinical Trial. *Frontiers in Nutrition*, *7*: 606124. https://doi.org/10.3389/fnut.2020.606124

Kang, C., Lee, H., Jung, E.-S., Seyedian, R., Jo, M. et al. (2012). Saffron (Crocus sativus L.) increases glucose uptake and insulin sensitivity in muscle cells via multipathway mechanisms. *Food Chemistry*, *135*(4): 2350–2358. https://doi.org/10.1016/j.foodchem.2012.06.092

Kianbakht, S. and Hajiaghaee, R. (2011). *Anti-hyperglycemic Effects of Saffron and its Active Constituents, Crocin and Safranal, in Alloxan-Induced Diabetic Rats. 39.*

Kosar, M., Demirci, B., Goger, F., Kara, I. and Baser, K.H.C. (2017). Volatile composition, antioxidant activity, and antioxidant components in saffron cultivated in Turkey. *International Journal of Food Properties*, *20*(sup1), S746–S754. https://doi.org/10.1080/10942912.2017.1311341

Lopresti, A.L. and Drummond, P.D. (2014). Saffron (*Crocus sativus*) for depression: A systematic review of clinical studies and examination of underlying antidepressant mechanisms of action. *Human Psychopharmacology: Clinical and Experimental*, *29*(6): 517–527. https://doi.org/10.1002/hup.2434

Maliha Sami, Abid Rashid, Muhammad Akram, Atheer Kadhim Ibadi and El Hadji Seydou Mbaye et al. (2022). Phytochemical analysis and therapeutic properties of medicinal plants Crocus sativus: Saffron. *World Journal of Advanced Research and Reviews*, *16*(2): 1117–1122. https://doi.org/10.30574/wjarr.2022.16.2.1277

Maqbool, Z., Arshad, M.S., Ali, A., Aziz, A., Khalid, W. et al. (2022). Potential Role of Phytochemical Extract from Saffron in Development of Functional Foods and Protection of Brain-Related Disorders. *Oxidative Medicine and Cellular Longevity*, *2022*: 1–14. https://doi.org/10.1155/2022/6480590

Mir, T.U.G., Wani, A.K., Singh, J. and Shukla, S. (2022). Therapeutic application and toxicity associated with Crocus sativus (saffron) and its phytochemicals. *Pharmacological Research - Modern Chinese Medicine*, *4*: 100136. https://doi.org/10.1016/j.prmcm.2022.100136

Mousavi, B, Bathaie, S.Z., Fadai, F., Ashtari, Z., Ali Beigi, N. et al. (2015). Safety evaluation of saffron stigma (Crocus sativus L.) aqueous extract and crocin in patients with schizophrenia. *Avicenna J Phytomed.;5: 413–419*. https://pmc.ncbi.nlm.nih.gov/articles/PMC4599113/

Mzabri, Addi and Berrichi. (2019). Traditional and Modern Uses of Saffron (Crocus Sativus). *Cosmetics*, *6*(4): 63. https://doi.org/10.3390/cosmetics6040063

Nabavizadeh, F., Salimi, E., Sadroleslami, Z., Karimian, S.M. and Vahedian, J. et al. (n.d.). *Saffron (Crocus sativus) increases gastric acid and pepsin secretions in rats: Role of nitric oxide (NO).*

Nair, S.C., Kurumboor, S.K. and Hasegawa, J.H. (1995). Saffron Chemoprevention in Biology and Medicine: A Review. *Cancer Biotherapy*, *10*(4): 257–264. https://doi.org/10.1089/cbr.1995.10.257

Nair, S.C., Pannikar, B. and Panikkar, K.R. (1991). Antitumour activity of saffron (Crocus sativus). *Cancer Letters*, *57*(2): 109–114. https://doi.org/10.1016/0304-3835(91)90203-T

Nam, K.N., Park, Y.-M., Jung, H.-J., Lee, J.Y., Min, B.D. et al. (2010). Anti-inflammatory effects of crocin and crocetin in rat brain microglial cells. *European Journal of Pharmacology*, *648*(1–3): 110–116. https://doi.org/10.1016/j.ejphar.2010.09.003

Negbi, M. (1999). *SAFFRON: Crocus sativus L.*

Pandita, D. (2021). Saffron (Crocus sativus L.): Phytochemistry, therapeutic significance and omics-based biology. In *Medicinal and Aromatic Plants* (pp. 325–396). Elsevier. https://doi.org/10.1016/B978-0-12-819590-1.00014-8

Popović-Djordjević, J.B., Kostić, A.Ž. and Kiralan, M. (2021). Antioxidant activities of bioactive compounds and various extracts obtained from saffron. In *Saffron* (pp. 41–97). Elsevier. https://doi.org/10.1016/B978-0-12-821219-6.00002-6

Rajaei, Z., Hadjzadeh, M.-A.-R., Nemati, H., Hosseini, M., Ahmadi, M. et al. (2013). Antihyperglycemic and Antioxidant Activity of Crocin in Streptozotocin-Induced Diabetic Rats. *Journal of Medicinal Food*, *16*(3): 206–210. https://doi.org/10.1089/jmf.2012.2407

Shakeri, M., Hashemi Tayer, A., Shakeri, H., Sotoodeh Jahromi, A., Moradzadeh, M. et al. (2020). Toxicity of Saffron Extracts on Cancer and Normal Cells: A Review Article. *Asian Pacific Journal of Cancer Prevention*, *21*(7): 1867–1875. https://doi.org/10.31557/APJCP.2020.21.7.1867

Shirali, S., Zahra Bathaie, S. and Nakhjavani, M. (2013). Effect of Crocin on the Insulin Resistance and Lipid Profile of Streptozotocin-Induced Diabetic Rats. *Phytotherapy Research*, *27*(7): 1042–1047. https://doi.org/10.1002/ptr.4836

Siddiqui, M., Saleh, M.M., Basharuddin, S.B.B., Zamri, S.B., Mohd Najib, M.B. et al. (2018). Saffron (*Crocus sativus* L.): As an antidepressant. *Journal of Pharmacy and Bioallied Sciences*, *10*(4): 173. https://doi.org/10.4103/JPBS.JPBS_83_18

Sotoodeh Jahromi, A., Kargar, M., Kafilzadeh, F., Jamalidoust, M. and Moradzadeh, M. et al. (2021). Crocin Promotes Apoptosis in Human EBV-Transformed B-Lymphocyte via Intrinsic Pathway:

Crocin Promotes Apoptosis in EBV-Transformed LCL. *Mediterranean Journal of Hematology and Infectious Diseases*, *13*(1): e2021049. https://doi.org/10.4084/MJHID.2021.049

Srivastava, R., Ahmed, H., Dixit, R., Dharamveer, and Saraf, S. et al. (2010). Crocus sativus L.: A comprehensive review. *Pharmacognosy Reviews*, *4*(8): 200. https://doi.org/10.4103/0973-7847.70919

Środa-Pomianek, K., Palko-Łabuz, A., Poła, A., Ferens-Sieczkowska, M., Wesołowska, O. et al. (2020). TMPE Derived from Saffron Natural Monoterpene as Cytotoxic and Multidrug Resistance Reversing Agent in Colon Cancer Cells. *International Journal of Molecular Sciences*, *21*(20): 7529. https://doi.org/10.3390/ijms21207529

Trease and Evans' pharmacognosy, 14th edition. (1997). *General Pharmacology: The Vascular System*, *29*(2): 291. https://doi.org/10.1016/S0306-3623(97)89724-8

Wang, Y., Han, T., Zhu, Y., Zheng, C.-J., Ming, Q.-L. et al. (2010). Antidepressant properties of bioactive fractions from the extract of Crocus sativus L. *Journal of Natural Medicines*, *64*(1): 24–30. https://doi.org/10.1007/s11418-009-0360-6

Wenger, T. (2022). History of saffron. *Longhua Chinese Medicine*, *5*: 15–15. https://doi.org/10.21037/lcm-21-67

Xi, L., Qian, Z., Shen, X., Wen, N. and Zhang, Y. et al. (2005). Crocetin Prevents Dexamethasone-Induced Insulin Resistance in Rats. *Planta Medica*, *71*(10): 917–922. https://doi.org/10.1055/s-2005-871248

Xi, L., Qian, Z., Xu, G., Zheng, S., Sun, S. et al. (2007). Beneficial impact of crocetin, a carotenoid from saffron, on insulin sensitivity in fructose-fed rats. *The Journal of Nutritional Biochemistry*, *18*(1): 64–72. https://doi.org/10.1016/j.jnutbio.2006.03.010

Zeinali, M., Zirak, M.R., Rezaee, S.A., Karimi, G. and Hosseinzadeh, H. et al. (2019). Immunoregulatory and anti-inflammatory properties of Crocus sativus (Saffron) and its main active constituents: A review. *Iranian Journal of Basic Medical Sciences*, *22*(4): https://doi.org/10.22038/ijbms.2019.34365.8158

Index

For Product Safety Concerns and Information please contact our EU representative GPSR@taylorandfrancis.com Taylor & Francis Verlag GmbH, Kaufingerstraße 24, 80331 München, Germany